Clifford Analysis and Its Applications

NATO Science Series

A Series presenting the results of scientific meetings supported under the NATO Science Programme.

The Series is published by IOS Press, Amsterdam, and Kluwer Academic Publishers in conjunction with the NATO Scientific Affairs Division

Sub-Series

I. Life and Behavioural Sciences	IOS Press
II. Mathematics, Physics and Chemistry	Kluwer Academic Publishers
III. Computer and Systems Science	IOS Press
IV. Earth and Environmental Sciences	Kluwer Academic Publishers

The NATO Science Series continues the series of books published formerly as the NATO ASI Series.

The NATO Science Programme offers support for collaboration in civil science between scientists of countries of the Euro-Atlantic Partnership Council. The types of scientific meeting generally supported are "Advanced Study Institutes" and "Advanced Research Workshops", and the NATO Science Series collects together the results of these meetings. The meetings are co-organized bij scientists from NATO countries and scientists from NATO's Partner countries – countries of the CIS and Central and Eastern Europe.

Advanced Study Institutes are high-level tutorial courses offering in-depth study of latest advances in a field.
Advanced Research Workshops are expert meetings aimed at critical assessment of a field, and identification of directions for future action.

As a consequence of the restructuring of the NATO Science Programme in 1999, the NATO Science Series was re-organized to the four sub-series noted above. Please consult the following web sites for information on previous volumes published in the Series.

http://www.nato.int/science
http://www.wkap.nl
http://www.iospress.nl
http://www.wtv-books.de/nato-pco.htm

Series II: Mathematics, Physics and Chemistry – Vol. 25

Clifford Analysis and Its Applications

edited by

F. Brackx
Department of Mathematical Analysis,
Ghent University, Ghent, Belgium

J.S.R. Chisholm
Institute of Mathematics and Statistics,
University of Kent,
Canterbury, United Kingdom

and

V. Souček
Mathematical Institute,
Charles University,
Prague, Czech Republic

Kluwer Academic Publishers

Dordrecht / Boston / London

Published in cooperation with NATO Scientific Affairs Division

Proceedings of the NATO Advanced Research Workshop on
Clifford Analysis and Its Applications
Prague, Czech Republic
October 30–November 3, 2000

A C.I.P. Catalogue record for this book is available from the Library of Congress.

ISBN 0-7923-7044-9

_____

Published by Kluwer Academic Publishers,
P.O. Box 17, 3300 AA Dordrecht, The Netherlands.

Sold and distributed in North, Central and South America
by Kluwer Academic Publishers,
101 Philip Drive, Norwell, MA 02061, U.S.A.

In all other countries, sold and distributed
by Kluwer Academic Publishers,
P.O. Box 322, 3300 AH Dordrecht, The Netherlands.

Printed on acid-free paper

_____

Table of Contents

Preface

Clifford Analysis offers a function theory for the solutions of the Dirac equation for spinor fields and is as such a direct generalization to higher dimensions of the classical function theory of holomorphic functions in the complex plane. The functions considered take their values in a Clifford algebra, a graded non-commutative and associative algebra in which arbitrary dimensional real or complex Euclidean space may be embedded in a natural way. Not only from the theoretical point of view but also with respect to applications, in particular to theoretical physics, Clifford Analysis has proven to be a valuable counterpart to the theory of several complex variables.

After a few preliminary attempts in the 1930's and the 1940's, Clifford analysis really took off in the 1960's through the effort of three research groups in Bucarest (Romania), Ghent (Belgium) and Arizona (USA). Meanwhile it has developed into an autonomous discipline in mathematical analysis with research groups all over the world. Moreover Clifford Algebra and Clifford Analysis turned out to play an important role in applications to theoretical physics. A series of conferences to stimulating the interaction between Clifford algebras, Clifford analysis and mathematical physics, was launched at Canterbury in 1985, JSR Chisholm organizing a NATO ASI-ARW, "Clifford Algebras and Their Applications to Mathematical Physics".

Fifteen years and many conferences later we have had the pleasure of organizing in Praha (Czech Republic) the NATO Advanced Research Workshop "Clifford Analysis and Its Applications". The present volume is devoted to this conference and by presenting a selection of the talks delivered, it aims at bringing a state-of-the-art view of the actual research in Clifford Analysis and its applications.

The field of Clifford analysis is indeed going through a period of quick evolution in directions belonging to the main classical stream of research in the field as well as to applications in various other parts of mathematics and physics. All traditional and new fields are very well represented in this volume, and a lot of papers contribute in a substantial way to a clarification of new important directions for further research.

Let us now go for a quick guided tour through the book. The classical topics of the main stream and their recent evolution are described in lectures on the use of the Rellich inequalities in a study of boundary operators (A Axelsson, R Grognard, J Hogan and A McIntosh) singular integral operators (V Kokilashvili and A Meskhi), Bergman and Hardy spaces (N Vasilevski), the Robin boundary value problem (L Lanzani), various alternatives for monogenicity equations in dimension eight and basic properties of their solutions (I Sabadini and D Struppa), double covers of pseudo-orthogonal groups (A Trautman), Clifford analysis on spheres and hyperbolic spaces (J Ryan), a geometric characterization of monogenic functions (H Malonek), the Beltrami equation and its relation to locally quasiconformal maps (P Cerejeiras and U Kähler), quaternionic generalization of the Riccati equation (V Kravchenko, V Kravchenko, B Williams), a study of the Riemann-Hilbert problem in the setting of Clifford analysis (S Bernstein) and wavelet theory (F Brackx and F Sommen). For several years the solutions of the Hodge operator on hyperbolic spaces have been studied by H Leutwiler and his coworkers (papers by H Leutwiler and SL Eriksson-Bique).

During the recent decade there has been a growing interest in a deeper study of Clifford analysis on Riemannian manifolds (or on manifolds with a given conformal structure) and of the relations to the broad and interesting field of invariant differential operators on manifolds with a given geometrical structure. These topics are discussed in the contributions by T Branson and V Souček; the contributions of J Bures, L Krump and P Somberg treat closely related subjects. The importance of these topics lies in the fact that they introduce very powerful geometrical methods in Clifford analysis. The theory of several Clifford variables has grown a lot already. In Clifford analysis, the analogues of the Dolbeault complex are complexes of differential operators; they have been studied for many years by D Struppa and his coworkers. The paper by I Sabadini and F Sommen reports on new results in this direction and on their relation to finite geometries and Platonic bodies. Moreover a geometrical and representational theoretical interpretation of these complexes and their relations to the Baston complexes are given in the papers of L Krump, P Somberg and V Souček. Generalizations of holomorphic cells were studied for some time in symplectic geometry, similar objects have appeared under the name of D-branes in topological field theory and string theory. A clifford version of such objects is studied in the paper by G Khimshiashvili.

In recent years Clifford analysis has broken out of its classical bounds and several very interesting generalizations have emerged. The first one is still inside the framework of invariant operators on conformal manifolds and their solutions. Classical Clifford analysis is devoted to the study of properties of the (Euclidean) Dirac operator. There exists a family of first order invariant differential operators for fields with values in more complicated spinor representations, the first one in the series being the Rarita-Schwinger operator. The properties of their solutions form a new topic in the field having very close interaction with both the geometrical methods in Clifford analysis and the theory of two Clifford variables. In such questions the role of the representation theory of simple Lie groups is becoming more and more important; it forms a new very important tool in the subject. These topics are treated in the contributions by P Van Lancker and J Bures. The second generalization is based on a different underlying geometry. It is a supersymmetric analogue of classical Clifford analysis based on the symplectic Clifford algebra. The corresponding analogue of the Dirac operator acts on fields with values in the infinite-dimensional analogue of the spinor representation introduced by B Kostant. The first properties of its solutions are formulated by F Sommen in his paper on the abstract variable approach, and its relation to the field of parabolic geometries is described in contributions by L Kadlčáková and V Souček. The third direction concerns the role of nilpotent Lie groups in possible generalizations of Clifford analysis (V Kisil).

The applications of Clifford analysis to other fields of mathematics and physics form a traditional part of Clifford analysis. Applications to hyperbolic systems of equations are treated by G Kaiser, relations to integral geometry (in the sense of I M Gelfand) are given in the paper by H Schaeben, W Sprößig and G van den Boogaart. Applications to boundary value problems in PDE's and their numerical description are further developed by W Sprößig and K Gürlebeck. Generalizations of Clifford algebras and their applications are described by A Kwasniewski. Applications to image recognition and the computational complexity of the corresponding algorithms are treated by V Labunets, E Labunets-Rundblad and J Astola. A large field of applications to problems in theoretical physics is reported on in contributions by JSR Chisholm, B Jancewicz and N Marchuk.

The NATO Advanced Research Workshop 976052 *Clifford Analysis and Its Applications* was organized with the financial support of NATO, Charles University Prague and Ghent University. The organizers seize the opportunity to thank heartily these three Institutions and especially the NATO Scientific and Environmental Affairs Division and the NATO Science Committee for their generous support.

Last but not least let us mention that the whole ARW with its scientific content and the presence and contributions of so many friends, was offered to Richard Delanghe (Ghent University) at the occasion of his 60th birthday. In this way all participants wanted to pay tribute to the founding father of Clifford Analysis. In particular the Ghent participants wanted to honour him as the driving force, already for more than thirty years, of the Ghent Clifford research group.

Riemann-Hilbert Problems in Clifford Analysis

Swanhild Bernstein (sbernstein@sfb.uni-weimar.de)*
Bauhaus-Universität Weimar

Abstract. We give an overview of Riemann-Hilbert problems and related topics in Clifford analysis as generalizations of the classical complex problem and its basic properties.

Keywords: Clifford analysis, Riemann-Hilbert problems

1991 Mathematics Subject Classification: 30G35

Dedicated to Richard Delanghe on the occasion of his 60th birthday.

1. Preliminaries

We denote by e_j, $j = 1, 2, \ldots n$ the generating elements of the Clifford algebra $\mathcal{Cl}_{p,q}$, $0 \leq p, q \leq n$, $p + q = n$, i.e. $e_j^2 = +1$, $1 \leq j \leq p$ and $e_j^2 = -1$, $p + 1 \leq j \leq p + q = n$, with multiplication rule

$$e_i e_j + e_j e_i = 0, \quad i \neq j \ 1 \leq i, j \leq n.$$

In case of the Clifford algebra $\mathcal{Cl}_{0,n}$ or $\mathcal{Cl}_{n,0}$ the Dirac operator is defined as

$$D = \sum_{j=1}^{n} e_j \frac{\partial}{\partial x_j}.$$

A function f which fulfills $Df = 0$ in a domain G of $\mathbb{R}^n$ will be called (left-)monogenic. The Cauchy kernel $e(x)$ is a fundamental solution of the Dirac operator. By the aid of the Cauchy kernel we define the integral operators

$$(T_G f)(x) := - \int_G e(x - y) f(y) \, dy, \quad x \in \mathbb{R}^n, \qquad \text{Teodorescu transform}$$

$$(F_\Gamma f)(x) := \int_\Gamma e(x - y) n(y) f(y) \, dy, \quad x \notin \Gamma, \qquad \text{Cauchy-type operator}$$

$$(S_\Gamma f)(x) := \int_\Gamma 2e(x - y) n(y) f(y) \, dy, \quad x \in \Gamma, \qquad \text{singular Cauchy operator}$$

---

* This research was carried out while being supported by DFG in the Collaborative Research Center 524

F. Brackx et al. (eds.), Clifford Analysis and Its Applications, 1–8.
© 2001 *Kluwer Academic Publishers. Printed in the Netherlands.*

where $n(y) = \sum_{j=1}^{n} n_j(y)e_j$ denotes the outward pointed unit vector to $\Gamma = \partial G$ at the point y. This leads to

THEOREM 1. *Let $f \in C^1(G) \cap C(\bar{G})$ then the Borel-Pompeiu formula holds:*

$$(F_\Gamma f)(x) + (T_G Df)(x) = \begin{cases} f(x), & x \in G \\ 0, & x \in \mathbb{R}^n \setminus \bar{G} \end{cases}$$

If $f \in C^{0,\alpha}(\Gamma), 0 < \alpha < 1$. Then the Plemelj-Sokhotzkij's formulae hold:

$$\lim_{G \ni x \to \xi \in \Gamma} (F_\Gamma f)(x) = \tfrac{1}{2}(I + S_\Gamma)f(\xi),$$

$$\lim_{\mathbb{R}^n \setminus \bar{G} \ni x \to \xi \in \Gamma} (F_\Gamma f)(x) = -\tfrac{1}{2}(I - S_\Gamma)f(\xi).$$

Meanwhile there are many good books on Quaternionic and Clifford analysis. It all starts with (Brackx, Delanghe, Sommen, 1982) and the reader will find there a lot of ideas. Several fields of Clifford analysis are also considered in (Gürlebeck, Sprößig, 1997), (Delanghe, Sommen, Souček, 1992) and (Gilbert, Murray, 1991).

2. The beginning

It starts in 1851 with Bernhard Riemann's thesis "Grundlagen für eine allgemeine Theorie der Funktionen einer veränderlichen complexen Größe." When studying conditions which uniquely define a holomorphic function in a region D the problem he stated might be formulated in the following way:

Let $\{M_t\}_{t \in \partial D}$ be a prescribed family of curves in the complex plane. Find all functions w holomorphic in the domain D and continuous in the closure $\bar{D}$ such that the boundary condition

$$w(t) \in M_t \quad \forall t \in \partial D$$

is satisfied.

This formulation stresses the **geometrical nature** of the problem. A less geometrical, but common utilized form of writing the boundary condition is

$$F(t, u(t), v(t)) = 0,$$

where u and v are the real and imaginary part of the unknown function w.

The linear **boundary value problem** is given by

$$a(t)u(t) + b(t)v(t) = c(t), \tag{1}$$

where $a(t), b(t), c(t)$ are given real-valued functions.

Not regarding the connection with the problem of conformal mapping, Riemann treated the problem only heuristically. In 1904 at the 3rd International Mathematical Congress David Hilbert presented a method for solving the linear problem. He introduced the **singular integral operator** with the cotangent kernel and transformed the problem into a Fredholm integral equation. He also was first to consider the index. Hilbert further considered the (homogeneous) **transmission problem** of finding two functions Φ^+ and Φ^- holomorphic inside, respectively outside a closed curve, which satisfies the transmission condition

$$\Phi^+(t) = G(t)\Phi^-(t) \text{ on } \partial D,$$

with a given complex-valued function G.

F. Noether's paper "Über eine Klasse singulärer Integralgleichungen." is commonly known as the beginning of the **functional analytic index**. When studying singular integral equations involving the Hilbert operator, F. Noether utilized the RHP (Riemann-Hilbert Problem) to determine the index. The functional analytic index can be expressed by the **winding number** of the curve $a(t) + ib(t)$ in the complex plane. The classical period found its end with papers by N.I. Muskhelishvili (1992), F.D. Gakhov (1990) and I.N. Vekua (1959). More about (non)-linear RHP can be found in (Wegert, 1992) where these notes are taken from.

All the basic ideas led to their own theories. We will show what are the analogues in Clifford analysis.

3. Boundary Value Problems

The linear boundary value problem (1) expresses a linear relation between the real and imaginary part of a holomorphic function. A generalization of this problem in Clifford analysis would be:

Determine a monogenic function $f = \sum_\alpha e_\alpha f_\alpha$ in G such that

$$\sum_\alpha a_\alpha(t) f_\alpha(t) = c(t) \text{ on } \Gamma = \partial G,$$

where $a_\alpha(t)$, $c(t)$ are given real-valued functions.

This problem is much more complicated than the complex problem, even with smooth $a_\alpha(t)$ the problem need NOT to be Fredholmean. The main point is that this single equation need not to contain enough information about the function f. This problem is well-discussed in (Stern, 1991).

4. The Hilbert operator

The classical Hilbert operator connects the boundary values of real and imaginary part of a holomorphic function. The operator connects so-called conjugated harmonic functions because real and imaginary parts of a holomorphic function are harmonic functions. One possible way to generalize this problem is the following:

Let W be a monogenic function in the upper half space $\mathbb{R}_+^{n+1}$ with values in the Clifford algebra $\mathcal{C}\ell_{n+1,0}$. Then there exist U, V with values in the Clifford algebra $\mathcal{C}\ell_{n,0}$ such that $W = U + e_{n+1}V$ and $\Delta U = \Delta V = 0$.

If we define the *Hilbert operator* by

$$(Hu)(x) = \int_{\mathbb{R}^n} 2e(x-y)u(y)\,dy$$

then we have the following analogue to "real" and "imaginary" part of a holomorphic function:

Let u be a function defined on $\mathbb{R}^n$ with values in $\mathcal{C}\ell_{n,0}$ then there exists a v defined on $\mathbb{R}^n$ with values in $\mathcal{C}\ell_{n,0}$ such that $w = u + e_{n+1}v$ are the boundary values of a monogenic function in the upper half space $\mathbb{R}_+^{n+1}$ and $v = Hu$.

This is based on $H^2 = -I$. Due to the lack of commutativity of the multiplication there is up to now no analogue to the Riemann mapping theorem and we don't know how to generalize the result to arbitrary domains. Nevertheless, it is also true for the unit ball which can be proved directly. The operator H is also a bounded and monotone operator which is also true for more general domains. Therefore monotonicity principles can be used to solve non-linear integral equations involving the Hilbert operator. More about these can be found in (Bernstein, 1999).

5. A singular integral equation and the transmission problem

Now we will have another view on the problem. The transmission problem connects the boundary values of a monogenic function inside the domain with boundary values of the considered monogenic function outside. To write it down we can use the Cauchy integral which describes monogenic functions and Plemelj-Sokhotzkij's formulae which give the connection to the boundary values. More precisely we have

Let Φ be a monogenic function in $\mathbb{R}^n \backslash \Gamma$ and Φ_- and Φ_+ its boundary values from the inside and the outside, then the generalized transmission problem

$$a(t)\Phi_-(t) = b(t)\Phi_+(t) + c(t) \text{ on } \Gamma$$

is equivalent to the singular integral equation involving the Cauchy integral

$$\frac{a(t) - b(t)}{2}\varphi(t) + \frac{a(t) + b(t)}{2}(S_\Gamma\varphi)(t) = c(t) \text{ on } \Gamma,$$

where $(S_\Gamma\varphi)(t) = \int_\Gamma 2e(t - \tau)n(\tau)\varphi(\tau)\,d\Gamma = H(n\varphi)(t)$.
In case of the lower (upper) half space and the unit sphere we have $H(n\varphi)(t) = -n(t)(H\varphi)(t)$ which allows a direct connection with the Hilbert operator.

The singular Cauchy integral on Lipschitz surfaces is well-discussed in (McIntosh, Li, Semmes, 1992).
Due to the non-commutativity of the multiplication we have in fact a much general RHP:

$$a_1(t)\varphi(t) + \varphi(t)a_2(t) + b_1(t)(S_\Gamma\varphi)(t) + (S_\Gamma\varphi)(t)b_2(t) = c(t) \text{ on } \Gamma$$

Also the transmission problem can be written as a right, left or mixed one, which means that the multiplication is done from the right, left or one from the left and one from the right. This problem and special cases are studied first by M.V. Shapiro and N.L. Vasilevski (1995), later in (Bernstein, 1996) and (Bernstein, 1999). We only want to mention the following result

THEOREM 2. *Let G be a bounded domain in $\mathbb{R}^n$ with Lipschitz boundary Γ. If $c(x), d(x) \in C(\Gamma, Cl_{0,n})(\mathbb{R})$ and $c(x)\overline{c(x)}$, $d(x)\overline{d(x)} \in \mathbb{R}$ for all $x \in \Gamma$ and $\inf_{x \in \Gamma} c(x)^2 - d(x)^2 > 0$ then:*
The singular integral equation

$$c(x)f(x) + d(x)(S_\Gamma f)(x) = h(x)$$

is uniquely solvable in $L_p(\Gamma, Cl_{0,n}(\mathbb{R}))$ for any $f \in L_p(\Gamma, Cl_{0,n}(\mathbb{R}), 1 < p < \infty$.

6. The Index

The most fascinating feature of the RHP is the connection between topological and analytical index. On one hand we have the generalization of the winding number to higher dimensions which is also called

wrapping number and on the other hand we have the Fredholm theory
of operators which involves the analytical index.

6.1. Wrapping number

An analogue to the winding number is described in several papers,
in a Clifford context it is done in (Sudbery, 1979), (Sommen, 1984),
(Habetha, 1986) and in connection with physics in
(Hestenes, Sobczyk, 1985).

DEFINITION 1. *Let $\mathcal{N}$ be a closed hypersurface in $\mathbb{R}^n$ then the wrapping number of $\mathcal{N}$ around the point $y \notin \mathcal{N}$ is defined by*

$$\#_y(\mathcal{N}) = \frac{\Gamma\left(\frac{n}{2}\right)}{2\pi^{n/2}} \int_{\mathcal{N}} \frac{(x-y) \cdot n(x)}{|x-y|^n} d\mathcal{N}(x).$$

where $n(x)$ is the outward-pointed unit normal at $x \in \mathcal{N}$.

If f is a one-to-one mapping of a closed hypersurface $\mathcal{N}'$ onto $\mathcal{N}$ in $\mathbb{R}^n$, we can define a wrapping number for f by identifying it with the wrapping number of $\mathcal{N}$. We have (see (Hestenes, Sobczyk, 1985))

$$\#_y(f) = \#_y(\mathcal{N}) = \deg\left[\frac{f(x')-y}{|f(x')-y|}\right]$$

and especially with $y = 0$ and $|f(x')| = 1$

$$\#_0(f) = \#_0(\mathcal{N}) = \deg f$$

where $\deg f$ denotes the mapping degree of f. Thus we have a connection between wrapping number and mapping degree. If we are able to connect mapping degree and singular integral operators we will have the analogue to the Noetherian theorems.

6.2. Symbol and Fredholmness

DEFINITION 2. *Let A be a mapping between Banach spaces. If at least one of the dimensions*

$$\alpha(A) = \dim \ker A \quad and \quad \beta(A) = \dim \operatorname{coker} A$$

is finite, their difference is called the index of A and denoted by

$$\operatorname{ind} A = \alpha(A) - \beta(A).$$

A normally solvable operator A is called a Fredholm operator if it has finite index.

The singular integral operator

$$(Au)(x) = a(x)u(x) + \int_{\mathbb{R}^n} K(x, x - y)u(y)\, dy,$$

where $K(x, x - y) = f(x, \frac{x-y}{|x-y|})|x - y|^{-n}$ has the symbol

$$\Phi(x, \xi) = a(x) + \hat{K}(x, \xi).$$

A is a Fredholm operator iff its symbol $\Phi(x, \xi)$ is invertible. From now on we only consider singular integral operators with paravector-valued symbols $\Phi(x, \xi)$. Thus the invertibility of the symbol is equivalent to

$$\inf_{x, \xi} |\Phi(x, \xi)\overline{\Phi(x, \xi)}| > 0.$$

Therefore, we set

$$\Psi(x, \xi) = \frac{\Phi(x, \xi)}{\sqrt{\Phi(x, \xi)\overline{\Phi(x, \xi)}}}$$

Now the symbol $\Psi(x, \xi) = \Psi_R(x, \xi) + i\Psi_I(x, \xi)$ is homotopic to

$$\chi(x, \xi) := \frac{\Psi_R(x, \xi)}{|\Psi_R(x, \xi)|}$$

and we have

$$\chi(x, \xi)\overline{\chi(x, \xi)} = 1, \quad \mathbb{R}^n \times S^{n-1} \to S^{2n-1}.$$

Due to a famous theorem of Hopf, see (Alexandroff, Hopf, 1935), such mappings classify homotopy classes. It is possible to show the mapping degree of this last mapping and the index of the singular integral operator with symbol $\chi(x, \xi)$ are homomorphisms of the Spoin-group into the integers such that if for some element of the Spoin-group with mapping degree zero the index of the associated singular integral operator is also zero. From this we get

THEOREM 3. *Let $\chi(x, \theta) \in \text{Spoin}_{2n-1}(\mathbb{R})$ be the symbol of singular integral operator. Then : The index of the singular integral operator with symbol $\chi(x, \theta)$ denoted by $\operatorname{ind}\chi$, is equal to μ-times the mapping degree of $\chi(x, \theta)\overline{\chi(x, \theta)} = 1 : \mathbb{R}^n \times S^{n-1} \to S^{2n-1}$, which is denoted by $\deg \chi$, i.e.*

$$\operatorname{ind}\chi = \mu \deg \chi = \mu \#_0(\chi\bar{\chi} = 1),$$

where μ is an integer.

References

Alexandroff, P., Hopf, H.: *Topology*. Springer Verlag, Berlin, 1935.

Bernstein, S.: 1996, 'On the left linear Riemann problem in Clifford analysis', *Bull. Belg. Math.Soc., Simon Stevin* Vol. no. 2, pp. 557–576

Bernstein, S.: 1999, 'The Quaternionic Riemann problem', in: Function Spaces, Proc. of the Third Conference on Function Spaces, Edwardsville, 1998, ed. by K. Jarosz, *Contemporary Mathematics* Vol. no. 232, pp. 69–84

Bernstein, S.: 1999, 'Nonlinear Singular Integral Equations Involving the Hilbert Transform in Clifford Analysis', *Zeitschrift für Analysis und ihre Anwendungen* Vol. no. 18(2), pp. 379–392

Brackx, F., Delanghe, R. and Sommen, F.: *Clifford Analysis*. Research Notes in Mathematics 76, Pitman Adv. Publ. Program, 1982.

Delanghe, R., Sommen, F. and Souček, V.: *Clifford algebra and Spinor-valued functions*. Kluwer Acad. Publ., Dordrecht, Boston, London, 1992

Gilbert, J. E. and Murray, M. A. M.: *Clifford Algebras and Dirac Operators in harmonic Analysis*. Cambridge studies in adv. math. Vol. 26, Cambridge University Press, Cambridge 1991

Gürlebeck, K. and Sprößig, W.: *Quaternionic and Clifford calculus for Engineers and Physicists*. Wiley & Sons Publ., 1997

Habetha, K.: 1986,'Eine Definition des Kronecker Indexes in $\mathbb{R}^{n+1}$ mit Hilfe der Clifford Analysis', *Zeitschrift für Analysis und ihre Anwendungen* Vol. no. 5(2), pp. 133–137

Gakhov, F. D.: *Boundary Value Problems*. Dover Publ. Inc., New York, 1990

Hestenes, D. and Sobczyk, G.: *Clifford algebra to Geometric Calculus*. D. Reidel Publ. Comp., Dordrecht, Holland, 1985

McIntosh, A., Li, C. and Semmes,S.: 1992, 'Convolution singular integrals on Lipschitz surfaces', *Journal of the American Mathematical Society* Vol. no. 5, pp.455–481

Muskhelishvili, N.I.: *Singular Integral Equations*. Dover Publ., Inc., New York, 1992

Shapiro, M.V. and Vasilevski, N.L.: 1995, 'Quaternionic ψ-holomorphic functions, singular integral operators and boundary value problems, I, II' *Complex Variables, Theory and Applications* Vol. no. 27, pp. 14–46 and 67–96

Stern, I.: 'Boundary value problems for generalized Cauchy–Riemann systems in the space.' *In: Kühnau, R. and Tutschke, W. (eds.), Boundary value and initial value problems in complex analysis*. Pitman Res. Notes in Math., pp.159–183

Sudbery, A.: 1979, 'Quaternionic analysis', *Math. Proc. Cambr. Phil. Soc.* Vol. no. 85, pp.199–225

Sommen, F.: 1984, 'Monogenic differential forms and homology theory.' *Proc. Royal Irish Academy* Vol. no. 84 A, pp.87–109

Vekua, I.N.: *Generalizes analytic functions*. Nauka, Moskow, 1959, (Russian)

Wegert, E.: *Nonlinear Boundary Value Problems for Holomorphic Functions and Singular Integral Equations*. Mathematical Research Vol. 65, Akademie Verlag Berlin, 1992

The Continuous Wavelet Transform in Clifford Analysis

Fred Brackx (fb@cage.rug.ac.be) and Frank Sommen
(fs@cage.rug.ac.be)*
Department of Mathematical Analysis, Ghent University

Abstract. A theory of higher dimensional continuous wavelet transforms in Clifford analysis is presented. The construction of Clifford-Hermite wavelets generates new generalized Hermite polynomials.

Keywords: continuous wavelet transform, Clifford analysis

Mathematics Subject Classification: 42B10, 44A15, 30G35

Dedicated to Richard Delanghe on the occasion of his 60th birthday.

1. Introduction

Wavelet analysis is a particular time- or space-scale representation of functions. In the last decade it has found numerous applications in mathematics and physics. Two of the main themes in wavelet theory are the continuous wavelet transform [CWT] and discrete orthonormal wavelets generated by multiresolution analysis . They enjoy more or less opposite properties and both have their specific field of application. The CWT plays an analogous rôle as the Fourier Transform and is a successful tool for the analysis of signals and feature detection in signals. The discrete wavelet transform (see e.g. [9]) is the analogue of the Discrete Fourier Transform and provides a powerful technique for e.g. data compression and signal reconstruction.

Most signals in practice are non-stationary and cover a wide range of frequencies; in general low frequency segments of a signal last a relatively long time, whereas high frequencies occur for a short moment. Classical Fourier analysis is inadequate for analysing such kind of signals: the information about the time localization of a given frequency is lost, it is unstable with respect to perturbation due to its global character and for reproducing even an almost flat signal it requires an infinite series.

For such non-stationary, inhomogeneous signals a time-frequency or a time-scale representation is far more adequate. Two parameters are

---

* Senior Research Associate, FWO, Belgium

9

F. Brackx et al. (eds.), Clifford Analysis and Its Applications, 9–26.

introduced, one indicating the time or the position in the signal, the second characterizing the scale. If in addition the transform has to be linear, which is a nontrivial assumption since there are also sesquilinear time-frequency representations (see [6]), it takes the form:

$$f(x) \mapsto F(a,b) = \int_{-\infty}^{+\infty} \overline{g^{a,b}(x)} \, f(x) \, dx,$$

where f is the signal and $g^{a,b}$ is the analysing function. Two specific transforms of this kind are particularly successful: the Windowed Fourier Transform, sometimes called the Gabor Transform, and the CWT.

The Windowed Fourier Transform was first introduced by Gabor in signal analysis on the real line (see [11]). For the family of analysing functions he used

$$g^{a,b}(x) = \exp(ix/a) \exp(x-b)^2 \quad ,$$

the window thus having a constant width and the number of oscillations in the window being governed by the parameter a.

The CWT on the real line is defined as the scalar product of the signal f with a transformed basic wavelet function g:

$$F(a,b) = T_g^{wav} f(a,b) = \frac{1}{\sqrt{a}} \int_{-\infty}^{+\infty} \overline{g(\frac{x-b}{a})} \, f(x) \, dx \quad .$$

The parameter $b \in \mathbb{R}$ indicates the time or the position of the wavelet, while the action of the parameter $a > 0$ is a dilation ($a > 1$) or a contraction ($a < 1$), the shape of the analysing function remaining unchanged. The equivalent expression in the frequency domain reads:

$$F(a,b) = \sqrt{a} \int_{-\infty}^{+\infty} \overline{\hat{g}(a\xi)} \, \hat{f}(\xi) \, \exp(ib\xi) \, d\xi,$$

where $\hat{f}$ denotes the Fourier transform of the function f.

The basic wavelet function g is a quite arbitrary L_2-function [a finite energy signal] which is well localized both in the time domain and in the frequency domain. In order to have the CWT satisfy a Parseval formula, this mother wavelet has to satisfy the so-called admissibility condition:

$$C_g = \int_{-\infty}^{+\infty} \frac{|\hat{g}(\xi)|^2}{|\xi|} \, d\xi < +\infty \quad .$$

In the case where g is also L_1, this admissibility condition implies that g has mean value zero, is oscillating, and decays to zero at infinity;

these properties explain the qualification as "wavelet" of this function g. Mostly the basic wavelet function also has a number of vanishing moments:

$$\int_{-\infty}^{+\infty} x^n \, g(x) \, dx = 0 \quad , \quad n = 0, 1, \ldots, N \quad ,$$

which means that the corresponding CWT will filter out polynomial behaviour of the signal up to degree N, making it more adequate at detecting singularities.

This filtering property of the CWT becomes more apparent when its definition is rephrased as the convolution

$$F(a, b) = h^{a,0} * f(b) \quad ,$$

where we have put

$$h(x) = \overline{g(-x)} \quad .$$

This shows that only those transforms subsist for which the analysing wavelet matches the signal. So the CWT performs a local filtering around the time b and the scale a.

Considering two finite energy signals f and g with respective CWT F and G, the following inner product may be introduced:

$$[F, G] = \frac{1}{C_g} \int_{-\infty}^{+\infty} \int_0^{+\infty} \overline{F(a,b)} G(a,b) \, \frac{da}{a^2} \, db \quad .$$

Taking into account the above mentioned admissibility condition of the wavelet function, the corresponding Parseval formula is readily obtained :

$$[F, G] = \langle f, g \rangle.$$

The CWT maps $L_2(\mathbb{R})$ into $L^2(\mathbb{R}_+ \times \mathbb{R}, C_g^{-1} a^{-2})$ but it is by no means a surjection onto this space. So there is a lot of freedom in constructing inversion formulae. However from the above Parseval formula it follows that if $f \in L_2(\mathbb{R})$ then:

$$f(x) = \frac{1}{C_g} \int_{-\infty}^{+\infty} \int_0^{+\infty} g^{a,b}(x) F(a,b) \, \frac{da}{a^2} \, db$$

$$= [\overline{g^{a,b}(x)}, \, F(a,b)]$$

to hold weakly in $L_2(\mathbb{R})$.

This means that the signal $f(x)$ may be reconstructed from its transform $F(a,b)$ or, in other words, that the CWT decomposes the signal $f(x)$ in terms of the analysing wavelets $g^{a,b}(x)$ with coefficients $F(a,b)$.

Moreover the image of $L_2(\mathbb{R})$ under the CWT may be characterized. It turns out to be a closed subspace of $L^2(\mathbb{R}_+ \times \mathbb{R}, C_g^{-1} a^{-2})$ and hence a Hilbert space with reproducing kernel or autocorrelation function given by

$$K(a, b; \tilde{a}, \tilde{b}) = T_g^{wav} g^{a,b}(x)(\tilde{a}, \tilde{b})$$

$$= \int_{-\infty}^{+\infty} \overline{g^{\tilde{a},\tilde{b}}(x)} \, g^{a,b}(x) dx \ .$$

Hence a function $F(a, b) \in L^2(\mathbb{R}_+ \times \mathbb{R}, C_g^{-1} a^{-2})$ is the CWT of a signal $f(x) \in L_2(\mathbb{R})$ iff it satisfies the reproducing property:

$$F(a, b) = [K(a, b; \tilde{a}, \tilde{b}), \ F(\tilde{a}, \tilde{b})]$$

$$= \frac{1}{C_g} \int_{-\infty}^{+\infty} \int_0^{+\infty} \overline{K(a, b; \tilde{a}, \tilde{b})} F(\tilde{a}, \tilde{b}) \, \frac{d\tilde{a}}{\tilde{a}^2} \, d\tilde{b} \ .$$

2. The Continuous Wavelet Transform in higher dimensions

The CWT may be extended to higher dimensions while still enjoying the same properties as in the one-dimensional case. Many wavelets are available for practical applications, often linked to a specific problem. A typical example is the m-dimensional Mexican hat or Marr wavelet:

$$g(\underline{x}) = -\Delta_{\underline{x}} \exp(-\frac{|\underline{x}|^2}{2}), \quad \underline{x} \in \mathbb{R}^m,$$

where $\Delta_{\underline{x}}$ denotes the Laplacian. This wavelet was originally introduced by Marr [8] and is used in image processing; also higher order Laplacians of the Gaussian function are used as wavelets.

Special attention should be paid to the two-dimensional CWT

$$F(a, \theta, \underline{b}) = \int_{\mathbb{R}^2} \overline{g(\frac{r_{-\theta}(\underline{x} - \underline{b})}{a})} \, f(\underline{x}) \, d\underline{x}$$

where the analysing wavelet g is translated by $\underline{b} \in \mathbb{R}^2$, dilated by $a > 0$ and rotated by an angle $\theta \in [0, 2\pi[$ (see [2]); it is an efficient tool for detecting oriented features of the signal provided the basic wavelet contains itself an intrinsic orientation.

In two previous papers ([4] and [5]) specific higher dimensional wavelet kernel functions and the corresponding CWT were constructed in the framework of Clifford analysis.

Using the radial Hermite polynomials introduced by F Sommen in [10]:

$$H_n(\underline{x}) = (-1)^n \exp\left(\frac{|\underline{x}|^2}{2}\right) \partial_{\underline{x}}^n \exp\left(-\frac{|\underline{x}|^2}{2}\right),$$

the resulting Clifford-Hermite wavelets obtained in [4]:

$$\psi_n(\underline{x}) = \exp\left(-\frac{|\underline{x}|^2}{2}\right) H_n(\underline{x})$$

$$= (-1)^n \partial_{\underline{x}}^n \exp\left(-\frac{|\underline{x}|^2}{2}\right)$$

offer a refinement of the Marr wavelets; they are $SO(m)$-invariant and real or vector valued and show vanishing moments of any order.

Making use of the generalized Hermite polynomials defined in [10] for $n = 0, 1, 2, \ldots$ and $k = 0, 1, 2, \ldots$ by the relation

$$\exp\left(\frac{|\underline{x}|^2}{2}\right)(-\partial_{\underline{x}})^n \left(\exp\left(-\frac{|\underline{x}|^2}{2}\right) P_k(\underline{x})\right) = H_{n,k}(\underline{x}) P_k(\underline{x})$$

where $P_k(\underline{x})$ is a monogenic homogeneous polynomial of degree k, the generalized Clifford-Hermite wavelets obtained in [5] read:

$$\psi_{n,k}(\underline{x}) = \exp\left(-\frac{|\underline{x}|^2}{2}\right) H_{n,k}(\underline{x}) P_k(\underline{x})$$

$$= (-1)^n \partial_{\underline{x}}^n \left(\exp\left(-\frac{|\underline{x}|^2}{2}\right) P_k(\underline{x})\right) .$$

The corresponding generalized Clifford-Hermite Continuous Wavelet Transform (GCHCWT) in $I\!R^m$ offers the possibility of a pointwise and a directional analysis of signals.

In this paper we present a more unifying theory of the CWT in Clifford analysis.

3. Clifford analysis

Clifford analysis (see e.g.[3]) offers a function theory which is a higher dimensional analogue of the function theory of the holomorphic functions of one complex variable.

Consider functions defined in $I\!\!R^m$ $(m > 1)$ and taking values in the Clifford algebra $I\!\!R_m$ or its complexification C_m. If $(e_1, \ldots, e_m)$ is an orthonormal basis of $I\!\!R^m$, the non-commutative multiplication in the Clifford algebra is governed by the rule:

$$e_j e_k + e_k e_j = -2\delta_{jk}, \quad j, k = 1, \ldots, m.$$

Two anti-involutions on the Clifford algebra are important.
Conjugation is defined as the anti-involution for which

$$\overline{e_j} = -e_j, \quad j = 1, \ldots, m$$

with the additional rule

$$\bar{i} = -i$$

in the case of C_m.
Inversion is defined as the anti-involution for which

$$e_j^\dagger = e_j, \quad j = 1, \ldots, m.$$

The Euclidean space $I\!\!R^m$ is embedded in the Clifford algebras $I\!\!R_m$ and C_m by identifying $(x_1, \ldots, x_m)$ with the vector variable $\underline{x}$ given by

$$\underline{x} = \sum_{j=1}^m e_j x_j.$$

Notice that

$$\underline{x}^2 = - <\underline{x}, \underline{x}> = -|\underline{x}|^2.$$

If $I\!\!R_m^k$ denotes the subspace of k-vectors, i.e. the space spanned by the products of k different basis vectors, then the even subalgebra $I\!\!R_m^+$ of $I\!\!R_m$ is defined by

$$I\!\!R_m^+ = \sum_{k \text{ even}} \oplus I\!\!R_m^k.$$

The Clifford group $\Gamma(m)$ of $I\!\!R_m$ consists of those invertible elements m in $I\!\!R_m$ for which the action $\overline{m} \, \underline{x} \, m$ on a vector $\underline{x}$ is again a vector. Its subgroup Γ^+ is the intersection of Γ with the even subalgebra $I\!\!R_m^+$. The Spin-group $\text{Spin}(m)$ is the subgroup of Γ^+ of the elements $m \in \Gamma^+$ for which $m \, m^\dagger = 1$. The Spin-group is a two-fold covering group of the rotation group $SO(m)$.

An $I\!\!R_m$- or C_m-valued function $F(x_1, \ldots, x_m)$ is called (left-) monogenic in an open region Ω of $I\!\!R^m$, if in Ω :

$$\partial_{\underline{x}} F = 0.$$

Here $\partial_{\underline{x}}$ is the Dirac operator

$$\partial_{\underline{x}} = \sum_{j=1}^{m} e_j \partial_{x_j} ,$$

which splits the Laplacian in $\mathbb{R}^m$:

$$\Delta_{\underline{x}} = -\partial_{\underline{x}}^2 .$$

In the sequel the monogenic homogeneous polynomials, or spherical monogenics for short, will play an important rôle. We call $\mathcal{M}_k$ the (right) module of the (left) monogenic homogeneous polynomials of degree k. Its dimension is given by

$$dim \mathcal{M}_k = \frac{(m+k-2)!}{(m-2)! \, k!} .$$

For $R_k \in \mathcal{M}_k$ we have the following fundamental formulae:

$$(-\partial_{\underline{x}})(\underline{x} \, R_k) = (m+2k)R_k$$

and

$$(-\partial_{\underline{x}})(R_k \, \underline{x}) = -(m-2)R_k \quad .$$

That Clifford analysis is a refinement of harmonic analysis is nicely illustrated by the well-known fact that the classical space $\mathcal{H}_{k+1}$ of spherical harmonics, i.e. harmonic homogeneous polynomials of degree $(k+1)$, can be decomposed as

$$\mathcal{H}_{k+1} = \mathcal{M}_{k+1} \oplus \underline{x} \, \mathcal{M}_k \quad .$$

Indeed, putting for each $S_{k+1} \in \mathcal{H}_{k+1}$:

$$P_k = \partial_{\underline{x}} S_{k+1} \quad ,$$

this P_k is left and right monogenic and moreover

$$S_{k+1} = R_{k+1} - \frac{1}{m+2k} \underline{x} \, P_k$$

for a certain $R_{k+1} \in \mathcal{M}_{k+1}$. By conjugation we get

$$S_{k+1} = \overline{R}_{k+1} - \frac{1}{m+2k} P_k \, \underline{x}$$

where $\overline{R}_{k+1}$ is a right spherical monogenic. In view of

$$\underline{x} \, P_k = \underline{x} \, \partial_{\underline{x}} S_{k+1} = - < \underline{x}, \partial_{\underline{x}} > S_{k+1} + \underline{x} \wedge \partial_{\underline{x}} S_{k+1}$$

and

$$P_k \, \underline{x} = \partial_{\underline{x}} \, S_{k+1} \, \underline{x} = - <\underline{x}, \partial_{\underline{x}}> S_{k+1} - \underline{x} \wedge \partial_{\underline{x}} \, S_{k+1}$$

we have that

$$\underline{x} \, P_k + P_k \, \underline{x} = -2(k+1) \, S_{k+1}$$

is scalar valued, which leads to the formula:

$$R_{k+1} = -\frac{1}{2k+2} \left(\frac{m-2}{m+2k} \underline{x} \, P_k + P_k \, \underline{x} \right) \quad .$$

This means that the $\mathcal{M}_{k+1}$-component in the decomposition of $\mathcal{H}_{k+1}$ may be expressed as a linear combination of $\underline{x} \, P_k$ and $P_k \, \underline{x}$. Moreover, as is the case for $\underline{x} \, P_k$ and $P_k \, \underline{x}$, R_{k+1} is the sum of a scalar and a bivector part:

$$R_{k+1} = S_{k+1} - \frac{1}{m+2k} <\underline{x}, \partial_{\underline{x}}> S_{k+1} + \frac{1}{m+2k} \underline{x} \wedge \partial_{\underline{x}} \, S_{k+1}$$

or

$$R_{k+1} = \frac{m+k-1}{m+2k} S_{k+1} + \frac{1}{m+2k} \underline{x} \wedge \partial_{\underline{x}} \, S_{k+1} \quad .$$

4. Clifford wavelets

We consider a Clifford algebra valued function ψ which for the time being we only assume to be in $L_1 \cap L_2(\mathbb{R}^m)$. The associated continuous family of wavelets

$$\psi^{a,\underline{b},s}(\underline{x}) = \frac{1}{a^{m/2}} \, s \, \psi \left(\frac{\overline{s}(\underline{x} - \underline{b})s}{a} \right) \overline{s} \quad ,$$

where $a \in \mathbb{R}_+$, $\underline{b} \in \mathbb{R}^m$ and $s \in \mathrm{Spin}(m)$, originates from the basic wavelet function ψ by dilation, translation and spinor-rotation. By a series of easily proved properties their Fourier transform is obtained as follows.

Property 1

$$\psi^{a,\underline{b},s}(\underline{x}) = \psi^{a,0,s}(\underline{x} - \underline{b})$$

Property 2

$$\widehat{\psi}^{a,\underline{b},s}(\underline{\xi}) = \exp(-i <\underline{\xi}, \underline{b}>) \, \widehat{\psi}^{a,0,s}(\underline{\xi})$$

Property 3

$$\widehat{\psi}^{a,0,s}(\underline{\xi}) = \widehat{\psi}^{a,0,1}(\overline{s}\underline{\xi}s)$$

Property 4

$$\widehat{\psi}^{a,0,1}(\overline{s}\underline{\xi}s) = a^{m/2}\, s\, \widehat{\psi}(a\overline{s}\underline{\xi}s)\, \overline{s}$$

Property 5

$$\widehat{\psi}^{a,\underline{b},s}(\underline{\xi}) = \exp(-i < \underline{\xi}, \underline{b} >)\, a^{m/2}\, s\, \widehat{\psi}(a\overline{s}\underline{\xi}s)\, \overline{s}$$

The corresponding Clifford Continuous Wavelet Transform [CCWT] applies to functions $f \in L_2(I\!\!R^m)$ by

$$
\begin{aligned}
T_\psi f(a, \underline{b}, s) &= F_\psi(a, \underline{b}, s) \\
&= \langle \psi^{a,\underline{b},s}, f \rangle \\
&= \int_{I\!\!R^m} \overline{\psi}^{a,\underline{b},s}(\underline{x})\, f(\underline{x})\, d\underline{x}
\end{aligned}
$$

This definition can be rewritten in terms of the Fourier transform as

$$
\begin{aligned}
F_\psi(a, \underline{b}, s) &= \frac{1}{(2\pi)^m}\langle\, \widehat{\psi}^{a,\underline{b},s}(\underline{\xi})\, ,\, \widehat{f}(\underline{\xi})\, \rangle \\
&= \frac{1}{(2\pi)^m}\langle\, a^{m/2}\, \exp(-i\langle \underline{b}, \underline{\xi}\rangle)\, s\, \widehat{\psi}(a\overline{s}\underline{\xi}s)\, \overline{s}\, ,\, \widehat{f}(\underline{\xi})\, \rangle \\
&= \frac{a^{m/2}}{(2\pi)^m}\, s\, \overline{\widehat{\psi}(a\overline{s}\underline{\xi}s)}\, \overline{s}\, \widehat{\widehat{f}(\underline{\xi})}(-\underline{b})\, .
\end{aligned}
$$

It is clear that the CCWT will map $L_2(I\!\!R^m)$ into a weighted L_2-space on $I\!\!R_+ \times I\!\!R^m \times \text{Spin}(m)$ for some weight function still to be determined. This weight function has to be chosen in such a way that the CCWT is an isometry, or in other words that the Parseval formula should hold.

Introducing the inner product

$$[F_\psi, G_\psi] = \frac{1}{C_\psi} \int_{\text{Spin}(m)} \int_{I\!\!R^m} \int_0^{+\infty} \overline{F_\psi(a, \underline{b}, s)}\, G_\psi(a, \underline{b}, s)\, \frac{da}{a^{m+1}}\, d\underline{b}\, ds \quad ,$$

where ds stands for the Haar measure on $\text{Spin}(m)$, we search for the constant C_ψ in order to have the Parseval formula

$$\langle f, g \rangle = [F_\psi, G_\psi]$$

fulfilled.

We have consecutively $[F_\psi, G_\psi] =$

$$= \frac{1}{C_\psi \, (2\pi)^{2m}} \int_{\mathrm{Spin}(m)} \int_{\mathbb{R}^m} \int_0^{+\infty} s \, \overline{\widehat{\psi}(a\bar{s}\underline{\xi}s)} \, \bar{s} \widehat{\overline{f}(\underline{\xi})(-\underline{b})}$$

$$s \, \overline{\widehat{\psi}(a\bar{s}\underline{\xi}s)} \, \bar{s} \, \widehat{g}(\underline{\xi})(-\underline{b}) \, \frac{da}{a} \, d\underline{b} \, ds$$

$$= \frac{1}{C_\psi \, (2\pi)^{2m}} \int_{\mathrm{Spin}(m)} \int_{\mathbb{R}^m} \int_0^{+\infty} \overline{\widehat{f}(\underline{\xi})} \, s \, \widehat{\psi}(a\bar{s}\underline{\xi}s) \, \bar{s}$$

$$s \, \overline{\widehat{\psi}_{n,k}(a\bar{s}\underline{\xi}s)} \, \bar{s} \, \widehat{g}(\underline{\xi}) \, \frac{da}{a} \, d\underline{\xi} \, ds \, .$$

Now making the assumption that

$$\widehat{\psi} \, \overline{\widehat{\psi}} \quad \text{is real valued},$$

and taking into account that $\bar{s} \, s = 1$, and putting

$$\int_{\mathrm{Spin}(m)} \int_0^{+\infty} s \, \widehat{\psi}(a\bar{s}\underline{\xi}s) \, \overline{\widehat{\psi}(a\bar{s}\underline{\xi}s)} \, \bar{s} \, \frac{da}{a} \, ds = C_\psi,$$

we get the desired result:

$$[F_\psi, G_\psi] = \frac{1}{(2\pi)^m} \int_{\mathbb{R}^m} \overline{\widehat{f}(\underline{\xi})} \, \widehat{g}(\underline{\xi}) \, d\underline{\xi}$$

$$= \frac{1}{(2\pi)^m} \langle \widehat{f}, \widehat{g} \rangle.$$

$$= \langle f, g \rangle.$$

This means that if the function ψ satisfies all assumptions made, the CCWT will be an isometry between two L_2-spaces. The above relation defining the constant C_ψ is called the admissibility condition for the Clifford wavelets, and the constant C_ψ involved is called the admissibility constant.

By means of the substitution

$$\underline{\xi} = \frac{r}{a} \, \underline{\omega} \quad , \quad \underline{\omega} \in S^{m-1}$$

and taking into account that $\overline{s}\underline{\omega}s = \eta \in S^{m-1}$ for all $\underline{\omega} \in S^{m-1}$, the admissibility condition may be simplified to

$$C_\psi = \int_0^{+\infty} \int_{S^{m-1}} \widehat{\psi}(r\underline{\eta}) \, \overline{\widehat{\psi}}(r\underline{\eta}) \, d\underline{\eta} \, \frac{dr}{r}$$

$$= \int_{I\!\!R^m} \widehat{\psi}(\underline{u}) \, \overline{\widehat{\psi}}(\underline{u}) \, \frac{d\underline{u}}{|\underline{u}|^m}$$

$$= \int_{I\!\!R^m} \frac{|\widehat{\psi}(\underline{u})|^2}{|\underline{u}|^m} \, d\underline{u} \quad .$$

So we have proved

THEOREM

A Clifford algebra valued function $\psi \in L_2 \cap L_1(I\!\!R^m)$ is a basic wavelet function if it satisfies in frequency space the following conditions:

(i)

$$\widehat{\psi} \, \overline{\widehat{\psi}} \text{ is real valued;}$$

(ii)

$$\int_{I\!\!R^m} \frac{|\widehat{\psi}(\underline{u})|^2}{|\underline{u}|^m} \, d\underline{u} \ < \ +\infty .$$

Notice that the admissibility condition (ii) implies that the wavelet function ψ has zero momentum:

$$\int_{I\!\!R^m} \psi(\underline{x}) \, d\underline{x} \ = \ 0 \quad .$$

It should be emphasized that T_ψ is no surjection onto $L^2(I\!\!R_+ \times I\!\!R^m \times \text{Spin}(m), C_\psi^{-1} \, a^{-(m+1)} \, da \, d\underline{b} \, ds)$. So as in the one-dimensional case, there is a lot of freedom in constructing inversion formulae. However from the above Parseval formula it follows that if $f \in L_2(I\!\!R^m)$ and $F_\psi(a, \underline{b}, s) = T_\psi f(\underline{x})$ then:

$$f(\underline{x}) = \frac{1}{C_\psi} \int_{\text{Spin}(m)} \int_{I\!\!R^m} \int_0^{+\infty} \overline{\psi}^{a,\underline{b},s}(\underline{x}) \, F_\psi(a, \underline{b}, s) \, \frac{da}{a^{m+1}} \, d\underline{b} \, ds$$

$$= [\psi^{a,\underline{b},s}(\underline{x}) \, , \, F_\psi(a, \underline{b}, s)]$$

to hold weakly in $L_2(I\!\!R^m)$.
This means that the CCWT decomposes the signal $f(\underline{x})$ in terms of the

analyzing wavelets $\psi^{a,\underline{b},s}(\underline{x})$ with coefficients $F_\psi(a,\underline{b},s)$. The CCWT thus offers an analysis of a signal in higher dimensional space which is pointwise as as well directional.

5. Two-sided Hermite wavelets

Now we introduce a new family of Clifford wavelet functions which encompasses the notions of Marr wavelets on the one side and the Clifford-Hermite wavelets on the other, both mentioned in section 1.

We define for $p = 0,1,2,\ldots$, $q = 0,1,2,\ldots$ and $k = 0,1,2,\ldots$, excluding the case where $p = q = k = 0$:

$$\psi_{p,q,k}(\underline{x}) = (-\partial_{\underline{x}})^p \left(\exp\left(-\frac{|\underline{x}|^2}{2}\right) P_k(\underline{x})\right) (-\partial_{\underline{x}})^q$$

where $P_k(\underline{x}) = \partial_{\underline{x}} S_{k+1}$ is a two-sided monogenic homogeneous polynomial of degree k, S_{k+1} being a spherical harmonic of degree $(k+1)$. Notice that $P_k(\underline{x})\overline{P}_k(\underline{x})$ is indeed real valued.

We now discuss the four cases depending upon the parity of the indices p and q, bearing in mind that $\partial_{\underline{x}}^2 = -\Delta_{\underline{x}}$ is a real differential operator.

Case A : p and q both even

If $p = 2p'$ and $q = 2q'$ then

$$\begin{aligned}
\psi_{2p',2q',k} &= \psi_{2p'+2q',0,k} \\
&= \psi_{0,2p'+2q',k} \\
&= (-\partial_{\underline{x}})^{2p'+2q'} \left(\exp\left(-\frac{|\underline{x}|^2}{2}\right) P_k\right) \\
&= \exp\left(-\frac{|\underline{x}|^2}{2}\right) H_{2p'+2q',k}(\underline{x}) P_k(\underline{x})
\end{aligned}$$

where $H_{2p'+2q',k}$ are the well known scalar valued generalized Hermite polynomials of even degree.

Case B : p odd and q even

If $p = 2p' + 1$ and $q = 2q'$ then

$$
\begin{aligned}
\psi_{2p'+1,2q',k} &= \psi_{2p'+2q'+1,0,k} \\
&= \psi_{1,2p'+2q',k} \\
&= (-\partial_{\underline{x}})\, \psi_{0,2p'+2q',k} \\
&= \exp\left(-\frac{|\underline{x}|^2}{2}\right)(\underline{x} - \partial_{\underline{x}})\, H_{2p'+2q',k}\, P_k \\
&= \exp\left(-\frac{|\underline{x}|^2}{2}\right) H_{2p'+2q'+1,k}\, P_k \\
&= \exp\left(-\frac{|\underline{x}|^2}{2}\right) \widetilde{H}_{2p'+2q',k}\, \underline{x}\, P_k
\end{aligned}
$$

where we rediscover the vector valued generalized Hermite polynomials of odd degree:

$$
H_{2p'+2q'+1,k} = (\underline{x} - \partial_{\underline{x}})\, H_{2p'+2q',k}
$$

or

$$
H_{2p'+2q'+1,k} = \widetilde{H}_{2p'+2q',k}\, \underline{x} \quad ,
$$

having introduced the scalar polynomials $\widetilde{H}_{2p'+2q',k}$ of even degree.

Case C : p even and q odd

If $p = 2p'$ and $q = 2q' + 1$ then

$$
\begin{aligned}
\psi_{2p',2q'+1,k} &= \psi_{2p'+2q',1,k} \\
&= \psi_{0,2p'+2q'+1,k} \\
&= \psi_{0,2p'+2q',k}\, (-\partial_{\underline{x}}) \\
&= \left(\exp\left(-\frac{|\underline{x}|^2}{2}\right) H_{2p'+2q',k}\, P_k\right)(-\partial_{\underline{x}}) \\
&= \exp\left(-\frac{|\underline{x}|^2}{2}\right) P_k\, H_{2p'+2q'+1,k} \\
&= \exp\left(-\frac{|\underline{x}|^2}{2}\right) \widetilde{H}_{2p'+2q',k}\, P_k\, \underline{x}
\end{aligned}
$$

Case D : p and q both odd

If $p = 2p' + 1$ and $q = 2q' + 1$ then $\psi_{2p'+1,2q'+1,k} =$

$$= \psi_{2p'+2q'+1,1,k}$$

$$= \psi_{1,2p'+2q'+1,k}$$

$$= (-\partial_{\underline{x}})\,\psi_{0,2p'+2q'+1,k}$$

$$= (-\partial_{\underline{x}}) \left(\exp\left(-\frac{|\underline{x}|^2}{2}\right) \tilde{H}_{2p'+2q',k}\,P_k\,\underline{x} \right)$$

$$= \exp\left(-\frac{|\underline{x}|^2}{2}\right) \left((\underline{x} - \partial_{\underline{x}})\,\tilde{H}_{2p'+2q',k}\,P_k\,\underline{x} - (m-2)\,\tilde{H}_{2p'+2q',k}\,P_k \right)$$

which leads to the introduction of the vector valued polynomials of odd degree:

$$\tilde{H}_{2p'+2q'+1,k} = (\underline{x} - \partial_{\underline{x}})\,\tilde{H}_{2p'+2q',k} \quad .$$

Summarizing, the new Clifford wavelets lead to three families of special functions coinciding with or closely related to the generalized Hermite polynomials introduced by F Sommen in [10]:

(i) the scalar valued $H_{2n,k}(\underline{x})$

(ii) the vector valued $H_{2n+1,k}(\underline{x}) = (\underline{x} - \partial_{\underline{x}})\,H_{2n,k}(\underline{x}) = \underline{x}\,\tilde{H}_{2n,k}(\underline{x})$

(iii) the vector valued $\tilde{H}_{2n+1,k}(\underline{x}) = (\underline{x} - \partial_{\underline{x}})\,\tilde{H}_{2n,k}(\underline{x})$

We have e.g.:

$$H_{0,k}(\underline{x}) = 1$$
$$H_{2,k}(\underline{x}) = \underline{x}^2 + 2k + m$$
$$H_{4,k}(\underline{x}) = \underline{x}^4 + 2(2k + m + 2)\underline{x}^2 + (2k + m)(2k + m + 2)$$
$$\text{etc.}$$

and

$$H_{1,k}(\underline{x}) = \underline{x}\,\tilde{H}_{0,k}(\underline{x}) = \underline{x}$$
$$H_{3,k}(\underline{x}) = \underline{x}\,\tilde{H}_{2,k}(\underline{x}) = \underline{x}\,(\underline{x}^2 + (2k + m + 2))$$
$$H_{5,k}(\underline{x}) = \underline{x}\,\tilde{H}_{4,k}(\underline{x}) = \underline{x}\,(\underline{x}^4 + 2(2k + m + 2)\underline{x}^2$$
$$+ (2k + m + 2)(2k + m + 4))$$
$$\text{etc.}$$

and also

$$\tilde{H}_{1,k}(\underline{x}) = \underline{x}$$
$$\tilde{H}_{3,k}(\underline{x}) = \underline{x}\,(\underline{x}^2 + (2k + m + 4))$$

etc.

Taking into account that

$$\exp\left(-\frac{|\underline{x}|^2}{2}\right) H_{2n+2,k}(\underline{x})\,P_k(\underline{x})$$

$$= (-\partial_{\underline{x}})^{2n+2}\left(\exp\left(-\frac{|\underline{x}|^2}{2}\right) P_k\right)$$

$$= (-\partial_{\underline{x}})\left(\exp\left(-\frac{|\underline{x}|^2}{2}\right) \tilde{H}_{2n,k}\,\underline{x}\,P_k\right)$$

$$= \exp\left(-\frac{|\underline{x}|^2}{2}\right)\left((\underline{x} - \partial_{\underline{x}})\,\tilde{H}_{2n,k}\,\underline{x}\,P_k + (m + 2k)\,\tilde{H}_{2n,k}P_k\right)$$

we also find the relation

$$H_{2n+2,k} = \tilde{H}_{2n+1,k}\,\underline{x} + (m + 2k)\,\tilde{H}_{2n,k}$$

which is nothing else but the recurrence relation in disguise for the generalized Hermite polynomials of even degree :

$$H_{2n+2,k} = (\underline{x} - \partial_{\underline{x}})\,H_{2n+1,k} - 2k\frac{\underline{x}}{|\underline{x}^2|}\,H_{2n+1,k} \quad.$$

6.   The generalized Marr wavelets

The Clifford wavelets introduced in the foregoing section are now shown to be related to the Marr wavelets, or better to their generalization defined for all $N \in I\!N$ by:

$$\phi_{N,k+1}(\underline{x}) = (-\Delta_{\underline{x}})^N\left(\exp\left(-\frac{|\underline{x}|^2}{2}\right) S_{k+1}\right)\quad,$$

where S_{k+1} is a spherical harmonic of degree $(k + 1)$.
In view of the observations made in section 3 on spherical harmonics, we have consecutively:

$$\phi_{N,k+1} = (-\partial_{\underline{x}})^{2N} \left(\exp\left(-\frac{|\underline{x}|^2}{2}\right) S_{k+1} \right)$$

$$= -\frac{1}{2(k+1)} (-\partial_{\underline{x}})^{2N} \left(\exp\left(-\frac{|\underline{x}|^2}{2}\right) (\underline{x} P_k + P_k \underline{x}) \right)$$

$$= -\frac{1}{2(k+1)} (\psi_{2N+1,0,k} + \psi_{2N,1,k})$$

thus meaning that the Clifford two-sided Hermite wavelets and the associated Clifford special functions are really fundamental for the higher dimensional CWT.

7.   The wavelet differential equation

If $p = q = k = 0$ it is well known that the Gaussian function

$$\psi_{0,0,0}(\underline{x}) = \exp\left(-\frac{|\underline{x}|^2}{2}\right)$$

has a non-vanishing zero momentum and thus cannot be used as a basic wavelet function.

As was already pointed out in [5], by introducing the time variable $t > 0$, the function

$$h(\underline{x}) = \frac{1}{(2\pi)^m} \psi_{0,0,0}(\underline{x})$$

generates the family of functions

$$h_t(\underline{x}) = \frac{1}{(2t)^{m/2}} h\left(\frac{\underline{x}}{\sqrt{2t}}\right)$$

which are basic in the theory of the higher dimensional windowed Fourier or Gabor transform (see [4]).

Moreover the function $h_t(x)$ is the fundamental solution of the heat operator in $\mathbb{R}^m$:

$$\partial_t + \partial_{\underline{x}}^2 = \partial_t - \Delta_{\underline{x}} .$$

This fact inspires the following calculation. First observe that

$$(\partial_t + \partial_{\underline{x}}^2) \left(\exp\left(-\frac{|\underline{x}|^2}{4t}\right) P_k\left(\frac{\underline{x}}{\sqrt{2t}}\right) \right) =$$

$$\frac{m-k}{2t} \left(\exp\left(-\frac{|\underline{x}|^2}{4t}\right) P_k\left(\frac{\underline{x}}{\sqrt{2t}}\right) \right)$$

holds, in view of the homogeneous and monogenic, and thus harmonic, character of the inner spherical monogenics P_k . Hence

$$(\partial_t + \partial_{\underline{x}}^2)\left(t^{-\frac{m-k}{2}}\,\exp\left(-\frac{|\underline{x}|^2}{4t}\right)\,P_k\left(\frac{\underline{x}}{\sqrt{2t}}\right)\right) = 0$$

and also

$$(\partial_t + \partial_{\underline{x}}^2)\left(t^{-\frac{p+q+m-k}{2}}\,\psi_{p,q,k}\left(\frac{\underline{x}}{\sqrt{2t}}\right)\right) = 0$$

and consequently

$$(\partial_t + \partial_{\underline{x}}^2)\left(t^{-\frac{2p+2q+m-2k}{4}}\,\psi_{p,q,k}^{\sqrt{2t},\underline{b},s}\left(\frac{\underline{x}}{\sqrt{2t}}\right)\right) = 0$$

which leads to the modified heat equation

$$\left(\partial_t - \Delta_{\underline{b}} - \frac{m + 2(p+q-k)}{4t}\right)\,\psi_{p,q,k}^{\sqrt{2t},\underline{b},s}(\underline{x}) = 0\,.$$

Notice that this differential equation still depends upon the degree of the polynomial factor in the wavelet $\psi_{p,q,k}^{\sqrt{2t},\underline{b},s}$. Introducing an extra variable the umbrella wavelets

$$\Psi_{p,q,k}^{\sqrt{2t},\underline{b},s}(x_0, \underline{x}) = \exp((2(p+q-k)+m)x_0)\,\psi_{p,q,k}^{\sqrt{2t},\underline{b},s}(\underline{x})$$

satisfy the so-called wavelet differential equation

$$\left(\partial_t - \Delta_{\underline{b}} - \frac{1}{4t}\partial_{x_0}\right)\,\Psi_{p,q,k}^{\sqrt{2t},\underline{b},s}(x_0, \underline{x})\,.$$

REFERENCES

1. L. Andersson, B. Jawerth and M. Mitrea : *The Cauchy singular integral operator and Clifford wavelets*, in: J. J. Benedetto and M. W. Frazier (eds.) : Wavelets: mathematics and applications, CRC Press, 1994, 525 - 546

2. J.-P. Antoine, R. Murenzi and P. Vandergheynst : *Two-dimensional directional wavelets in image processing*, Int. J. of Imaging Systems and Technology, 7, 152-165

3. F. Brackx, R. Delanghe and F. Sommen : *Clifford analysis*, Pitman Publ., 1982

4. F. Brackx and F. Sommen : *Clifford-Hermite Wavelets in Euclidean Space*, Journal of Fourier Analysis and Applications, volume 6, no 3, 2000, 299-310

5. F. Brackx and F. Sommen : *Clifford-Hermite Wavelets in Euclidean Space*, to appear in the Proceedings of the Conference on *Dirac Operators and Applications*, Cetraro (Italy), October 4-10, 1998

6. L. Cohen : General phase-space distribution functions, J. Math. Phys., 7, 781-786, 1966

7. I. Daubechies : *Ten Lectures on Wavelets*, SIAM, Philadelphia, 1992

8. D. Marr : *Vision*, Freeman, San Francisco, 1982

9. O. Rioul and M. Vetterli : Wavelets and signal processing, IEEE SP Magazine, October 1991, 14-38

10. F. Sommen : *Special Functions in Clifford analysis and Axial Symmetry*, Journal of Math. Analysis and Applications, 130, no 1, 1988, 110-133

11. D. Walnut : *Application of Gabor and wavelet expansions to the Radon transform*, Probabilistic and Stochastic Methods in Analysis, with Applications, NATO ASI Series, J. Byrnes et al., eds., Kluwer Academic Publishers, The Netherlands, 1992, 187-205

Automated Symbolic Computation in Spin Geometry

Thomas Branson (branson@math.uiowa.edu)
Department of Mathematics, University of Iowa

Abstract. We describe a program of computing identities in tensor-spinor calculus which are beyond the range of hand calculation, but are nevertheless worth knowing.

Keywords: Symbolic computation, spinors

Mathematics Subject Classification: 53A30, 53C27, 68W30

Dedicated to Richard Delanghe on the occasion of his 60th birthday.

1. Introduction

Symbolic computation in tensor-spinor calculus has produced identities and formulas that play central roles in physical field theories, and in the global geometric and harmonic analysis of manifolds. These abstract index computations are closely related to weight-theoretic considerations in group representation theory, and the two approaches can often be used effectively in tandem. For example, representation-theoretic arguments often reveal which tensor-spinor calculations are likely to yield useful information, as well as the qualitative form of this information. This is the case not only for linear problems, but also for nonlinear problems which arise from the study of linear differential operators (Branson, 1996). The strategy behind an important calculation often requires a certain inspiration, but each calculation also has a purely mechanical side which can sometimes be quite daunting, even to an expert practitioner.

We claim that many worthwhile results are lurking at and beyond the frontier of reasonable hand calculations. We would like to describe a program to pursue the identities and formulas that will support future results. We are interested in several types of symbolic calculations, but all are motivated by a core list of problems in fundamental Physics and Physical Mathematics.

F. Brackx et al. (eds.), Clifford Analysis and Its Applications, 27–38.

2. Tensor-spinor calculus

The calculations in question require a tensor calculus package with correct handling of dummy indices, covariant derivatives, curvature identities, symmetries, and dependence on an arbitrary dimension n. We have used the Ricci package created by Jack Lee (Lee, 1992-2000) (which runs under *Mathematica*), together with some collections of rules and relations which we have designed to handle issues like conformal variation and the Weyl tensor (Weyl.m), and spinors and Clifford algebra (Clifford.m).

These calculations are *interactive*; that is, one does not attempt to write a program which will do a given calculation without intervention. Instead, the idea is to look at intermediate results and make decisions about what the package should try next. Ideally, this is done without micromanagement: One gets an idea of which sequences of steps are likely to manipulate an expression in a certain direction. But ideally, one would like to avoid very specific instructions, like an instruction to interchange certain covariant derivatives in a certain term. An alternative explanation of the same philosophy is that one really *is* producing a program that does not require intervention, and that the interactive session is merely a debugging phase. Indeed, after stumbling toward a conclusion in a truly interactive session, it is desirable to then produce a clean session which is well documented and aims itself directly at the ultimate answer.

A feature of computations with spinors is that one is always dealing with the Clifford section γ of the bundle $TM \otimes \text{End}(\Sigma M)$ on a manifold M. Here TM is the tangent and ΣM the spinor bundle. (The Clifford section is the natural extension of Dirac's concept of gamma matrices to the general curved space setting.) The principal properties of γ are the *Clifford relations*

$$\gamma^i \gamma^j + \gamma^j \gamma^i = -2g^{ij}, \tag{1}$$

and the *compatibility* of the spin connection with γ:

$$\nabla \gamma = 0.$$

The indices in (1) are tensor indices. Here, as in most hand calculations, spinor indices are suppressed. In automated computation, however, it is useful to make spinor indices explicit, since the interlocking of indices affords a natural way to keep track of the order in iterated noncommutative products. For example,

$$\gamma_k{}^T{}_U \gamma^{iP}{}_Q \gamma^{jS}{}_T \gamma^{kQ}{}_S = (\gamma^i \gamma^k \gamma^j \gamma_k)^P{}_U, \tag{2}$$

by the interlocking of the repeated dummy spinor indices Q, S, and T. It is symbolic drudgery for a human to pick out this sort of interlocking, but an ideal task for an automaton. This suggests the tactic of *packing*, a kind of instruction which takes the left side of (2) and returns the right side, expressed as something like

$$\mathrm{Clif}^{ikj}{}_{k}{}^{P}{}_{U}.$$

Gammas are constantly being created and destroyed in actual computation, since application of the Clifford relations reduces the number of gammas, and the interchange of covariant derivatives produces terms based on the *spin curvature*

$$\sigma_{ij} = -\frac{1}{4}R_{klij}\gamma^{k}\gamma^{l},$$

and thus increases the number of gammas. As a result, packing and unpacking go on constantly, and are important connective material in the fabric of more substantial instructions.

Another fundamental tactic is the use of a normal form for Clifford algebra elements: each has a unique expression as a linear combination of the

$$\alpha_{i_1...i_p} := \gamma_{[i_1} \cdots \gamma_{i_p]}.$$

(As usual in tensor calculus, the square brackets denote antisymmetrization.) Clifford.m supports γ-to-α and α-to-γ operations. The payoff here is much more than just the normal form: α allows us to play off the near-anticommutativity of the gammas against the near-commutativity of covariant derivatives. More precisely, the Clifford relations allow us to *anticommute* gammas in any Clifford expression, modulo terms with fewer gammas, while the definition of curvature,

$$[\nabla_i, \nabla_j] = \mathrm{Curv}_{ij}$$

allows us to *commute* covariant derivatives, modulo terms with fewer differentiations. This has *extremely* powerful computational consequences when one can anticommute or commute to a normal form with a single instruction. The transition from gammas to alphas provides the turbo-anticommutation, while Ricci's OrderCovD instruction provides the turbo-commutation in this strategy. What OrderCovD does is to interchange covariant derivatives until a certain lexicographical order is attained, shooting out the appropriate curvature terms all the while. This lexicographical order is generally independent of index choices that have been made by the human in the picture, since previous simplification instructions have renamed dummy indices in a more or less uniform way.

A side benefit of the α-normal form is its smooth cooperation with the Bianchi identities, which are usually quite troublesome from an automated computational viewpoint. For example,

$$R_{ijkl}\alpha^{iklm} = 0, \qquad (\nabla_m R_{ijkl})\alpha^{ikm} = 0. \tag{3}$$

More generally, whenever 3 dummy indices in an α pair with 3 in an R or ∇R, the result vanishes. This really just reflects the fact that the 3-tensors break up, under $GL(n)$, into symmetric, antisymmetric, and Bianchi-like summands.

A practical application of the above is to the computation of Bochner-Weitzenböck (henceforth, BW) formulas on tensor-spinors, and similar formulas, involving different source and target bundles, which we shall call *mixed BW* formulas. Indeed, the first equation of (3) is the fundamental calculation behind the *Lichnerowicz formula*, the most famous BW formula involving spinors. Up to now, there has been very little study of mixed BW formulas; and indeed, it is not well known that the sporadic examples in the literature can be linked by an overall theory.

The Lichnerowicz formula says that for the Dirac operator $\nabla\!\!\!\!/ = \gamma^i \nabla_i$ on spinors,

$$\nabla\!\!\!\!/^{\,2} = -\nabla^i \nabla_i + \frac{1}{4}K, \tag{4}$$

where K is the scalar curvature. Since $\nabla\!\!\!\!/$ is self-adjoint and $-\nabla^i \nabla_i = \nabla^* \nabla$, this formula expresses the difference of two manifestly nonnegative second-order operators as an order zero curvature operator. This is, in fact, the common theme of BW formulas. Another such is the differential form BW formula

$$\Delta = d^* d + d d^* = \nabla^* \nabla - R^{ijkl}\varepsilon_i \iota_j \varepsilon_k \iota_l, \tag{5}$$

relating the de Rham and Bochner Laplacians on differential forms. Here ε is exterior and ι interior multiplication. Many results on the non-existence of harmonic spinors and forms (*vanishing theorems*) are based directly on these formulas.

In (Branson, 1997), it is shown that the number of independent BW formulas on a given irreducible Spin(n)-bundle $\mathbf{V}$ is approximately half the number $N(\mathbf{V})$ of different natural first-order differential operators originating in $\mathbf{V}$. (See (Branson, 1997) for precise statements.) $N(\mathbf{V})$ may be used as a kind of index of complexity for the bundle. Most familiar bundles have a low $N(\mathbf{V})$; for example, 2 for spinor bundles and 3 for form bundles. A by-product of the result is an expression for the left side of such a BW formula which resides, in its mode of expression, somewhere between representation theory and tensor calculus. *Left* and *right* sides are meant here in the sense of $\nabla\!\!\!\!/^{\,2} - \nabla^* \nabla$ and $d^* d + d d^* - \nabla^* \nabla$

as the *left* sides of (4) and (5), and the curvature terms as the right sides. The job, given a bundle of interest, is thus to make the left side of such a formula explicit, and to compute the right side.

To illustrate, let us say more about mixed BW formulas, and then more about a specific one. If **V** and **W** are non-isomorphic natural bundles, some representation theory shows that there are either 0, 1, or 2 paths from **V** to **W** via natural first-order operators. If there are two paths,

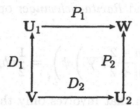

then there is one linear relation between the principal parts of $P_1 D_1$ and $D_2 P_2$:

$$P_1 D_1 + a D_2 P_2 = (\text{a curvature action } \mathbf{V} \to \mathbf{W})$$

for some real number a. If there is one path,

$$\mathbf{V} \xrightarrow{\;D\;} \mathbf{U} \xrightarrow{\;P\;} \mathbf{W},$$

then PD is either second order, or an order 0 action of the Weyl tensor. It is always possible to say which situation we are in, based on the *highest weights* of the bundles relative to the structure group Spin(n).

For example, consider the bundle **Tw** of *twistors*, or spinor-one-forms φ with $\gamma^i \varphi_i = 0$, and consider the diagram

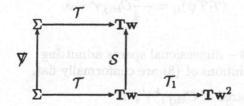

in which $\mathbf{T}\mathbf{w}^k$ is the bundle of spinor-k-forms $\Phi_{a\dots b}$ with

$$\gamma^b \Phi_{a\dots b} = 0, \tag{6}$$

and

$$
\begin{aligned}
(\mathcal{T}\psi)_i &= \nabla_i \psi + \frac{1}{n}\gamma_i \nabla\!\!\!/\,\psi, \\
(\mathcal{S}\varphi)_i &= n\gamma^j \nabla_j \varphi_i - 2\gamma_i \nabla^j \varphi_j, \\
(\mathcal{T}_1 \varphi)_{ij} &= \nabla_i \varphi_j - \nabla_j \varphi_i + \frac{1}{n-2}\left(\gamma_i \gamma^k \nabla_k \varphi_j - \gamma_j \gamma^k \nabla_k \varphi_i\right).
\end{aligned}
\tag{7}
$$

$\mathcal{T}$ and $\mathcal{S}$ are the *twistor* and *Rarita-Schwinger* operators. The relation from the 2-path principle,

$$\left(\left(\mathcal{S}\mathcal{T} - \frac{n-2}{n+2}\mathcal{T}\nabla\!\!\!/\,\right)\psi\right)_i = \frac{1}{2}nb_{ij}\gamma^j\psi$$

features a curvature action that involves only the Einstein (trace-free Ricci) tensor b. Since

$$\langle b_{ij}\gamma^j\psi, b^i{}_k\gamma^k\psi\rangle = b_{ij}b^{ij}\langle\psi,\psi\rangle = |b|^2|\psi|^2,$$

the existence, on a compact space, of a nonzero solution of the overdetermined system

$$
\begin{cases}
\gamma^i \nabla_i \psi = \kappa\psi, \\
\nabla_j \psi + \frac{1}{n}\gamma_j \underbrace{\gamma^i \nabla_i \psi}_{\kappa\psi} = 0
\end{cases}
\tag{8}
$$

(κ any real number) leads to $b = 0$, since any such solution has constant length:

$$\nabla_j |\psi|^2 = 2\operatorname{Re}\langle\nabla_j\psi,\psi\rangle = -\frac{2\kappa}{n}\operatorname{Re}\underbrace{\langle\gamma_j\psi,\psi\rangle}_{\text{imaginary}} = 0.$$

That is, the existence of a solution forces the background to be an Einstein manifold. From the one-path principle, we have

$$(\mathcal{T}_1 \mathcal{T}\psi)_{ij} = -\frac{1}{4}C_{klij}\gamma^k\gamma^l\psi. \tag{9}$$

As a result,

$$4 - \text{dimensional spaces admitting}$$
$$\text{solutions of (8) are conformally flat,}$$

since the norm-squared of $C_{klij}\gamma^k\gamma^l\psi$ is

$$4|C|^2|\psi|^2$$

in dimension 4. These results work together with the Lichnerowicz formula and the *Friedrich identity*

$$\nabla^2 = \frac{n}{n-1}\mathcal{T}^*\mathcal{T} + \frac{nK}{4(n-1)}$$

to provide a by now well-developed theory relating eigenvalue problems, vanishing conditions, and local geometry.

The bundles and formulas just above are not too difficult to compute with by hand. But without venturing too far into the realm of unfamiliar bundles, we encounter calculations that are somewhat beyond the range of reasonable pen and paper calculation. Furthermore, these calculations are potentially useful, as they have a chance to connect with information on familiar bundles to produce vanishing theorems and eigenvalue estimates. For example, we might want to compute generalizations of (9) to spinor-forms of higher degree; the next case is

$$\mathbf{Tw} \xrightarrow{\mathcal{T}_1} \mathbf{Tw}^2 \xrightarrow{\mathcal{T}_2} \mathbf{Tw}^3.$$

Here, for the unique (and suitably normalized) natural first-order operator $\mathcal{T}_2$, we have, after a Ricci calculation,

$$(\mathcal{T}_2\mathcal{T}_1\varphi)_{i_0i_1i_2} = 6\left\{-\frac{1}{8}C_{kl[i_0i_1}\gamma^k\gamma^l\varphi_{i_2]} + \frac{1}{n-4}C^k{}_{[i_2i_1l}\gamma_{i_0]}\gamma^l\varphi_k \right.$$
$$\left. + \frac{1}{4(n-4)(n-3)}C_{kl}{}^j{}_{[i_2}\gamma_{i_0}\gamma_{i_1]}\gamma^k\gamma^l\varphi_j\right\}.$$

A more substantial calculation goes as follows. Let **Z** be the bundle of spinor-symmetric-2-tensors ψ which are annihilated by interior Clifford multiplication (condition (6) again, with ψ for Φ). Let **A** be the bundle of spinor-3-tensors κ which are alternating in their last two arguments, Bianchi-like in all three arguments, and annihilated by interior Clifford multiplications. In symbols,

$$\kappa_{ijk} = -\kappa_{ikj}, \quad \kappa_{ijk} + \kappa_{jki} + \kappa_{kij} = 0, \quad \gamma^i\kappa_{ijk} = 0, \quad \gamma^j\kappa_{ijk} = 0.$$

The interior Clifford conditions on **Z** and **A** imply that tensor traces vanish, since for example

$$0 = (\gamma^i\gamma^j + \gamma^j\gamma^i)\kappa_{ijk} = -2\kappa^j{}_{jk}.$$

Corresponding to the diagram

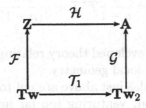

in which $\mathcal{F}$, $\mathcal{G}$, $\mathcal{H}$ are the essentially unique natural first-order operators between the bundles in question, we have, by a Ricci calculation,

$$\mathcal{H}\mathcal{F} - \mathcal{G}\mathcal{T}_1 = \frac{2}{n-2}\mathcal{B}(b) + \mathcal{C}(C),$$

where

$$
\begin{aligned}
(\mathcal{B}(b)\Phi)_{kij} = &-\frac{n^2 - 2n - 2}{2(n+2)(n-2)}\left\{\Phi_i b_{jk} - \Phi_j b_{ik}\right\} \\
&+\frac{1}{2(n+2)}\left\{\alpha_{lk}(\Phi_i b_j{}^l - \Phi_j b_i{}^l) - g_{jk}\Phi_l b_i{}^l + g_{ik}\Phi_l b_j{}^l\right\} \\
&-\frac{n}{2(n+2)(n-2)}\left\{\alpha_{jl}\Phi_i b^l{}_k - \alpha_{il}\Phi_j b^l{}_k\right\} \\
&+\frac{1}{2(n+2)(n-2)}\left\{\alpha_{jk}\Phi_l b_i{}^l - \alpha_{ik}\Phi_l b_j{}^l + g_{ik}\alpha_{jl}\Phi_m b^{lm} - g_{jk}\alpha_{il}\Phi_m b^{lm}\right\} \\
&-\frac{1}{(n+2)(n-2)}\left\{\alpha_{ij}\Phi_l b^l{}_k + \alpha_{jl}\Phi_k b_i{}^l - \alpha_{il}\Phi_k b_j{}^l\right\},
\end{aligned}
$$

and

$$
\begin{aligned}
(\mathcal{C}(C)\Phi)_{kij} = &\frac{1}{6}\alpha_{lm}\left(-\Phi_k C_{ij}{}^{lm} + \Phi_{[i}C_{j]k}{}^{lm}\right) \\
&+\frac{1}{3n(n+2)(n-2)}\left(\alpha_{lk}\Phi_m C_{ij}{}^{lm} - \alpha_{l[i}C_{j]k}{}^{lm}\Phi_m\right) \\
&-\frac{2n^3 - 4n^2 - 7n + 8}{2n(n+2)(n-2)}\Phi_l C_{ij}{}^l{}_k + \frac{2n^2 - 9n - 6}{6n(n+2)(n-2)}\alpha_{lk}\Phi_m C_{ij}{}^{ml} \\
&+\frac{4n^2 - 6n - 15}{3n(n+2)(n-2)}\Phi_m \alpha_{l[i}C_{j]}{}^l{}_k{}^m + \frac{(n+3)(2n-3)}{3n(n+2)(n-2)}\Phi_m \alpha_{l[j}C_{i]}{}^{ml}{}_k \\
&+\frac{5n+8}{6n(n+2)(n-2)}\left(\alpha_{lkm[j}C_{i]}{}^{plm} + \alpha_{ijlm}C_k{}^{plm}\right)\Phi_p \\
&+\frac{n^2 - n - 8}{2n(n+2)(n-2)}\alpha_{lm}\Phi_p g_{k[j}C_{i]}{}^{plm}.
\end{aligned}
$$

The whole potential catalogue of such relations, including relations on fairly simple tensor types (no spin structure involved), is largely untouched, but the technology is there to compute them.

3. Conformally invariant differential operators

A *conformally invariant differential operator* (henceforth CIO) is a natural operator between natural bundles, $D : \mathbf{V} \to \mathbf{W}$, with the property that

$$\bar{g} = \Omega^2 g, \ \Omega \text{ a positive function} \ \Rightarrow \ \bar{D}\varphi = \Omega^{-b} D(\Omega^a \varphi) \qquad (10)$$

on any section φ of $\mathbf{V}$. The best-known CIO are the conformal Laplacian

$$L = \Delta + \frac{n-2}{4(n-1)} K$$

on scalar functions, where $\Delta = -\nabla^i \nabla_i$, and the Dirac operator. When expressed in terms of density bundles of the correct weights, (10) can be rewritten as $\bar{D}_1 = D_1$ for a related operator D_1, justifying the use of the term *invariant*. The question of when CIO exist, and efforts at their construction, have produced a large literature in the 1980's and 1990's. The central motivation is physical: massless particles move along null geodesics, which are preserved by conformal change of metric. Therefore massless field equations must be conformally invariant.

One approach to the automated computation of CIO is as follows. First, there is the question of whether a CIO of a given order and between given bundles exists. There is a general theory of *embeddings of Verma modules* and *Bernstein-Gelfand-Gelfand* (BGG) *resolutions* which resolves almost all such questions. (In the conformally flat case, it is a complete answer. There are still, however some values of the relevant weight parameters where the question is open in the arbitrarily conformally curved case.) Next, using spectral data from (Branson, 1997), one can use the leading asymptotics of the operator on the sphere to write the operator's principal part. This gives an exact formula for the operator in flat space, and the conformal covariance relation can then be used to get an expression for arbitrary conformally flat metrics, by repeated application of the conformal change equation for the Ricci tensor:

$$\eta_{ij} = -V_{ij} - \eta_i \eta_j + \frac{1}{2} \eta_k \eta^k g_{ij},$$

where $g = e^{2\eta} g_{\text{flat}}$. Here

$$J = \frac{K}{2(n-1)}, \qquad V = \frac{r - Jg}{n-2}, \qquad (11)$$

and we have employed the usual notational abuse: for a scalar function, $\eta_{j\ldots i} := \nabla_i \cdots \nabla_j \eta$.

Given that we are searching for an operator that is really there in the conformally flat case, and that we have the correct principal part, the first derivatives η_i will wash out of the final formula, which is then polynomial in ∇ and the Ricci tensor. Now three things can happen:

- 1. The expression is a CIO in the arbitrarily conformally curved case;

- 2. After adding a nontrivial adjustment expression, each summand of which has a factor of the form

$$C_{ijkl|(m_1 \cdots m_k)_0},$$

we have a CIO in the arbitrarily conformally curved case;

- 3. There is no CIO in the arbitrarily conformally curved case with the original parameters.

We can often rule out possibility (3) by general theory (Eastwood-Slovak, 1997) – this is sometimes called the *curved translation principle*, or CTP. We can sometimes conclude that (1) is the case because there are no terms of the form described in (2), or because such terms must be separately invariant. This is always the situation for second-order operators. An example of a higher-order operator of this type is the following third-order operator on spinors:

$$D_3\psi := -\gamma^i \nabla^j \nabla_j \nabla_i \psi - \frac{n+4}{2(n-2)} R_i{}^j \gamma^i \psi_j$$
$$+ \frac{n^2 - 2n - 12}{8(n-1)(n-2)} R \nabla\!\!\!/\, \psi - \frac{n+4}{8(n-1)} (\nabla_i R) \gamma^i \psi. \tag{12}$$

By the Lichnerowicz formula (4), this has principal part $\nabla\!\!\!/^{\,3}$. By weight considerations, there are no actions of ∇C on spinors, nor actions of C taking $T^*M \otimes \Sigma$ (where $\nabla\psi$ lives) to Σ. As a result, (12) must fully conformally invariant, since its parameters pass the CTP test. Another example is the fourth-order Paneitz operator

$$\Delta^2 + \delta T d + \frac{n-4}{2} Q,$$

which is important in the study of the functional determinant in dimension 4; see (Branson-Ørsted, 1991), (Branson-Chang-Yang, 1992), (Chang-Yang, 1995), and (Connes, 1994), Sec. IV.4.γ. Here (recall (11))

$$T = (n-2)J - 4V\cdot,$$
$$Q = \frac{n}{2} J^2 - 2|V|^2 + \Delta J,$$

where $(V \cdot \varphi)_j = V^i{}_j \varphi_i$ is the natural action of V on one-forms. The automated algorithm quickly produces this formula upon input of the flat Δ^2. The only available Weyl term of the type described above is $|C|^2 = C^{ijkl}C_{ijkl}$, which is separately invariant. (This pair of examples is somewhat deceptive, however, in that powers of principal parts of CIO are almost never themselves principal parts of CIO.)

For a calculation similar to, but more difficult than (12), consider once again the bundle $\mathbf{Tw}$ of twistors, and the twistor and Rarita-Schwinger operator from (7). BGG resolutions predict a third-order conformally invariant operator S_3 on twistors (i.e., between different orders of twistor densities). Spectral data predicts that the principal part of S_3 should be

$$\frac{n+2}{4}n^2 S^3 - \frac{4}{n(n-2)}\mathcal{T}\mathcal{T}^*S.$$

A `Ricci` calculation then gives

$$
\begin{aligned}
(S_3\varphi)_i = {} & \left(\left\{ \frac{n+2}{4}n^2 S^3 - \frac{4}{n(n-2)}\mathcal{T}\mathcal{T}^*S \right\} \varphi \right)_i \\
& - \frac{n+2}{n} J\gamma_i \nabla^j \varphi_j + V_i{}^j \gamma^k \nabla_k \varphi_j + (n+2)V_i{}^k \gamma_k \nabla^j \varphi_j \\
& + (n+1)V^{jk}\gamma_i \nabla_k \varphi_j - \frac{n(n+2)}{2} V^{jk}\gamma_k \nabla_j \varphi_i \\
& + (n-1)V^{jk}\gamma_k \nabla_i \varphi_j + V^{jl}\alpha_i{}^k{}_l \nabla_k \varphi_j + \frac{n}{2}(\nabla^j J)\gamma_i \varphi_j \\
& - \frac{n(n+2)}{4}(\nabla^j J)\gamma_j \varphi_i + n(\nabla^k V_i{}^j)\gamma_k \varphi_j.
\end{aligned}
$$

The curved translation principle predicts that there will be an arbitrarily conformally curved generalization of S_3, since it is not a so-called *longest arrow*. Possible correction terms take the schematic forms $C\nabla A$ and $(\nabla C)A$. An alternative to using correction terms with undetermined coefficients is a direct (and automated) computation of the operator using the conformally invariant *tractor* calculus (Bailey-Eastwood-Gover, 1994). In joint work with Rod Gover, we have developed code for use with `Ricci` which allows such calculations to be automated.

References

Bailey, T., M. Eastwood, and A.R. Gover *The Thomas structure bundle for conformal, projective and related structures.* Rocky Mountain J. Math. **24** (1994) 1–27.

Branson, T. *Nonlinear phenomena in the spectral theory of geometric linear differential operators.* Proc. Symp. Pure Math. **59** (1996) 27–65.

Branson, T. *Stein-Weiss operators and ellipticity*. J. Funct. Anal. **151** (1997) 334–383.

Branson, T., S.-Y.A. Chang, and P. Yang *Estimates and extremals for zeta function determinants on four-manifolds*. Commun. Math. Phys. **149** (1992) 241–262.

Branson, T. and B. Ørsted *Explicit functional determinants in four dimensions*. Proc. Amer. Math. Soc. **113** (1991) 669–682.

Chang, S.-Y.A., and P. Yang *Extremal metrics of zeta function determinants on 4-manifolds*. Annals of Math. **142** (1995) 171–212.

Connes, A. *Noncommutative Geometry*. Academic Press, San Diego, 1994.

Eastwood, M. and J. Slovak *Semi-holonomic Verma modules*. J. Alg. **197** (1997) 424–448.

Lee, J.M. *Ricci: A Mathematica package for doing tensor calculations in differential geometry*. Archived at http://www.math.washington.edu/ lee/Ricci/.

Stein, E. and G. Weiss *Generalization of the Cauchy-Riemann equations and representations of the rotation group*. Amer. J. Math. **90** (1968) 163–196.

Monogenic Forms of Polynomial Type [*]

Jarolím Bureš (jbures@karlin.mff.cuni.cz)
Mathematical Institute of Charles University

Abstract. The classical Dirac equation for spin 1/2-fields is an example of (first order) equation for spin λ-fields related with irreducible representation spaces of the group $Spin(m)$ with weight λ on an oriented riemannian spin manifold M. For the flat space $M = R^m$ there is the Clifford analysis as a natural method for the study of properties of these fields. Their Taylor series are composed from polynomial-type fields, the elements of some finite-dimensional representation space of the group $Spin(m)$. The representation character (decomposition into irreducible components) of polynomial-type fields were studied for Rarita Schwinger fields in [4], for symmetric analogies of Rarita-Schwinger fields in two related papers [7, 8]. In this paper the representation character of monogenic s-forms of polynomial-type with $s < m/2$ is described in details.

Keywords: conformally invariant differential operators, twisted Dirac operators, monogenic forms

Mathematics Subject Classification: 50G25,53C25

Dedicated to Richard Delanghe on the occasion of his 60th birthday.

1. Introduction

Let M be an oriented spin manifold endowed with a conformal structure. There is a set of conformally invariant differential operators naturally defined on M. These operators are defined for any conformal spin structure on M and their geometrical properties (solutions, eigenvalues, index, etc.) depends on geometry as well as topology of the manifold. The simplest first order conformal invariant operator is the Dirac operator, other first order conformal invariant operators can be constructed from twisted Dirac operators. Harmonic spinors (solutions of the Dirac equation) were studied intensively, only few facts are known about the properties of solutions of the other equations.

A special system $\mathcal{EL}$ of first order conformally invariant operators consists of elliptic operators of the first order

$$D_\lambda : \Gamma(M, V_\lambda) \to \Gamma(M, V_\lambda)$$

[*] This work was supported by grant GAUK No. 247/2000 an CEZ J13/98113200007

F. Brackx et al. (eds.), Clifford Analysis and Its Applications, 39–48.
© *2001 Kluwer Academic Publishers. Printed in the Netherlands.*

which have source and target in the same space of sections and which can be uniquely defined (together with some normalization). The system $\mathcal{EL}$ can be indexed by a special set Λ^{el} of weights of $spin(n)$, dominant weights for n odd, and sums of the related dominant weights for n even (for details see [4, 6]).

The set Λ^{el} is ordered by the order defined on the weight space. For any $\lambda \in \Lambda^{el}$ the set

$$B(\lambda) := \{\lambda' \in \Lambda^{el}; \lambda' < \lambda\}$$

is finite. The study of the space of solutions of equations $D_\lambda \psi = 0$, as well as eigenvalues of the operators from $\mathcal{EL}$ is an interesting and open problem. There is a possibility to find a relation between solutions (eigenvalues) of D_λ and $D'_\lambda, \lambda' \in B(\lambda)$ on M. using other (twistor type) conformally invariant operators ([6, 5])

$$D_{\lambda',\lambda} : \Gamma(M, V_{\lambda'}) \to \Gamma(M, V_\lambda)$$

and properties of Bernstein-Gelfand-Gelfand sequences. Let us call solutions ψ of the equation

$$D_\lambda \psi = 0$$

spin λ-fields on M. Let us consider as a basic example the flat space $M = R^m$ with the standard euclidean metric and flat conformal structure. There is a natural method of study of analytic properties of solutions of conformally invariant operators on the flat space using Clifford analysis. All needed facts from Clifford analysis in our case are contained in [12].

One of main related analytic problems is to define and study analogy of Taylor (Fourier) series for the corresponding fields, which are given by polynomial solutions of corresponding equations. Polynomial solutions (of the fixed degree) of the equations on flat space (called polynomial λ-fields) are finite-dimensional representation spaces of the group $Spin(n)$, and can be decomposed into irreducible pieces. There is an idea of existence of an inductive procedure how to get all polynomial λ-fields mainly from the polynomial λ'-fields, with $\lambda' \leq \lambda$ and the rest from the eigenspaces of the induced operators on the unit sphere $S^{m-1} \subset R^m$. This problem is solved for the Rarita-Schwinger fields ($\lambda = (3/2, 1/2, ..., 1/2)$) in the paper [5] and for the case of the solutions of symmetric analogues of Rarita-Schwinger equations ($\lambda = (k + 1/2, 1/2..., 1/2)$ by induction on k in two related papers [7, 8]. In the present paper the solution of the problem is given by (finite) induction procedure for the case of monogenic differential s-forms with $s < n/2$. A classification of conformally invariant first order operators was given in [13], see also [9, 6]. The list of (strong) elliptic first order operators in even, resp. odd dimensions is given in [6]. We shall use the notations and results from the [6, 4] in all the following parts of the paper.

2. Spinor-valued forms.

There is a family of conformally invariant elliptic operators of the first order defined on spinor-valued differential forms. The Dirac operator and the Rarita-Schwinger operator $\mathcal{R}_1$ both belong to this family, but there are other elliptic different operators $Q_j; j < m/2$ there, in our notations $Q_j = D_{\lambda_j}$ with $\lambda_j = (3/2, ..3/2, 1/2, ..1/2)$, $3/2$ appearing j-times in the bracket. So Q_0 is the Dirac operator, $Q_1 = \mathcal{R}_1$ is the Rarita-Schwinger operator.

The study of the properties of the operators Q_j and the structure of spaces of their polynomial-type solutions on he flat space R^n is the main topic of this section. The main reference for this section is [14].

2.1. The spinor valued forms on the flat space.

Denote by $\Omega^s(\mathbf{S}_{\frac{1}{2}}) := \Gamma(\mathbf{S}_{\frac{1}{2}} \otimes \Lambda^s(\mathbf{R}^{m*}))$ the space of spinor valued s-forms on the flat space $\mathbf{R}^m$. We have defined covariant derivative and its extension in the classical way for any s

$$\nabla : \Omega^s(\mathbf{S}_{\frac{1}{2}}) \to \Omega^{s+1}(\mathbf{S}_{\frac{1}{2}})$$

which is conformally invariant first order operator.

Let us suppose $s < \frac{m}{2}$ in the whole paper, because for $s > \frac{m}{2}$ the corresponding spaces are isomorphic, the remaining case $s = \frac{m}{2}$, for m even do not fit into our scheme.

There is a decomposition of $\Omega^s(\mathbf{S}_{\frac{1}{2}})$ into irreducible components with respect to representation of the group $Spin(n)$ in the following way:

$$\Omega^s(\mathbf{S}_{\frac{1}{2}}) \simeq \mathcal{E}^{(s,0)} \oplus \mathcal{E}^{(s,1)} \oplus ... \oplus \mathcal{E}^{(s,s)}.$$

The spinor-valued forms from $\mathcal{E}^s := \mathcal{E}^{(s,s)}$ are called primitive and they are sitting on the side of the triangle of decomposition of spinor valued forms.

LEMMA 2.1. *([14]) For any (s, k) we have*

$$\nabla : \mathcal{E}^{s,k} \to \mathcal{E}^{s+1,k-1} \oplus \mathcal{E}^{s+1,k} \oplus \mathcal{E}^{s+1,k+1}$$

Let us denote by $X : \Omega^s(\mathbf{S}_{\frac{1}{2}}) \to \Omega^{s+1}(\mathbf{S}_{\frac{1}{2}})$ an operator defined by

$$X(\psi \otimes \omega) = \sum_i e_i.\psi \otimes \epsilon^i \wedge \omega$$

There is also second invariant operator $Y : \Omega^{s+1}(\mathbf{S}_{\frac{1}{2}}) \to \Omega^s(\mathbf{S}_{\frac{1}{2}})$ defined on spinor valued forms by

$$Y(\psi \otimes \omega) = -\sum_i e_i.\psi \otimes \iota(e_i)\omega$$

with

$$[X,Y] = (2s - m)Id$$

on $\Omega^s(\mathbf{S}_{\frac{1}{2}})$

The operators X, Y are invariant operators (determined by the algebraic operators $\mathbf{X}, \mathbf{Y}$).

LEMMA 2.2. *The operator X maps the space $\mathcal{E}^{s,k}$ isomorphically onto the space $\mathcal{E}^{(s+1,k)}$ for $k < s$. The operator Y maps the space $\mathcal{E}^{s,k}$ isomorphically onto the space $\mathcal{E}^{s-1,k}$ for $k < s$.*

The twisted Dirac operator (for a fixed s)

$$D_T : \Omega^s(\mathbf{S}_{\frac{1}{2}}) \to \Omega^s(\mathbf{S}_{\frac{1}{2}})$$

is given by the formula $D_T = -(Y\nabla + \nabla Y)$

For any fixed s the twisted Dirac operator is given by the formula $D_T(\psi) = \sum_I D\psi_I \otimes dx^I$. where D is the Dirac operator.

Let us denote for fixed s by $\tilde{\mathcal{E}}^k := X^k(\mathcal{E}^k)$

LEMMA 2.3. *If $\psi \in \mathcal{E}^s$ then*

$$D_T\psi \in \mathcal{E}^s \oplus \tilde{\mathcal{E}}^{s-1}$$

There is the following diagram:

$$\Omega^s(\mathbf{S}_{\frac12}) \xrightarrow{\quad D_T \quad} \Omega^s(\mathbf{S}_{\frac12})$$

$$
\begin{array}{ccc}
\| & & \| \\
\mathcal{E}^s & & \mathcal{E}^s \\
\oplus & & \oplus \\
\tilde{\mathcal{E}}^{s-1} & & \tilde{\mathcal{E}}^{s-1} \\
\oplus & & \oplus \\
\vdots & & \vdots \\
\oplus & & \oplus \\
\mathcal{E}^1 & \xrightarrow{\ Q'_1\ } & \mathcal{E}^1 \\
\oplus & & \oplus \\
\tilde{\mathcal{E}}^0 & \xrightarrow{\ Q'_0\ } & \tilde{\mathcal{E}}^0
\end{array}
$$

with operators Q_s, T'_s, $T^{*\prime}_s$, Q'_{s-1}, Q'_1, Q'_0.

Let us call the operator Q_s higher spin (HS) Dirac operator and equation $Q_s\psi = 0$ higher spin Dirac equation (HS-Dirac equation).

REMARK 2.4. *There are the following natural relations*

$$\mathcal{E}^s = \{\psi \in \Omega^s(\mathbf{S}_{\frac12}), Y\psi = 0\}$$

$$Q_s = Y \circ \nabla$$
$$T = \nabla - Y\nabla - Y^2\nabla, \quad T^* = XY^2\nabla$$

and

$$T^*_s \circ T^*_{s+1} = 0, \quad T_s \circ T_{s-1} = 0$$

Immediately from the definition follows

LEMMA 2.5. ψ *is a solution of the HS-Dirac equation* $Q_s\psi = 0$ *if and only if there exists an element* $\phi \in \mathcal{E}^{s-1}$ *with*

$$D_T\psi = X(\phi)$$

Let us denote $\phi := \delta_s(\psi)$.

There is a nonzero constant c_s such that

$$Q'_s = c_s.X \circ Q_s \circ Y.$$

LEMMA 2.6. *Let ψ be a solution of higher spin Dirac equation $Q_s\psi = 0$, then*

$$Q_{s-1}(\delta_s(\psi)) = 0.$$

Let us denote by $\mathcal{P}_k(s)$ the space of polynomial k-homogeneous spinor-valued forms in $\mathcal{E}^s$.

We would like to describe the space $\mathcal{M}(s)$ of all solutions of the equation $Q_s\psi = 0$ on $\mathbf{R}^n$ and its subspace $\mathcal{M}_k(s)$ of solutions, which are elements of $\mathcal{P}_k(s)$.

The space $\mathcal{M}_k(s)$ is a finite dimensional representation space of the group $Spin(m)$, it can be decomposed into irreducible pieces and it is possible to find their representation types, determined by its highest weight.

For the further computations let us denote $\{e_1, ..., e_m\}$ the standard (orthonormal) basis and also basic fields on R^m, $e_i = \frac{\partial}{\partial x_i}$ and $\{\epsilon^1, ..., \epsilon^m\}$ the dual basis $\epsilon^i = dx_i$. Then the Dirac operator has a form $D = \sum_1^m e_i \frac{\partial}{\partial x_i}$

Let us introduce the map: $\mathcal{L} : \mathcal{P}_k(s) \rightarrow \mathcal{P}_{k+1}(s-1)$ by

$$\mathcal{L}(\psi \otimes \omega) = \sum_i x_i . \psi \otimes \iota(e_i)\omega$$

LEMMA 2.7. *We have $\mathcal{L}^2 = 0$, let ψ be a polynomial solution of $Q_s\psi = 0$, then*

$$Q_{s-1}(\mathcal{L}(\psi)) = \mathcal{L}(T^*\psi)$$

2.2. THE CASE OF DIRAC AND RARITA-SCHWINGER EQUATIONS.

The problem of description of the spaces $\mathcal{M}_s^k$ was solved completely for the Dirac operator ($s = 0$) and the Rarita-Schwinger operator ($s = 1$) (see e.g [16, 5]). Let us recall it briefly.

Q_0 is the Dirac operator D and

$$\mathcal{M}_k(0) = \{\psi \in \mathcal{P}_k; | Q_0\psi = 0\} \simeq (\frac{2k+1}{2}, \frac{1}{2}, ..., \frac{1}{2})$$

Q_1 is the Rarita-Schwinger operator.

The space $\mathcal{M}_k(1)$ is a direct sum

$$\mathcal{M}_k(1) = \mathcal{M}_k^A(1) \oplus \mathcal{M}_k^{B1}(1) \oplus \mathcal{M}_k^{B2}(1)$$

of three spaces (representation types) of solutions (see[5]), namely

Type A.: It is the space of all solutions ψ of equation $D_T\psi = \phi$ with $\phi \neq 0$ and moreover satisfying the additional condition (for the uniqueness)

$$\psi \in x \bullet \mathcal{P}_{k-1}(0)$$

It will be denoted by $\mathcal{M}_k^A(1)$ and can be identified in a canonical way with $\mathcal{M}_{k-1}(0)$ which is of type $(\frac{2k-1}{2}, \frac{1}{2}, ..., \frac{1}{2})$.

Type B: This type is characterized by condition $\phi = 0$ (also by $Q_0\mathcal{L}(\psi) = 0$) and consists of solutions not only of RS-equation but of the whole twisted Dirac equation.

There are two subtypes:

Type B1: The solutions of this type satisfy the condition $\mathcal{L}(\psi) = 0$ The corresponding space of k- homogeneous solutions $\mathcal{M}_k^{B1}(1)$ is in one-to-one correspondence with the eigenspace of RS-equation on the unit sphere of type $(\frac{2k+1}{2}, \frac{3}{2}, \frac{1}{2}, ..., \frac{1}{2})$. The correspondence is done by restriction of solution to the sphere.

Type B2: There is $\mathcal{L}(\psi) \neq 0$ for $\psi \neq 0$ and $D_0\mathcal{L}(\psi) = 0$. The space of solutions of this type is denoted by $\mathcal{M}_k^{B2}(1)$ and it is the image of $\mathcal{M}_{k+1}(0)$ by the twistor operator T_1. and has the type $(\frac{2k+3}{2}, \frac{1}{2}, ..., \frac{1}{2})$.

2.3. GENERAL CASE.

Let us consider the case for general $s < m/2$. We would like to find a procedure how to find a structure of homogeneous solutions of $Q_s = 0$. Examples presented in preceding section give us a suggestion, that is can be possible to get several types of polynomial solutions of $Q_s = 0$ from the polynomial solutions of the equation $Q_{s-1} = 0$ and one type of solutions from special eigenspace of induced operator on the sphere.

Let us mention that the results are different from the case of higher spin Rarita-Schwinger equations [7, 8].

Let us introduce the following short notation for the representation spaces of $Spin(m)$:

$$[k,q] = (\frac{2k-1}{2}, \frac{3}{2}, ...\frac{3}{2}, ..., \frac{1}{2})$$

where $\frac{3}{2}$ appears q-times starting from the second place, and $\frac{1}{2}$ on the remaining places. For $q < 0$ we set $[k,q] = 0$.

Let us remark that there exists an invariant scalar product on the space of spinor-valued differential forms, denote $U^\perp$ the orthogonal complement to U with respect to this product.

THEOREM 2.8. *Suppose in the following $k \geq s$. The space of polynomial solutions of the equation $Q_s\psi = 0$ degree k has a decomposition* *as representation space of Spin(m) of the form*

$$\mathcal{M}_k(s) \simeq [k+1, s-2] \oplus [k+2, s-1] \oplus [k, s-1] \oplus [k+1, s]$$

Remark : If we need to consider $[p, q]$ as subspace of $\mathcal{M}_k^s$ we shall use the more precise notation $[p, q]_s^k$.

Sketch of proof: The construction is done by induction, it follows the method used in the first step $s = 0 \to s = 1$.

There are again two types of the homogeneous solutions of degree k of $Q_s \psi = 0$.

Type A : The space $\mathcal{M}_k^A(s)$ of solutions of the equation

$$D_T \psi = X(\phi)$$

with $\phi \neq 0$. Then $Q_{s-1}(\phi) = 0$ and candidates for ϕ are elements of the space of (k-1)-homogenic solutions of $Q_{s-1}\phi = 0$:

$$\mathcal{M}_{k-1}(s-1) \simeq [k, s-3]_{s-1}^{k-1} \oplus [k+1, s-2]_{s-1}^{k-1} \oplus [k-1, s-2]_{s-1}^{k-1} \oplus [k, s-1]_{s-1}^{k-1}.$$

which are in the kernel of the operator T_{s-1}^*, namely $U^A(k, s) = [k+1, s-2]_{s-1}^{k-1} \oplus [k, s-1]_{s-1}^{k-1}$.

LEMMA 2.9. *The map T^* is an isomorphism from the space $\mathcal{M}_k^A(s) := (Ker T_s^*)^\perp$ into $U^A(k, s)$.*

Let us denote $T^{*-1} : U^A(k, s) \to \mathcal{M}_k^A(s)$ the inverse mapping on the orhogonal complement.

Type B. This type is characterized by the condition $\phi = 0$ and consists of solutions of the whole twisted Dirac equation.

There are still two subcases available :

Type B1: This type is characterized by conditions $\mathcal{L}(\psi) = 0$ and the corresponding space of k-homogeneous solutions $\mathcal{M}_k^{B1}(s)$ corresponds to the eigenspace of induced operator Q_s^S on the unit sphere and induces solutions of $Q_s \psi =)$ of the type $[k+1, s]$. The correspondence is done by restriction of the solution ψ to the unit sphere.

Type B2: There is $\mathcal{L}(\psi) \neq 0$, $Q_{s-1}\mathcal{L}(\psi) = 0$, The corresponding space $\mathcal{M}_k^{B1}(s)$ is constructed from the space of polynomial solutions of degree $k+1$ of the equation $Q_{s-1}\phi = 0$ as the image of the twistor operator T_s. Candidates for the preimage are the elements of

$$\mathcal{M}_{k+1}(s-1) \simeq$$

$$[k+2, s-3]_{s-1}^{k+1} \oplus [k+3, s-2]_{s-1}^{k+1} \oplus [k+1, s-2]_{s-1}^{k+1} \oplus [k+2, s-1]_{s-1}^{k+1}$$

which are not in the image of T_{s-1}, namely

$$[k+1, s-2]_{s-1}^{k+1} \oplus [k+2, s-1]_{s-1}^{k+1}.$$

LEMMA 2.10. *The map T_s maps the space $[k+2, s-1]_{s-1}^{k+1}$ isomorphically onto $\mathcal{M}_k^{B2}(s)$ and the space $[k+1, s-2]_{s-1}^{k+1}$ isomorphically onto $[k+1, s-2]_s^k$.*

We can introduce the following schema for $\mathcal{M}_k(s))$:

$$[k+1, s-2]_s^k$$

$$[k+2, s-1]_s^k \qquad\qquad [k, s-1]_s^k$$

$$[k+1, s]_s^k$$

and we have the following picture:

Polynomial solutions of $Q_s = 0$ of degree k

$$[k+1, s-2]_s^k = T^{*-1}([k+1, s-2]_{s-1}^{k-1}) = T([k+1, s-1]_{s-1}^{k+1})$$

$$[k+2, s-1]_s^k = T([k+2, s-1]_{s-1}^{k+1})$$

$$[k, s-1]_s^k = T^{*-1}([k, s-1]_{s-1}^{k-1})$$

$$[k+1, s]_s^k \text{spherical}$$

The structure of the space of homogeneous solutions of Q_s

1. Let $\psi \in \mathcal{M}_k(s)$ be a solution. If $T^*\psi = \phi \neq 0$ we can find a solution ψ_1 of the type A with

$$Q_s(\psi_1) = \phi$$

Then $\psi' = \psi - \psi_1$ satisfies the equation $Q_s\psi' = 0$ and $T^*\psi' = 0$.

2. If $\mathcal{L}\psi' \neq 0$ there is possible to find a solution ψ_2 of the equation $Q_s\psi_2 = 0$ of the form $\psi_2 = T\phi$ with $\mathcal{L}\psi_2 = \mathcal{L}\psi'$.

3. We have for $\psi'' = \psi - \psi_1 - \psi_2$ the condition $\mathcal{L}(\psi'') = 0$ and we can find a solution ψ_3 of $Q_s\psi_3 = 0$ which can be constructed from its restriction to sphere and $\psi''|S^n$ is an eigenvector of the spherical Q_s operator. For k-homogeneous ψ'' we have

Together we have

$$\psi = \psi_1 + \psi_2 + \psi_3$$

and this decomposition can be defined in a unique way.

J Bureš

References

1. Baston R.J., Eastwood M.G. : The Penrose transform. Its interaction with representation theory, Oxford University Press, New York, 1989 .
2. Brackx, F., Delanghe, R., Sommen, F.: Clifford analysis, Pitman, 1982
3. Branson T. : Stein-Weiss operators and ellipticity, J.Funct. Anal. 151 (1997), 334-383 .
4. Bureš J.: The higher spin Dirac operators, Proc. conf DGA, Brno 1998
5. Bureš, J.: The Rarita-Schwinger operator and spherical monogenic forms, to be published in Complex Variables:Theory and Applications.
6. Bureš, J., Souček, V.: Eigenvalues of conformally invariant operators on spheres, Proc.Winter School 1988, Suppl.di Rendiconti Palermo,II. 109-122, No 59, 1999 to be published
7. Bureš, J., Van Lancker, P., Sommen, F., Souček, V.: Symmetric analogies of Rarita-Schwinger equations, submitted for publication
8. Bureš, J., Van Lancker, P., Sommen, F., Souček, V.: Rarita-Schwinger type Operators in Clifford Analysis, submitted for publication
9. Čap A.,Slovák J.,Souček V. - Invariant operators with almost hermitean symmetric structures, III. Standard operators , to be published .
10. Čap A.,Slovák J.,Souček V. - Bernstein-Gelfand-Gelfand sequences , to be published .
11. Delanghe R.,Souček V. : On the structure of spinor-valued differential forms, complex Variables, 1992, vol.18, 223-236 .
12. Delanghe, R., Sommen, F., Souček, V.: Clifford algebras and spinor valued functions: Function theory for Dirac operator, Kluwer, 1992
13. Fegan H.D. : Conformally invariant first order differential operators, Quart. J. Math. 27 (1976), 371-378 .
14. V.Severa : Invariant Differential operators on Spinor-Valued Differential Forms, PhD thesis, Prague 1998
15. F. Sommen, N. Van Acker: *Invariant differential operators on polynomial valued functions*, F. Brackx et al. (eds.), Clifford Algebras and their Applications in Mathematical Physics, Kluwer Acad. Publ., 1993, pp. 203-212.
16. Souček V. : Higher spins and conformal invariance in Clifford analysis, Lecture in Seiffen, Proc. Conf. Seiffen 1996 .
17. Souček V. : Monogenic forms on manifolds, in Oziewicz et al. (Eds.) Spinors, Twistors,Clifford Algebras and Quantum Deformation, Kluwer 1993, 159-166 .

On Beltrami Equations in Clifford Analysis and Its Quasi-conformal Solutions

Paula Cerejeiras (pceres@mat.ua.pt) and Uwe Kähler
(uwek@mat.ua.pt)*
Departamento de Matemática, Universidade de Aveiro

Abstract. In this paper we investigate the problem of existence of locally quasiconformal solutions of Beltrami equations for functions with values in a Clifford algebra, based on a characterization of a monogenic homeomorphism by its derivative.

Keywords: Beltrami equation, quasiconformal mappings, local homeomorphisms

Mathematics Subject Classification 2000: 30G35, 30G20

Dedicated to Richard Delanghe on the occasion of his 60th birthday.

1. Introduction

Beltrami equations play an important role in complex analysis due to the fact that its solutions are quasiconformal mappings. Although there is a well-developed theory of quasiconformal mappings in higher dimensions, the same cannot be said about the theory of Beltrami equations, mainly due to the fact that it is not clear if the solutions are quasiconformal or not. In (CGKM, 2001) the authors investigated the problem of locally quasiconformal solutions of quaternionic Beltrami equations by means of a higher dimensional equivalent of the well-known theorem in complex analysis, that a holomorphic function $f(z)$ with $f'(z_0) \neq 0$ in some point z_0 realizes a locally conformal mapping in a neighbourhood of this point. Directly, the results obtained in (CGKM, 2001) lead to the question of the case of $\mathbb{R}^n$. Here, we will consider this problem in the setting of Clifford analysis, an area which was founded by R. Delanghe (D, 1970).

But almost immediately one realizes the problems in this setting, which make a direct translation almost impossible. First, we cannot use the embedding of $\mathbb{H}$ in $\mathbb{C}^2$. Second, we are dealing with function defined over $\mathbb{R}^{n+1}$ with values in a Clifford algebra, i.e. in a 2^n-dimensional

* This paper was done while the author was a recipient of a PRAXIS XXI-scholarship of the Fundação para a Ciência e a Tecnologia visiting the Universidade de Aveiro in Portugal.

F. Brackx et al. (eds.), Clifford Analysis and Its Applications, 49–58.
© 2001 *Kluwer Academic Publishers. Printed in the Netherlands.*

space and, finally, due to the problem of considering paravectors, we cannot simply consider rotations of the form $PZ\overline{P}$. Nevertheless, the embedding of the paravectors of $C\ell_n$ in the space of vectors in $C\ell_{n+1}$ allows us to obtain a generalization of the above mentioned complex theorem in case of the $\mathbb{R}^{n+1}$ and to transform the problem of the existence of a locally quasiconformal solution of the Beltrami equation into a problem of the positive definiteness of real symmetric matrix.

2. Preliminaries

We shall denote by $C\ell_n$ the universal real Clifford algebra generated by the vector space $\mathbb{R}^n$ together with its orthonormal basis $e_1, \ldots, e_n$ endowed with the multiplication rules $e_i e_j + e_j e_i = -2\delta_{ij}$, for all $i, j = 1, \ldots, n$. It is a 2^n-dimensional real associative algebra, with basis given by

$$\mathcal{B} = \{1, e_A = e_{h_1} \cdots e_{h_k}, A \subset N\}$$

where $A = \{h_1, \ldots, h_k\} \subset N = \{1, \ldots, n\}$, for $1 \leq h_1 < \cdots < h_k \leq n$. Hence, each element $X \in C\ell_n$ can be written as a linear combination of the elements of the basis. Any linear combination of basic elements with equal length k is designated a k–vector, and we shall denote by $[X]_k$ the k-vector part of $X \in C\ell_n$. We introduce an *involutory automorphism* in the algebra $C\ell_n$, denoted *conjugation* $X \to \overline{X}$, defined by its action on the basis elements as $\overline{1} = 1$ and $\overline{(e_{h_1} \cdots e_{h_s})} = (-1)^{s+[s/2]} e_{h_1} \cdots e_{h_s}$ for all $h_1, \ldots, h_s$ satisfying $1 \leq h_1 < \cdots < h_s \leq n$.

We define an arbitrary paravector $Z \in C\ell_n$ as the element

$$Z = x_0 + x_1 e_1 + \cdots + x_n e_n,$$

where all x_i are real.

As in the complex case, we denote by $\mathrm{Sc}\,(Z) = x_0$ the *scalar part* of the paravector Z and by $\mathrm{Vec}\,(Z) = x_1 e_1 + \cdots + x_n e_n$ its *vectorial part*. The *conjugate element* of Z is given by $\overline{Z} = \mathrm{Sc}\,(Z) - \mathrm{Vec}\,(Z) = x_0 - x_1 e_1 - \cdots - x_n e_n$ and satisfies $Z\overline{Z} = \overline{Z}Z = x_0^2 + x_1^2 + \cdots + x_n^2$, the Euclidean norm of Z considered as an element of the vectorial space $\mathbb{R}^{n+1}$. Moreover, each non-zero paravector Z has a unique inverse $Z^{-1} = \frac{\overline{Z}}{|Z|^2}$.

Furthermore, a Clifford-valued function $W = W(Z)$ will be written as

$$W(Z) = \sum_{A \subset N} w_A(Z) e_A$$

with all $w_A(Z)$ being real-valued functions.

The usual properties (as continuity, differentiability, and so on) are ascribed to a Clifford-valued function $W(Z)$ by imposing that all its real-valued components $w_A(Z)$, $A \subset N$ fulfil that same condition.

The *generalized Cauchy-Riemann operator*

$$D = \partial_{x_0} + \sum_{i=1}^{n} \partial_{x_i} e_i$$

factorizes the $n + 1$-dimensional Laplace operator Δ, in the sense that $\Delta = \overline{D}D = D\overline{D}$, where

$$\overline{D} = \partial_{x_0} - \sum_{i=1}^{n} \partial_{x_i} e_i$$

is the *conjugate* of the generalized Cauchy-Riemann operator.

A Clifford-valued function $W(Z) = \sum_{A \subset N} w_A(Z) e_A$ is said to be *(left-) monogenic* in Ω if $DW(Z) = 0$ for all $Z \in \Omega$.

3. Influence of the derivative of a monogenic function on the mapping realized by it

As already mentioned in (CGKM, 2001), the generalized derivative of a monogenic function $W(Z)$ is given by the term $\frac{1}{2}\overline{D}W(Z)$, a result proved in (GM, 1999).

We will start, in what follows, by presenting conditions on the rank of the Jacobi matrix that ensures the existence of a local homeomorphism realized by a monogenic function $W(Z)$, conditions which admit a generalization to $C^{1,\alpha}$-functions. Finally, we end this section by presenting a link between the Jacobi matrix of a paravector-valued function and its generalized derivative.

Let $J = \left(\frac{\partial w_A}{\partial x_i} \right)_{A \subset N, \, i=0,\cdots,n}$ denote the Jacobi matrix of a monogenic function $W(Z)$. Then the following result holds.

LEMMA 3.1. $\mathrm{rk}J|_{Z=Z_0} = n + 1$ *implies* $\overline{D}W(Z_0) \neq 0$

Taking into consideration that $D + \overline{D} = 2\partial_{x_0}$, we can show that whenever $\overline{D}W(Z_0) = 0$ for a monogenic function W, then the first row of the Jacobi matrix J will be zero, and its rank will be less than $n + 1$. As in the quaternionic case, this is not a sufficient condition (see (CGKM, 2001), for the example $W(Z) = x_1 - x_0 e_0$ at $Z_0 = 0$).

In what follows, we need to take into consideration the natural embedding of the Clifford algebra Cl_n into the even subalgebra Cl_{n+1}^+ of all $X = \sum_{k \ even}[X]_k \in Cl_{n+1}$.

Indeed, let $\epsilon_0, \epsilon_1, \cdots, \epsilon_n$ be the orthogonal basis for the vectorial space $\mathbb{R}^{n+1}$ associated with the universal Clifford algebra Cl_{n+1}, with multiplication rules $\epsilon_i\epsilon_j + \epsilon_j\epsilon_i = -2\delta_{ij}$, $i, j = 0, \ldots, n$. Then

i) Each paravector $Z = x_0 + x_1 e_1 \cdots x_n e_n$ can be identified with the $n + 1$ dimensional vector $\underline{Z} = x_0\epsilon_0 + x_1\epsilon_1 \cdots x_n\epsilon_n$.

ii) Setting

$$\mathbf{e}_j = -\epsilon_0\epsilon_j, \quad j = 1, \cdots, n, \tag{1}$$

then the *Dirac operator* $\partial = \sum_{i=0}^{n} \epsilon_i\partial_{x_i}$ relates to the Cauchy-Riemann operator D by $-\epsilon_0\partial = D$. Moreover, (1) establishes an isomorphism between Cl_n and Cl_{n+1}^+, its action on the Cl_n-valued function $W(Z) = \sum_{A\subset N} w_A(Z)\mathbf{e}_A$ being given by

$$\underline{W}(\underline{Z}) = \sum_{A\subset N, \#A \ even} w_A(Z)\epsilon_A - \sum_{A\subset N, \#A \ odd} w_A(Z)\epsilon_0\epsilon_A.$$

Using the property of conformal weights of first order for the Dirac operator ((Ry, 1982), (Cn, 1994)) we have that

$$\begin{aligned}
\partial\underline{W}(\underline{Z}) &= \partial\overline{\underline{P}}W(\underline{P}\underline{Z}\overline{\underline{P}}) \\
&= \overline{\underline{P}}\partial\underline{W}(\underline{P}\underline{Z}\overline{\underline{P}})
\end{aligned} \tag{2}$$

for every $\underline{P} \in \mathbf{Pin}(n+1) = \{\underline{Z} = \prod_{i=1}^{k} Z_i, Z_i \in \mathbb{R}^{n+1}, \underline{Z}_i^2 \neq 0, \forall i, \forall k \in \mathbb{N}, |\overline{\underline{Z}}\underline{Z}| = 1\}$.

We construct the corresponding action of $\underline{P}\underline{Z}\overline{\underline{P}}$ in terms of paravectors, denoted by $R_P(Z)$, and given by

$$R_P(Z) = -\sum_{s=1}^{n}(p_0 p_s x_s - p_s^2 x_0)$$

$$-\sum_{i=1}^{n}\left[\sum_{j=0}^{i-1}(p_j^2 x_i - p_i p_j x_j) + \sum_{j=i+1}^{n}(p_i p_j x_j - p_j^2 x_i)\right]\mathbf{e}_i$$

ensuring that $\partial\underline{W}(\underline{P}\underline{Z}\overline{\underline{P}}) = 0$ whenever $DW(R_P(Z)) = 0$.

So we obtain, for $Y = R_P(Z)$, the result

$$\begin{aligned}
\partial_{x_0}W(R_P(Z)) &= \sum_{j=0}^{n}\frac{\partial W}{\partial y_j}\frac{\partial y_j}{\partial x_0} \\
&= \left(\sum_{i=1}^{n}p_i^2\right)\frac{\partial W}{\partial y_0} + \sum_{j=1}^{n}p_j p_0 \frac{\partial W}{\partial y_j},
\end{aligned}$$

combined with the fact that $\mathrm{rk}J|_{Z=Z_0} < n+1$ implies that $\frac{\partial W}{\partial x_0}, \frac{\partial W}{\partial x_1},$
$\cdots, \frac{\partial W}{\partial x_n}$ to be linear dependent at Z_0, that is, it exists $\alpha_0, \alpha_1, \cdots, \alpha_n$ not
all zero such that $\sum_{j=0}^n \alpha_j \frac{\partial W}{\partial x_j}(Z_0) = 0$. Hence, we obtain a paravector
P satisfying $|\mathrm{Vec}\,(P)|^2 = \alpha_0$ and $p_i \mathrm{Sc}\,(P) = \alpha_i$, for $i = 1, \cdots, n$, which
can be normalized.

Moreover, if W is a monogenic function, the relation

$$\overline{D} = 2\partial_{x_0} - D \tag{3}$$

leads to the following statement:

THEOREM 3.1. *For* $W \in \ker D$ *then*

$$\mathbf{rk}J|_{Z=Z_0} < n+1 \Longleftrightarrow \exists P : |P| = 1 \wedge \overline{D}W(R_P(Z_0)) = 0.$$

Also, a criterion for the rank of the Jacobi matrix can be obtained.

$$\mathbf{rk}J|_{Z=Z_0} < n+1 \Longleftrightarrow \min_{P : |P|=1} |\overline{D}W(R_P(Z_0))| > 0.$$

THEOREM 3.2. *Let* $W \in C^{1,\alpha}(\Omega)$ *be a solution of*
$DW(Z) = Q(Z)\overline{D}W(Z)$ *in* Ω, *with* $Q(Z_0) = 0$. *Then*

$$\mathbf{rk}J|_{Z=Z_0} < n+1 \Longleftrightarrow \exists P : |P| = 1 \wedge \overline{D}W(R_P(Z_0)) = 0.$$

From the Beltrami system, we have $2\partial_{x_0}W(Z) = (Q(Z)+1)\overline{D}W(Z)$.
Obviously, $\mathbf{rk}J|_{Z=Z_0} < n+1$ is a necessary condition for $\partial_{x_0}W(R_P(Z_0))$
$= 0$ for some $P : |P| = 1$, hence to $\overline{D}W(R_P(Z_0)) = 0$.

For the sufficient condition, $\mathbf{rk}J|_{Z=Z_0} < n + 1$ is equivalent, by
similar reasoning as in theorem 3.1, to the existence of a normalized
paravector P such that $0 = \partial_{x_0}W(R_P(Z_0))$. Also, $Q(Z_0) = 0$ leads to
$DW(Z_0) = 0$ (Beltrami equation) and, by the conformal weight properties, to $DW(R_P(Z_0)) = 0$. Now, relation (3) gives $\overline{D}W(R_P(Z_0)) = 0$.

REMARK 3.1. *The above theorem remains true when the condition*
$Q(Z_0) = 0$ *is replaced by the weaker condition that* $Q(R_P(Z_0)) + 1$ *is
no zero divisor.*

Assume now $W(Z)$ to be a paravector-valued function. In these con-
ditions, $\mathbf{rk}J|_{Z=Z_0} = n + 1$ is equivalent to $\det J(Z_0) \neq 0$. We shall
discuss the quasiconformality induced by $W(Z)$ in the neighbourhood
of $Z = 0$.

Also, $W(Z)$ is expandable in terms of the *totally regular variables* $\theta_i = x_i - e_i x_0$ in the form

$$W(Z) = W(0) + \sum_{i=1}^{n} A_i \theta_i + O(|Z|^2)$$

or, in terms of its Taylor' series expansion,

$$W(Z) = W(0) + \sum_{i=0}^{n} x_i \partial_{x_i} W(0) + O(|Z|^2).$$

In order to estimate the term $|W(Z) - W(0)|$ on the sphere $|Z| = r$, we consider the Jacobian expressed as

$$J = \left(\frac{1}{2} \overline{D} W, A_1, \cdots, A_n \right),$$

$$J^t J = \begin{pmatrix} |\frac{1}{2} \overline{D} W|^2 & \frac{1}{2} \overline{D} W \cdot A_1 & \cdots & \frac{1}{2} \overline{D} W \cdot A_n \\ \frac{1}{2} \overline{D} W \cdot A_1 & |A_1|^2 & \cdots & A_1 \cdot A_n \\ \vdots & \vdots & & \vdots \\ \frac{1}{2} \overline{D} W \cdot A_n & A_1 \cdot A_n & \cdots & |A_n|^2 \end{pmatrix},$$

where the inner product is to be understood as the usual real inner product in $\mathbb{R}^{n+1}$.

Therefore, $\mathrm{tr} J^t J = |\frac{1}{2} \overline{D} W|^2 + \sum_{i=1}^{n} |A_i|^2$, and we get the following estimates for the greatest eigenvalue σ_{max},

$$|\frac{1}{2} \overline{D} W|^2 \leq \mathrm{tr} J^t J \leq (n+1)\sigma_{max},$$

while the lowest eigenvalue σ_{min} is

$$\sigma_{min} = \min_{|Z| \neq 0} \frac{J^t J Z \cdot Z}{Z \cdot Z}.$$

Assuming $Z = e_1$ we get $\sigma_{min} \leq |\frac{1}{2} \overline{D} W|^2$, that is

$$\sigma_{min} \leq |\frac{1}{2} \overline{D} W|^2 \leq (n+1)\sigma_{max}.$$

From this inequality, we obtain

$$\frac{\sigma_{min}}{\sigma_{max}} |\frac{1}{2} \overline{D} W|^2 \leq \sqrt{|\det J|} \leq \frac{\sigma_{max}}{\sigma_{min}} |\frac{1}{2} \overline{D} W|^2$$

and

$$\frac{\sigma_{min}}{\sigma_{max}} \sqrt{|\det J|} \leq |\frac{1}{2} \overline{D} W|^2 \leq \frac{\sigma_{max}}{\sigma_{min}} \sqrt{|\det J|},$$

which proves the equivalence between the Jacobian determinant and the generalized derivative of a paravector-valued function.

4. Solvability

Let us consider the Beltrami equation

$$DW = Q\overline{D}W$$

with $Q \in \mathcal{W}_p^1(\mathbb{R}^{n+1}), p > n+1$. Applying the ansatz

$$W = \Phi + TH,$$

where $TH = -\frac{1}{\omega} \int_{\mathbb{R}^{n+1}} \frac{\overline{\xi-Z}}{|\xi-Z|^{n+1}} H(\xi) d\mathbb{R}_\xi^{n+1}$ is the Teodorescu-transform, ω the area of the unit sphere in $\mathbb{R}^{n+1}$, and

$$\Phi(Z) = \begin{cases} x_0 + e_1 x_1 - e_2 x_2 + e_3 x_3 - \ldots + e_n x_n & n \text{ odd} \\ 2x_0 + e_1 x_1 - e_2 x_2 + e_3 x_3 - \ldots + e_{n-1} x_{n-1} + e_n x_n & n \text{ even} \end{cases}$$

we obtain the singular integral equation

$$H = Q + Q\Pi H, \tag{4}$$

where $\Pi H = -\frac{1}{\omega} \int_{\mathbb{R}^{n+1}} \frac{(n-1)+(n+1)\frac{\overline{\xi-Z}^2}{|\xi-Z|^2}}{|\xi-Z|^{n+1}} H(\xi) d\mathbb{R}_\xi^{n+1} + \frac{1-n}{1+n} H(Z)$ is the singular Π-operator investigated in detail in (GK1, 1996), (GK2, 1997). For $\|Q\|_{\mathcal{W}_p^1} \leq q_c < 1/\|\Pi\|_{\mathcal{W}_p^1}, p > n+1$, this singular integral equation can be solved using the Banach fixed point theorem. We only remark that the Sobolev spaces $\mathcal{W}_p^k(\mathbb{R}^{n+1})$ form a Banach algebra for $kp > n+1$. Moreover, according to the mapping properties of the aforementioned operators and the well-known Sobolev embedding theorem we obtain a solution $W \in \mathcal{W}_p^2, p > n+1$, which also lies in the space $C^{1,\alpha}(\mathbb{R}^{n+1}), \alpha = 1 - (n+1)/p$. Based on this property, we will investigate in the next section the question of a locally or globally quasi conformal solution.

5. Local homeomorphisms

Due to the two different forms of $\Phi(Z)$ we will restrict ourselves to the case n even without any loss of generality.

Suppose that $Q \in \mathcal{W}_p^1(\mathbb{R}), p > n+1, \|Q\| \leq q_c < 1/\|\Pi\|$. Then for each neighbourhood $U_\delta(0) = \{Z \in \mathbb{R}^{n+1}, |Z| < \delta\}$ we consider the function $Q_\delta(Z) = Q(Z)\varphi(Z)$ with

$$\varphi(Z) = \begin{cases} 1 & , |Z| < \frac{1}{2}\delta \\ 2\left(\frac{|Z|}{\delta}\right)^3 - \frac{9}{2}\left(\frac{|Z|}{\delta}\right)^2 + 3\left(\frac{|Z|}{\delta}\right) - \frac{1}{2} & , \frac{1}{2}\delta \le |Z| \le \delta \\ 0 & , |Z| > \delta \end{cases}$$

Then $\quad \|Q_\delta\|_{\mathcal{W}_p^1} \quad \le \quad \|Q\|_{\mathcal{W}_p^1}\|\varphi\|_{\mathcal{W}_p^1} \quad \le \quad C_\varphi \delta^n \|Q\|_{\mathcal{W}_p^1}, \quad$ with $C_\varphi = \left(\frac{2^n+15}{2^{n+1}(n+1)} + \frac{2}{n+4} + \frac{3}{2(n+3)} - \frac{6}{(n+2)}\right)\omega$, ω being the area of the unit sphere in $\mathbb{R}^{n+1}$.

Let us denote by $\overset{o}{\mathcal{W}_p^1}(U_\delta(0))$ the space of all $\mathcal{W}_p^1(\mathbb{R}^{n+1})$-functions with support in $U_\delta(0)$. Then we have for the operator Π_δ, defined by $\Pi_\delta H = Q_\delta \Pi H$, the mapping property $\Pi_\delta : \overset{o}{\mathcal{W}_p^1}(U_\delta(0)) \mapsto \overset{o}{\mathcal{W}_p^1}(U_\delta(0))$, $1 < p < \infty$. Moreover, from $\|Q\|_{\mathcal{W}_p^1} \le q_c < 1/\|\Pi\|_{\mathcal{W}_p^1}, p > n+1$, and $\|Q^\delta\|_{\mathcal{W}_p^1} \le C_\varphi \delta^n \|Q\|_{\mathcal{W}_p^1}$ it follows that for all $\delta \le \frac{1}{\sqrt[n]{C_\varphi}}$ the operator Π_δ is a contraction over $\mathcal{W}_p^1(U_\delta(0)), p > n+1$.

Therefore, we obtain that $W = \Phi + TH_\delta$, where H_δ is a solution of $H_\delta = Q_\delta + \Pi_\delta H_\delta$, is a solution of the Beltrami equation $DW = Q_\delta \overline{D}W$ over $\mathbb{R}^{n+1}$, and also a solution of the Beltrami equation $DW = Q\overline{D}W$ over $U_\delta(0)$. Furthermore, from Banach's fixed-point theorem we get the estimate

$$\|H_\delta\|_{\mathcal{W}_p^1} \le \frac{\|Q_\delta\|}{1 - \|Q_\delta\|_{\mathcal{W}_p^1}\|\Pi\|_{\mathcal{W}_p^1}} < \frac{C_\varphi \delta^n \|Q\|_{\mathcal{W}_p^1}}{1 - C_\varphi \delta^n \|Q\|_{\mathcal{W}_p^1}\|\Pi\|_{\mathcal{W}_p^1}}.$$

If $\delta < 1/\sqrt[n]{2C_\varphi}$ then we have

$$\|H_\delta\|_{\mathcal{W}_p^1} < 2C_\varphi \|Q\|_{\mathcal{W}_p^1}\delta^n < \frac{2C_\varphi}{\|\Pi\|_{\mathcal{W}_p^1}}\delta^n.$$

Therefore, we obtain for all $\epsilon > 0$, that there exists a δ such that

$$\|H_\delta\|_{\mathcal{W}_p^1} < \frac{\epsilon}{C_p\|\Pi\|_{\mathcal{W}_p^1}}, \tag{5}$$

where C_p denotes the embedding constant of $\mathcal{W}_p^1(\mathbb{R}^{n+1}), p > n+1$, in $\mathcal{C}(\mathbb{R}^{n+1})$. This allows us to investigate the problem of a locally quasiconformal solution based on section 3. For answering the question whether $\overline{D}_Z W(R_P Z) \ne 0$ at the point $Z = 0$ for all $P \in \mathbb{R}^{n+1}$, we can transform the term $\overline{D}_Z W(R_P Z) = \overline{D}_Z \Phi(R_P Z) + \overline{D}_Z TH(R_P Z)$ into $\sum_A \mathbf{e}_A \left(P^T Q_A P\right)$, where Q_A are real symmetric matrices and P is considered as a vector $P = (p_0, p_1, \ldots, p_n)^T$. If one of these matrices Q_A is positive (or negative) definite, we have solved our problem.

Therefore, it is enough to look only at the term $\operatorname{Sc} \overline{D}_Z W(R_P Z) = \operatorname{Sc} \overline{D}\Phi(R_P Z) + \operatorname{Sc} \overline{D}TH(R_P Z) = 3\sum_{i=0}^{n} p_i^2 + \operatorname{Sc} \overline{D}TH(R_P Z)$. For this term we obtain the matrix

$$Q_0 = \begin{pmatrix} 3 + C_{00} & C_{01} & \cdots & C_{0n} \\ C_{10} & 3 + C_{11} & \cdots & C_{1n} \\ C_{20} & C_{21} & \cdots & C_{2n} \\ \vdots & \vdots & \cdots & \vdots \\ C_{n0} & C_{n1} & \cdots & 3 + C_{nn} \end{pmatrix}$$

where C_{ij} can be considered as singular integral operators of H evaluated at the point zero. It can be observed, that the C_{ij} depend only on the scalar part, the vector part, and the bivector part of H. For the calculation of the C_{ij} we simply remark that $\partial_0 R_P Z = \sum_{i=1}^{n} p_i^2 + \sum_{i=1}^{n} e_i p_i p_0$ and $\partial_k R_P Z = -p_0 p_k + e_k \left(\sum_{i=0}^{k-1} p_i^2 - \sum_{i=k+1}^{n} p_i^2\right) - \sum_{i=1}^{k-1} e_i p_i p_k + \sum_{i=k+1}^{n} e_i p_i p_k$. By analogy with the quaternionic case, we find

$$|C_{ij}| \leq \|\tilde{C}_{ij} H_\delta\|_C \leq C_p \|\tilde{C}_{ij}\|_{\mathcal{W}_p^1} \|H_\delta\|_{\mathcal{W}_p^1} \leq C_p \|\Pi\|_{\mathcal{W}_p^1} \|H_\delta\|_{\mathcal{W}_p^1}$$

for all indices i and j, where $\tilde{C}_{ij}$ denotes the singular operator with $C_{ij} = \tilde{C}_{ij} H_\delta(0)$ and C_p is again the embedding constant of $\mathcal{W}_p^1(\mathbb{R}^{n+1})$, $p > n + 1$, in $C(\mathbb{R}^{n+1})$. Therefore, due to (5) we have $|C_{ij}| < \epsilon$. Furthermore, our matrix $Q_0 = (q_{ij})$ is positive definite if $q_{ii} > \sum_{j \neq i} |q_{ij}|$. By choosing some $\epsilon < 3/(n+1)$ we can proceed to establish:

THEOREM 5.1. *Suppose that $Q \in \mathcal{W}_p^1(\mathbb{R}^{n+1})$ for a certain $p > n + 1$ and $\|Q\|_{\mathcal{W}_p^1} \leq q_c < 1/\|\Pi\|_{\mathcal{W}_p^1}$. Suppose also that $Q(Z) + 1$ is not a zero divisor in any point $Z \in \mathbb{R}^{n+1}$. Then the generalized Beltrami equation*

$$DW(Z) = Q(Z)\overline{D}W(Z)$$

has a solution, which realizes a locally quasiconformal mapping at each point Z_0.

Moreover, we can state the following theorem:

THEOREM 5.2. *If $Q \in \mathcal{W}_p^1(\mathbb{R}^{n+1})$, $\|Q\|_{\mathcal{W}_p^1} \leq q_c < \frac{2}{(n-1)\|\Pi\|_{\mathcal{W}_p^1}}$ for some $p > n + 1$ and $Q(Z) + 1$ is not a zero divisor for all $Z \in \mathbb{R}^{n+1}$, then the function*

$$W = \Phi + TH,$$

where H satisfies the corresponding singular integral equation (4), is a solution of the Beltrami equation

$$DW(Z) = Q(Z)\overline{D}W(Z)$$

which realizes a locally quasiconformal mapping in each point Z_0.

The proof is simply an extension of the considerations above, using the affine transformation $\eta = Z - Z_0$ and the norm estimate $\|H\| < \dfrac{\|Q\|}{1 - \|Q\| \|\Pi\|}$

References

Brackx, F.,Delanghe, R., and Sommen, F. - *Clifford analysis*, Pitman Advanced Publishing Program, 1982.

Cerejeiras, P., Gürlebeck, K., Kähler, U., and Malonek, H. - A quaternionic Beltrami type equation and the existence of local homeomorphic solutions, *ZAA*, to appear.

Cnops, J. - *Hurwitz pairs and applications of Möbius transformations*, Habilitation Thesis - Ghent, 1994.

Delanghe, R.: 1970 - On regular-analytic functions with values in a Clifford Algebra, *Math. Ann.*, Vol. **185**, pp. 91-111

Gürlebeck, K. and Kähler, U.: 1996 - On a spatial generalization of the complex Π-operator, *ZAA* , Vol. **15** No. **2**, pp. 283-297.

Gürlebeck, K. and Kähler, U.: 1997 - On a boundary value problem of the biharmonic equation, *Math. Meth. in the Applied Sciences*, Vol. **20**, pp. 867-883.

Gürlebeck, K. and H. Malonek: 1999 - A Hypercomplex Derivative of Monogenic Functions in $\mathbb{R}^{n+1}$ and its Applications, *Complex Variables*, **39**, pp. 199 - 228.

Gürlebeck, K. and W. Sprößig - *Quaternionic and Clifford calculus for Engineers and Physicists*, Chichester, John Wiley &. Sons 1997.

Ryan, J.: 1982 - Properties of Isolated Singularities of some Function taking Values in Real Clifford Algebras, *Math. Proc. Camb. Philos. Soc.*, Vol. **95**, pp. 277-298.

Parallel Transport of Algebraic Spinors on Clifford Manifolds

J.S.Roy Chisholm (chisholmrandm@wcb.u-net.com)

Institute of Mathematics and Statistics, University of Kent

Abstract. A Clifford manifold is defined by a position-dependent frame field $\{e_\mu(x)\}$ and metric $g_{\mu\nu}(x)$, satisfying the anti-commutation rule $\{e_\mu, e_\nu\} = g_{\mu\nu}I$. At each point x, orthonormal basis vector sets define the tangent space and the spin group. The Riemannian and spin connections are defined by imposing covariance under coordinate and spin group transformations. Then the frame field is necessarily parallel transported as a spin vector, subject to both connections.

Contraction of parallel transported vector fields with the frame field defines 'spin elements' which are parallel transported with two-sided spin connection. Idempotents, and hence algebraic spinors, are defined in terms of spin elements, and it is shown that the idempotent property is preserved under parallel transport.

The two-sided spin connection gives rise to an extra interaction of the kind studied earlier, proportional to the algebraic unit. It is suggested that this term might be the source of a cosmological 'constant'. More generally, it is conjectured that the necessary asymmetry of spinor idempotents might be the source of asymmetries observed in nature.

Keywords: manifold, framefield, Clifford algebra, algebraic spinors, parallel transport, spin group, cosmological constant, electroweak interactions

Math Subject Classifications: 11E88, 15A63, 15A66, 53B15, 53B25, 53C07, 53C80, 53Z05, 83C60, 83E15

Dedicated to Richard Delanghe on the occasion of his 60th birthday.

1. Clifford Manifolds

In the development of spin gauge theory models of elementary particle interactions [1,2,3], Ruth Farwell and I evolved a description of a space-time manifold which combined a number of existing concepts within a Clifford algebraic formalism, and later we gave a non-rigorous account [4] of these ideas. Throughout our work, we have aimed to develop a single mathematical structure which accounts for the observed properties of elementary particles and their interactions, including gravitation. In order to include particle spin, we must go beyond Riemannian ideas, but we have found that our concepts do not easily fit into the standard fibre bundle formalism; this is because the basic Clifford algebraic relations (1.2) below, are related to both 'base space' and 'spin bundle'. Our

59

F. Brackx et al. (eds.), Clifford Analysis and Its Applications, 59–69.

concepts are less general, and therefore simpler, than those formulated in terms of fibre bundles.

We adopt the model of an n-dimensional manifold as a union of a countable number of patches, with each patch P corresponding to a continuous range of real coordinates $(x^1, x^2, \ldots, x^n) = x$ changing to a different coordinate system y, where the Jacobian $\partial(y)/\partial(x)$ is non-zero, introduces the group GL(n) of coordinate transformations. Increments of the coordinates in the two systems are related by

$$dy^\nu = (\partial y^\nu/\partial x^\mu)dx^\mu. \tag{1.1}$$

We can now introduce the first of two fundamental principles governing equations describing physical systems:

Principle 1: Any equation which describes physically observable quantities must be covariant under the group GL(n) of coordinate transformations.

This principle has governed the General Theory of Relativity in space-time. However, it also applies to models of physical systems in higher-dimensional spaces. For this reason, we shall develop the theory for manifolds of any dimensionality n and signature (p, q). I now introduce the concept of a Clifford manifold.

An n-dimensional Clifford manifold is defined by a Frame Field, consisting of a set of n basis vectors $\{e_\mu(x); \mu = 1, 2, \ldots, n\}$ at each point x, and a non-degenerate metric $g_{\mu\nu}(x)$, satisfying the basic Clifford algebraic anti-commutation relations

$$\{e_\mu, e_\nu\} \equiv 2Ig_{\mu\nu}, \tag{1.2}$$

where I is the algebraic unit, with the increment of distance on the manifold given by

$$ds = e_\mu(x)dx^\mu. \tag{1.3}$$

To ensure that ds is invariant, $\{e_\mu(x)\}$ must transform cogrediently under allowed coordinate transformations on P, and the non-degeneracy of the metric ensures that they linearly independent.

The vectors

$$e_\mu(x) = \partial s/\partial x^\mu \tag{1.4}$$

are tangent to the manifold at x, and span the tangent space at that point. At each point x, we can define from $\{e_\mu(x)\}$ an orthonormal set of vectors $\{f_r(x); r = 1, 2, \ldots, n\}$, satisfying the Clifford relations

$$\{f_r(x), f_s(x)\} = 2I\eta_{rs}, \tag{1.5}$$

where η_{rs} is the Minkowski metric corresponding to the signature of the coordinate metric. Then

$$e_\mu(x) = h_\mu^r(x)f_r(x), \tag{1.6}$$

where the real expansion coefficients $\{h^r_\mu(x)\}$ constitute the vielbein field. Substituting (1.6) into (1.2) and using (1.5) gives the standard relation

$$g_{\mu\nu} = \eta_{rs} h^r_\mu(x) h^s_\nu(x). \tag{1.7}$$

Weinberg [5], developing the work of Utiyama [6] and Kibble [7], has emphasised the physical importance of orthonormal frames in Relativity theories in space-time. Transformations between orthonormal frames constitute the Lorentz group of Special Relativity, and the choice of a local orthonormal frame is the basis of the Principle of Equivalence in General Relativity. *This physical significance is paralleled, and indeed generalised, by the mathematical significance of orthonormal frames in Clifford algebras: they are the set of frames which define the grading structure of an algebra.* Products of k of the vectors $\{f_r(x)\}$ define the set of k-vectors of the algebras, but products of the set $\{e_\mu(x)\}$ are in general of mixed grade. In particular, the bivectors of an algebra at a point x, products of pairs of an orthonormal set, generate the spin group $Spin(p,q)$, a subgroup of the Clifford-Lipschitz group [8]. We can transform continuously to another orthonormal frame by an operation of the spin group. If $\Lambda(x)$ is an element of the spin group, this effects the transformation

$$f_r(x) \rightarrow \Lambda(x) f_r(x) \Lambda^{-1}(x), \tag{1.8}$$

which preserves the basic relations (1.5). We can therefore choose the orthonormal frame at each point.

The second rule to be satisfied by physical quantities is their independence of the choice of local frame $\{f_r(x)\}$. This is formulated as:
Principle 2: Any equation describing a physical system must be covariant under a change (1.8) of local frame of reference.
Weinberg [5] has emphasised the importance and independence of *Principle 1* and *Principle 2*.

2. Connections; the Parallel Transport Identity

If $\{V_\nu(x)\}$ are the components of a vector field on a manifold, the coordinate derivatives $\partial_\mu V_\nu$ on the manifold do not transform as a tensor. Introducing the Riemannian connection coefficients

$$\Gamma^\lambda_{\mu\nu} = g^{\lambda\sigma}(\partial_\mu g_{\nu\sigma} + \partial_\nu g_{\mu\sigma} - \partial_\sigma g_{\mu\nu})/2, \tag{2.1}$$

ensures that the covariant derivative

$$D_\mu V_\nu \equiv \partial_\mu V_\nu - \Gamma^\lambda_{\mu\nu} V_\lambda \tag{2.2}$$

transforms as a tensor, satisfying *Principle 1*.

Shortly after Dirac had discovered his theory of the electron, based on spinors in flat space-time, Fock and Ivanenko [9] began the study of covariant derivatives of spinors on a space-time manifold, defining an incremental distance through (1.3). They interpreted the space-time basis vectors, denoted by $\{\gamma_\mu(x)\}$ instead of $\{e_\mu(x)\}$, as geometric objects, noting that (1.1) and (1.2) ensure that

$$ds^2 = I\, g_{\mu\nu} dx^\mu dx^\nu, \tag{2.3}$$

the standard formula for a Riemannian manifold. However, they also noted that $\{\gamma_\mu(x)\}$ were the generalisations of Dirac matrices, which have a physical meaning.

There is a long history of these two interpretations of this formalism in space-time. The geometric interpretation has been emphasised, for example, by H.S Green [10] in his study of teleparallelism, by Hestenes and Sobczyk [11], and in our own work [1,2,3]. The frame field concept, with vectors not necessarily satisfying the Clifford relation (1.1), has also been used by Chern [12] as a basic concept in differential geometry and topology.

The physical interpretation in terms of spin has also been widely studied. In 1932, Schrödinger [13] investigated the covariant derivative of the space-time basis elements $\{\gamma_\nu(x)\}$, defining their parallel transport, and showed that it had to be of the form

$$\Delta_\mu \gamma_\nu = \partial_\mu \gamma_\nu - \Gamma_{\mu\nu}^\lambda \gamma_\lambda + [\Gamma_\mu, \gamma_\nu], \tag{2.4}$$

where $\{\Gamma_\mu\}$ are described as a vector of matrices, now known as the spin connection. He also stated the necessity for this equation to satisfy both of *Principles 1 and 2*.

These invariance principles are certainly satisfied if the set $\{\Gamma_\mu\}$ are chosen so that $\Delta_\mu \gamma_\nu = 0$. This condition was studied by several authors, and Fletcher [14] gave a general evaluation of the spin connection derived from this assumption. The generalisation of this assumption to a general Clifford manifold is

$$\Delta_\mu e_\nu = \partial_\mu e_\nu - \Gamma_{\mu\nu}^\lambda e_\lambda + [\Gamma_\mu, e_\nu], \tag{2.5}$$

and it can then be shown [14,15,1,4] that Γ_μ must be of the form

$$\Gamma_\mu = [e_\nu D_\mu e^\nu]/8 + I S_\mu, \tag{2.6}$$

where $S_\mu(x)$ is an arbitrary field, with

$$e^\nu = g^{\mu\nu} e_\nu. \tag{2.7}$$

Although S_μ has sometimes been given a physical interpretation, we take $S_\mu \equiv 0$.

However, Weinberg [5] has shown that, for space-time, the form (2.6) of the spin connection can be derived in a more fundamental way, by postulating invariance under both *Principle 1* and *Principle 2*. His derivation can be adapted to any Clifford manifold, and a proof of this result will be given elsewhere. Then both the Riemannian curvature (2.1) and the spin curvature (2.6) are derived from the two invariance principles, which is more satisfactory than making the apparently arbitrary assumption (2.5). The standard relation between the two curvatures then follows unambiguously [16]. The derivation of the form of the two curvatures from invariance principles changes the status of (2.5): it now becomes an identity, the *Parallel Transport Identity*.

$$\Delta_\mu \mathbf{e}_\nu \equiv \partial_\mu \mathbf{e}_\nu - \Gamma^\lambda_{\mu\nu} \mathbf{e}_\lambda + [\Gamma_\mu, \mathbf{e}_\nu], \qquad (2.8)$$

Although we assume that $S_\mu \equiv 0$, (2.4) is true for any field S_μ. It is important to note that the proof of this identity depends upon the assumption of the basic Clifford relations (1.2). *The Parallel Transport Identity is therefore essential only for Clifford manifolds, or for manifolds which assume relation (1.2) between the basis vectors and the metric.*

The identity (2.4) has a direct geometric meaning: *the Frame Field* $\{\mathbf{e}_\mu(x)\}$ *is parallel transported on a Clifford manifold*. At first sight this seems to be a remarkable result. But since the manifold in essence *is* the frame field, the identity is better seen as a check on the consistency of the Clifford manifold concept. Since $\Gamma^\lambda_{\mu\nu}$ and Γ_μ are the Riemannian and spin connections, (2.11) is covariant under both coordinate and spingroup transformations, provided that these connections transform in the standard way [16]. *So it is generally true that the frame field is parallel transported as a 'spin vector'.*

3. Spin elements

There are two definitions of spinors. The standard theory of quantum mechanics uses column spinors ψ and conjugate row spinors $\overline{\varphi}$. The transformations of column and row spinors under the appropriate spin groups are of the form

$$\psi(x) \to \Lambda(x)\psi(x), \qquad \overline{\varphi}(x) \to \overline{\varphi}(x)\Lambda^{-1}(x) \qquad (3.1)$$

These one-sided transformations correspond to observable spin-$\frac{1}{2}$ properties of the electron and other fermions. However, the transformation of an element $\mathbf{E}(x)$ of a Clifford algebra is of the two-sided form

$$\mathbf{E}(x) \to \Lambda(x)\mathbf{E}(x)\Lambda^{-1}(x). \tag{3.2}$$

Likewise, the covariant derivative of a column spinor is [5]

$$\partial_\mu \psi + \Gamma_\mu \psi , \tag{3.3}$$

with left multiplication by Γ_μ instead of the commutation operation in (2.4) and (2.8).

The alternative definition of spinors is as minimal left ideals of an algebra, known as algebraic spinors. If $\mathbf{U}(x)$ is a primitive idempotent of the algebra and $\mathbf{E}(x)$ any element, spinors are of the form

$$\psi(x) = \mathbf{E}(x)\mathbf{U}(x). \tag{3.4}$$

The action of the spin group on algebraic spinors is normally taken to be given by (3.1), and the Lorentz transformations of special relativity must be of this form. However, we have proposed a form of gauge theory [17] in which the action of the gauge group on spinors is two-sided, as in (3.2). Invariance is then assured if gauge fields $A_\mu(x)$ are introduced with a two-sided interaction involving the commutator

$$[A_\mu(x), \psi(x)], \tag{3.5}$$

and we have shown that this modification leads to a simplification of the interaction Lagrangian in the theory of electroweak interactions.

Here, I am proposing a way of defining algebraic spinors which are covariant on a Clifford manifold, in terms of covariant vector fields. However, the spin connection appears in the commutator form (3.5), rather than as the left multiplier in (3.3). Since the derivative $\partial_\mu \psi$ satisfies Leibnitz' theorem, it can conveniently be written in commutator form $[\partial_\mu, \psi]$ when ∂_μ is operating on a product. Then the algebraic spinor covariant derivative appears in the form

$$[\partial_\mu + \Gamma_\mu, \psi], \tag{3.6}$$

with a connection term of the form (3.5).

Suppose that $W^\nu(x)$ is a vector field with Riemannian covariant derivative

$$\partial_\mu W^\nu - \Gamma^\nu_{\mu\lambda} W^\lambda . \tag{3.7}$$

Now contract $W^\nu(x)$ with the frame field $\mathbf{e}_\nu(x)$ to form the 'spin element'

$$\mathbf{W}(x) = \mathbf{e}_\nu(x)W^\nu(x) ; \tag{3.8}$$

then the 'two-sided covariant derivative' of $\mathbf{W}(x)$ is, using (2.8),

$$[\partial_\mu + \Gamma_\mu, \mathbf{W}(x)] = [\partial_\mu + \Gamma_\mu, \mathbf{e}_\nu]W^\nu + \mathbf{e}_\nu(\partial_\mu W^\nu)$$

$$= \mathbf{e}_\nu(\partial_\mu W^\nu + \Gamma^\nu_{\mu\lambda}W^\lambda). \qquad (3.9)$$

Since the bracket contains the covariant derivative of W^ν, the two-sided covariant derivative (3.9) is in fact covariant under both coordinate and spin transformations. Because (3.9) is of commutator form, it follows at once that, if $\{W^\nu_r(x);\ r = 1, 2, \ldots, k\}$ are contravariant vector fields, and $\{\mathbf{W}_r(x)\}$ are spin elements of the form (3.8), the commutator

$$[\partial_\mu + \Gamma_\mu, \prod_r \mathbf{W}_r(x)] \qquad (3.10)$$

is coordinate and spin covariant. *So any polynomial in spin elements $\{\mathbf{W}_r(x)\}$ has a covariant derivative which is a sum of terms containing factors of the form (3.9), and is thus coordinate and spin covariant.*
This property of spin elements will be used in the next section to construct idempotents, and hence algebraic spinors. But the effectiveness of this construction will depend upon a further property of spin elements. The scalar product of two spin elements is

$$\mathbf{W}_r(x).\mathbf{W}_s(x) \equiv \tfrac{1}{2}\{\mathbf{W}_r(x), \mathbf{W}_s(x)\}$$

$$= \tfrac{1}{2}\{\mathbf{e}_\mu(x), \mathbf{e}_\nu(x)\}W^\mu_r W^\nu_s = \mathbf{I}g_{\mu\nu}W^\nu_r W^\nu_s. \qquad (3.11)$$

Thus the Clifford scalar product of spin elements in the local vector space is essentially equal to the inner product of vectors on the manifold. In particular, an orthonormal set of spin elements $\{\mathbf{W}_r(x)\}$ in the Clifford algebra corresponds to an orthonormal set of vectors $\{W^\mu_r\}$ relative to the manifold metric.
This key result is particularly important for a set of vectors $\{W^\mu_r\}$ which are parallel transported with the Riemannian connection. Then the inner products $g_{\mu\nu}W^\mu_r W^\nu_s$ are preserved, and so are constant over the whole patch P. Therefore the Clifford scalar products $\mathbf{W}_r(x).\mathbf{W}_s(x)$, parallel transported with two-sided spin connection (3.9), are constant over the patch, since $\mathbf{I}$ is the universal algebraic identity element. In particular, an orthonormal set of spin elements $\{\mathbf{W}_r(x)\}$ are parallel transported as an orthonormal set, with two-sided spin connection (3.9).

4. Gravitating algebraic spinors

Instead of giving a general definition of gravitating spinors, I shall define them for a particular model. This model is based on the Clifford algebra $R_{1,6}$, and was used in Appendix B of paper [17] to show how 'two-sided interactions' of the form (3.5) can simplify the standard electroweak Lagrangian of the standard theory. The basis vectors $\{f_0, f_r; r = 1, \ldots, 6\}$ satisfy

$$f_0^2 = -f_r^2 = I, \qquad (4.1)$$

and $\{f_0, f_1, f_2, f_3\}$ correspond to space-time. The pseudoscalar of the algebra satisfies all the properties of the unit imaginary, and so is denoted by i. The idempotent chosen to define spinors is

$$U = \tfrac{1}{8}(I + f_{03})(I - ie_{12})(I - ie_{45}), \qquad (4.2)$$

where $\rho_3 = ie_{45}$ plays the role of the third isospin operator. We can take U to be the idempotent at one point, $x = x_0$ say, on a manifold. Now define vector fields $\{W_r^\mu\}$ to be the unit vectors at $x = x_0$, so that the corresponding spin elements are

$$W_r(x_0) = f_r. \qquad (4.3)$$

If we now assume that $\{W_r^\mu\}$ are parallel transported with Riemannian connection, then any polynomial in $\{W_r(x)\}$ is parallel transported with two-sided spin connection. So with this choice of vector fields, making the replacement

$$f_r \to W_r(x) \qquad (4.4)$$

defines the parallel transport of any element of the Clifford algebra. So writing idempotent (4.2) in the form

$$U(x) = \tfrac{1}{8}[I + W_0(x)W_3(x)][I - iW_1(x)W_2(x)][I - iW_4(x)W_5(x)] \qquad (4.5)$$

ensures that it is the parallel transport of (4.2). Further, since the idempotent property depends the orthonormality of the defining frame, the orthonormality of $\{W_r(x)\}$ *ensures that (4.5) is idempotent.*
Spinors at $x = x_0$ are of the form $\psi(x_0) = E[f_r(x_0)]U[f_r(x_0)]$, where $E[f_r(x_0)]$ is some polynomial in the basis vectors. The parallel transported spinor on the patch P is formed simply by making the replacement (4.4), giving

$$\psi(x) = E[W_r(x)]U[W_r(x)]. \qquad (4.6)$$

This is the principal result of this investigation.

So on a patch of a Clifford manifold, if we know the form of the vector fields $\{W_r^\mu\}$, *satisfying (4.3) and parallel transported with Riemannian connection, then any spinor defined in terms of the Clifford algebra basis* $\{f_r\}$ *is automatically parallel transported by making the replacements (4.4).*

5. An asymmetry hypothesis

The model based on the algebra $R_{1,6}$, with chosen idempotent (4.2), was used to simplify the electroweak interaction Lagrangian. This interaction is unsymmetrical in two ways. First, although the helicity-plus component of the interaction is symmetrical between the three isospin components, the helicity-minus component contains only one isospin component, $\mathbf{i}e_{45} \equiv \rho_3$, associated with the $W_{4\mu}$ field. Second, this single isospin component, defined in terms of elements of the algebra, is combined with a term proportional to the unit $\mathbf{I}$ of the algebra. The use of algebraic spinors, together with gauge transformations which give rise to two-sided interactions of the form (3.5), allowed us to eliminate the term proportional to $\mathbf{I}$ from the Lagrangian. This is because the extra 'right-hand' ρ_3 interaction term is absorbed by the term $(\mathbf{I} - \mathbf{i}e_{45}) \equiv (\mathbf{I} - \rho_3)$ in the idempotent (4.2), and is equivalent to $-\mathbf{I}$.

Although the use of algebraic spinors has enabled us to eliminate one of the asymmetries of the electroweak interaction from the Lagrangian, we have not succeeded in fully symmetrising it between positive and negative helicity components. Ideally, we would like to derive the observed asymmetric interactions from a perfectly symmetrical Lagrangian. However, our partial success suggests that there may be a necessary source for the asymmetries observed in nature. *Whatever we choose as a spinor idempotent, it is necessarily asymmetrical.* For example, the idempotent (4.2) selects out Clifford algebra elements f_{03}, $\mathbf{i}e_{12}$ and $\mathbf{i}e_{45} \equiv \rho_3$, which break various symmetries Then perfectly symmetrical two-sided Lagrangian terms will give rise to unsymmetrical interaction terms containing the unit $\mathbf{I}$.

If we now consider the two-sided spin curvature term $[\Gamma_\mu, \psi]$, where Γ_μ is the bivector part of (2.6), we see that, because of the spacetime bivectors in (4.2), the term $\psi\Gamma_\mu$ appearing in the Lagrangian will give rise to an interaction term proportional to $\mathbf{I}$, but with unsymmetrical coefficients proportional to $\eta^{rt}h_t^\sigma(D_\mu h_\sigma^s)$, with $(r, s) = (0, 3)$ and $(1, 2)$ only. The dependence on $\mathbf{I}$, and the fact that this term is of gravitational magnitude, suggests that it might play the role of a cosmological 'constant', but the spatially asymmetrical functional form is quite different from a universal constant. The question arises: could this extra spin curvature term, appearing as a particle interaction, give rise to a cosmological 'constant' on astronomical scales ? Let me make an even wilder speculation. The factor $(I + f_{03})$ in the idempotent (4.2) has a time asymmetry. Could a factor such as this be responsible for the overall time-asymmetry we experience in the universe?

ACKNOWLEDGMENTS

I have been helped by the interest shown by Ruth Farwell in the work on gravitating spinors, which is based upon our collaborative work over many years. I am also grateful to Carol Chisholm for her translation from German of Schrödinger's paper [13].

REFERENCES

1. J.S.R. Chisholm and R.S. Farwell : *Gravity and the Frame Field*, J. Phys. **A20**, 6561, 1987.

2. J.S.R. Chisholm and R.S. Farwell : *Unified Spin Gauge Theory of Electroweak and Gravitational Interactions*, J. Phys. **A22**, 1059, 1989.

3. J.S.R. Chisholm and R.S. Farwell : *Unified Spin Gauge Theories of the Four Fundamental Forces*, 'The Interface of Mathematics and Physics', eds. D.G.Quillan et al. (OUP), 193, 1990.

4. J.S.R. Chisholm and R.S. Farwell : *Clifford Approach to Metric Manifolds*, Supp. ai Rendiconti del Circ. Mat. di Palermo, **2**, 26, 123, 1991.

5. S. Weinberg : *"Gravitation and Cosmology"*, John Wiley and Sons, 365-373, 1972

6. R. Utiyama :*Invariant Theoretical Interpretation of Interaction*, Phys. Rev. **101**, 5, 1597, 1955.

7. T.W.B. Kibble : *Lorentz Invariance and the Gravitational Field*, J.Math.Phys. **2**, 2, 212, 1961.

8. I.R. Porteous : *Chapter 16, 'Clifford Algebras and the Classical Groups'* (CUP), 1995.

9. V. Fock and D. Iwanenko : *Geometrie Quantique Lineare et Deplacement Parallele*, Comptes Rendues **188**, 1470, 1929.

10. H.S. Green : *Spinor Fields in General Relativity*, Proc. Roy. Soc. **A245**, 521, 1958.

11. D. Hestenes and G. Sobczyk : *Chapter 6, "Clifford Algebra to Geometric Calculus"* (D.Reidel), 1984.

12. S. Chern : *Vector Bundles with a Connection*, "Global Differential Geometry" (Math. Assn. America), 1989.

13. E. Schrödinger : *Diracsches Elektron im Schwerefeld I*, Preuss. Akad. Wiss. (Berlin), Mitt. der Phys-Math., **105**, 1932.

14. J.G. Fletcher : *Dirac Matrices in Riemannian Space*, Il Nuovo Cimento **8**, 3, 451, 1958.

15. J.G. Loos : *Spin Connection in General Relativity*, Ann. Physics **25**, 91, 1963.

16. E.E. Fairchild : *Yang-Mills Formulation of Gravitational Dynamics*, Phys. Rev. **D 16**, 8, 2438, 1977.

17. J.S.R. Chisholm and R.S. Farwell : *Gauge Transformations of Spinors within a Clifford Algebraic Structure*, J. Phys. A **32**, 2805, 1999.

A Correspondence of Hyperholomorphic and Monogenic Functions in $\mathbb{R}^4$

Sirkka-Liisa Eriksson-Bique
(sirkka-liisa.eriksson-bique@joensuu.fi)
Department of Mathematics, University of Joensuu

Abstract. Let $\mathbb{H}$ be the algebra of quaternions generated by e_1, e_2 satisfying $e_i e_j + e_j e_i = -2\delta_{ij}$ for $i = 1, 2$. Consider the Clifford algebra $\mathbb{H} \oplus \mathbb{H}$ generated by e_1, e_2 and e_3. Any element x in $\mathbb{H} \oplus \mathbb{H}$ may be decomposed as $x = Px + Qxe_3$ for quaternions Px and Qx. The Dirac operator in $\mathbb{R}^4$ is defined by $D = \frac{\partial}{\partial x_0} + \frac{\partial}{\partial x_1} e_1 + \frac{\partial}{\partial x_2} e_2 + \frac{\partial}{\partial x_3} e_3$. Leutwiler noticed that the power function $(x_0 + x_1 e_1 + x_2 e_2 + x_3 e_3)^m$ is a solution of the modified Cauchy-Riemann system $x_3 Df + 2f_3 = 0$ which has connections to the hyperbolic metric. We study solutions of the equation $x_3 Df + 2Q'(f) = 0$, called hyperholomorhic functions, where $'$ is the main involution in $\mathbb{H} \oplus \mathbb{H}$. We prove that for any monogenic function g in $\mathbb{H} \oplus \mathbb{H}$ there exists locally a hyperholomorphic functions f with $\Delta f = g$. Moreover, if f is hyperholomorphic, then Δf is monogenic.

Keywords: hyperholomorphic functions, monogenic functions, (hyperbolic) harmonic functions, quaternions

Mathematics Subject Classification: 30G35

Dedicated to Richard Delanghe on the occasion of his 60th birthday.

1. Hyperholomorphic functions

Let $\mathbb{H}$ be the real associative algebra of quaternions generated by e_1, e_2 satisfying the usual relations $e_1^2 = e_2^2 = -1$ and $e_1 e_2 = -e_2 e_1$. Set $e_{12} = e_1 e_2$. The conjugation $\bar{q}$ of the quaternion $q = t + x e_1 + y e_2 + z e_{12}$ is defined by $\bar{q} = t - x e_1 - y e_2 - z e_{12}$. The main involution $' : \mathbb{H} \to \mathbb{H}$ is the isomorphism defined by $q' = t - x e_1 - y e_2 + z e_{12}$. The second involution $* : \mathbb{H} \to \mathbb{H}$, called **reversion**, is the anti-isomorphism defined by $q^* = t + x e_1 + y e_2 - z e_{12}$.

We embed the algebra of quaternions into the Clifford algebra $\mathbb{H} \oplus \mathbb{H}$ generated by the elements e_1, e_2 and e_3 satisfying the relation $e_i e_j + e_j e_i = -2\delta_{ij}$ where δ_{ij} is the usual Kronecker delta.

The elements $x = x_0 + x_1 e_1 + x_2 e_2 + x_3 e_3$ for $x_0, x_1, x_2, x_3 \in \mathbb{R}$ are called *paravectors* in $\mathbb{H} \oplus \mathbb{H}$. The space $\mathbb{R}^4$ is identified with the set of paravectors. We also denote $e_0 = 1$.

71

F. Brackx et al. (eds.), Clifford Analysis and Its Applications, 71–80.

The involution $'$ is extended to an isomorphism in $\mathbb{H} \oplus \mathbb{H}$ by

$$(q_1 + q_2 e_3)' = q_1' - q_2' e_3 \quad (q_i \in \mathbb{H}) . \tag{1}$$

Note that

$$e_3 q = q' e_3 \tag{2}$$

for any $q \in \mathbb{H}$.

The reversion $*$ is extended to an anti-isomorphism in $\mathbb{H} \oplus \mathbb{H}$ as follows

$$e_3^* = e_3, \ (ab)^* = b^* a^*.$$

and the conjugation by $\bar{a} = (a^*)' = (a')^*$. Note that $\overline{ab} = \overline{b}\,\overline{a}$.

The projection operators $P : \mathbb{H} \oplus \mathbb{H} \to \mathbb{H}$ and $Q : \mathbb{H} \oplus \mathbb{H} \to \mathbb{H}$ are defined by $P(q_1 + q_2 e_3) = q_1$ and $Q(q_1 + q_2 e_3) = q_2$ for $q_i \in \mathbb{H}$. Applying (1) we note that $P(a') = (Pa)'$ and $Q(a') = -(Qa)'$. By virtue of (2) we obtain

$$P(ab) = PaPb - Qa(Qb)', \tag{3}$$

$$Q(ab) = aQb + (Qa)b'. \tag{4}$$

The proofs are in [10, Lemma 3] and [11, Lemma 1].

The notation $f \in C^k(\Omega)$ for a function $f : \Omega \to \mathbb{H} \oplus \mathbb{H}$, defined in an open set $\Omega \subset \mathbb{R}^4$ means as usual that the real coordinate functions of f are k-times continuously differentiable on Ω.

The Dirac operator (also called generalized Cauchy-Riemann operator) is defined by

$$Df = \sum_{i=0}^{3} e_i \frac{\partial f}{\partial x_i}$$

for a mapping $f : \Omega \to \mathbb{H} \oplus \mathbb{H}$, defined in an open subset Ω of $\mathbb{R}^4$, whose components are partially differentiable. The operator $\overline{D}$ is defined by

$$\overline{D}f = \frac{\partial f}{\partial x_0} - \sum_{j=1}^{3} e_j \frac{\partial f}{\partial x_j}.$$

Note that $D\overline{D} = \overline{D}D = \Delta$, where Δ is the Laplace operator in $\mathbb{R}^4$. If $Df = 0$ the function f is called (left) *monogenic* (or regular). For the reference on properties of monogenic functions in the general case see e.g. [1] , [4] and for quaternions [20].

Using (3) and (4) we obtain

$$P(Df) = \sum_{i=0}^{2} e_i \frac{\partial Pf}{\partial x_i} - \frac{\partial Q'f}{\partial x_3} \tag{5}$$

$$Q(Df) = \sum_{i=0}^{2} e_i \frac{\partial Qf}{\partial x_i} + \frac{\partial P'f}{\partial x_3}. \tag{6}$$

The **modified Dirac operator** M is defined by

$$(Mf)(x) = (Df)(x) + \frac{2}{x_3}Q'f$$

and the operator $\overline{M}$ by

$$(\overline{M}f)(x) = (\overline{D}f)(x) - \frac{2}{x_3}Q'f.$$

Let $\Omega \subset \mathbb{R}^4$ be open. If $f \in C^2(\Omega)$ and $Mf(x) = 0$ for any $x \in \Omega \setminus \{x \mid x_3 = 0\}$, the function f is called **hyperholomorphic** in Ω. If f is paravector-valued hyperholomorphic in Ω, the function f is called an **H-solution**.

The H-solutions in $\mathbb{R}^3$ were introduced by H. Leutwiler ([16]). They are notably studied in $\mathbb{R}^n$ by H. Leutwiler ([17], [18], [19]), T. Hempfling ([12], [13], [14]), J. Cnops ([3]), P. Cerejeiras ([2]) and S.-L. Eriksson-Bique ([5], [9], [7], [8]). In $\mathbb{R}^3$ the hyperholomorphic functions are researched by W. Hengartner and H. Leutwiler ([15]) and in $\mathbb{R}^n$ by H. Leutwiler and S.-L. Eriksson-Bique ([10], [11]).

PROPOSITION 1. *If $f : \Omega \to \mathbb{H} \oplus \mathbb{H}$ is hyperholomorphic, then*

$$x_3 \Delta Pf - 2\frac{\partial Pf}{\partial x_3} = 0 \qquad (7)$$

and the function $\frac{Qf}{x_3}$ is harmonic on $\Omega \setminus \{x_3 = 0\}$. Moreover, if $f \in C^3(\Omega)$, then the mapping $h : \Omega \to \mathbb{R}$ defined by

$$h(x) = \begin{cases} \frac{Q(x)}{x_3}, & \text{if } x_3 \neq 0, \\ \frac{\partial Q}{\partial x_3}(x), & \text{if } x_3 = 0, \end{cases}$$

is harmonic on Ω.

The equation (7) is the Laplace-Beltrami equation associated with the hyperbolic metric $ds^2 = x_3^{-2}(dx_0^2 + dx_1^2 + dx_2^2 + dx_3^2)$. Functions satisfying the equation (7) are called *hyperbolic harmonic*.

PROPOSITION 2. *The function x^m is hyperholomorphic for all $m \in \mathbb{Z}$.*

PROPOSITION 3. *The space of hyperholomorphic functions in an open subset Ω of $\mathbb{R}^4$ forms a right quaternionic vector space.*

LEMMA 4. *Let Ω be an open subset of $\mathbb{R}^4$ and $f : \Omega \to \mathbb{H} \oplus \mathbb{H}$ be hyperholomorphic. Then $\frac{\partial f}{\partial x_l}$ is hyperholomorphic for $l = 0, 1, 2$.*

The proofs of the preceding results are in ([10]).

Hyperholomorphic functions are related to hyperbolic harmonic functions similarly as monogenic functions to harmonic functions. The both results can be proved the same way ([10, Theorem 17]).

THEOREM 5. *Let Ω be an open subset of $\mathbb{R}^4$ and $f : \Omega \to \mathbb{H} \oplus \mathbb{H}$ twice continuously differentiable. Then f is hyperholomorphic (resp. monogenic) if and only if for any $a \in \Omega$ and a ball $B(a,r)$ with $\overline{B(a,r)} \subset \Omega$ there exists a quaternionic valued hyperbolic harmonic (resp. harmonic) function H in $B(a,r)$ satisfying the equation $f = \overline{D}H$.*

THEOREM 6. *A mapping f is hyperholomorphic on a domain $\Omega \subset \mathbb{R}^4$ if and only if there exist locally H-solutions g_i such that*

$$f = g_0 + g_1 e_1 + g_2 e_2 + g_{12} e_{12}.$$

Checking the proof of the preceding theorem in [10, Theorem 4] we note that the corresponding result holds also for monogenic functions, i.e.,

THEOREM 7. *A mapping f is monogenic on a domain $\Omega \subset \mathbb{R}^4$ if and only if there exist locally paravector-valued monogenic functions g_i such that*

$$f = g_0 + g_1 e_1 + g_2 e_2 + g_{12} e_{12}.$$

2. A relation between hyperholomorphic and monogenic functions

It is possible to construct in two ways a hyperholomorphic function from a monogenic function, a result proved in ([11]).

PROPOSITION 8. *Let Ω be an open subset of $\mathbb{R}^4$ and $f : \Omega \to \mathbb{H}$ monogenic (and therefore $Qf = 0$). Then f is hyperholomorphic. Moreover the function F defined by*

$$F(x) = x_3^2 f(x) e_3$$

is hyperholomorphic.

Applying (5) and (6) for $x_3 M f = 0$ we obtain the following system.

PROPOSITION 9. *Let Ω be an open subset of $\mathbb{R}^4$ and $f : \Omega \to \mathbb{H}\oplus\mathbb{H}$ be a mapping with continuous partial derivatives. The equation $x_3 M f = 0$ is equivalent with the system of equations*

$$x_3 \left(D_2 \left(P f \right) - \frac{\partial Q' f}{\partial x_3} \right) + 2 Q' f = 0,$$
$$D_2 \left(Q f \right) + \frac{\partial P' (f)}{\partial x_3} = 0, \tag{8}$$

where $D_2 = \sum_{i=0}^{2} e_i \frac{\partial f}{\partial x_i}$.

The corresponding system for monogenic functions is the following.

THEOREM 10. *Let Ω be an open subset of $\mathbb{R}^4$ and $f : \Omega \to \mathbb{H}\oplus\mathbb{H}$ be a mapping with continuous partial derivatives. The equation $D f = 0$ is equivalent with the system of equations*

$$D_2 \left(P f \right) - \frac{\partial Q' f}{\partial x_3} = 0,$$
$$D_2 \left(Q f \right) + \frac{\partial P' f}{\partial x_3} = 0. \tag{9}$$

A relation between harmonic and hyperholomorphic functions is given next.

PROPOSITION 11. *Let $h : \Omega \to \mathbb{H}$ be harmonic. Then there exists locally a hyperholomorphic function f with values in $\mathbb{H} \oplus \mathbb{H}$ satisfying*

$$Q' f \left(x \right) = x_3 h \left(x \right).$$

Proof. Let $h : \Omega \to \mathbb{H}$ be harmonic. Assume that $a \in \Omega$ and $B(a,r)$ is a ball satisfying $\overline{B(a,r)} \subset \Omega$. Define a mapping $p : \mathbb{R}^4 \to \mathbb{R}^3$ by $p(x_0, ..., x_3) = (x_0, ..., x_2)$. Find a mapping g from $B(a,r) \cap \{x \,|\, x_3 = a_3\}$ into $\mathbb{H}$ satisfying

$$D_2 g \left(p \left(x \right) \right) = a_3 \frac{\partial h}{\partial x_3} \left(p \left(x \right), a_3 \right) - h \left(p \left(x \right), a_3 \right)$$

for all $x \in B(a,r)$. Set $Q f \left(x \right) = x_3 h' \left(x \right)$ and

$$P f \left(x \right) = - \int_{a_3}^{x_3} t \overline{D}_2 h \left(p \left(x \right), t \right) dt + g \left(p \left(x \right) \right). \tag{10}$$

We verify that the function $f = P f + Q f e_3$ satisfies the equations of (8). Using the definition of $P f$ and $Q f$ we obtain

$$\frac{\partial P' f}{\partial x_3} = - x_3 \left(\overline{D}_2 h \left(x \right) \right)' = - \left(\overline{D}_2 Q' f \left(x \right) \right)' = - D_2 Q f. \tag{11}$$

Hence the second equation of (8) holds. Since h is harmonic, the mapping $Q'f = x_3 h$ satisfies the equation $\triangle Q'f - 2\frac{\partial h}{\partial x_3} = 0$ and therefore

$$\triangle_2 Q'f = 2\frac{\partial h}{\partial x_3} - \frac{\partial^2 Q'f}{\partial x_3^2}.$$

Calculating $D_2 Pf$ from (10) using the above equation we find that

$$x_3\left(D_2 Pf - \frac{\partial Q'f}{\partial x_3}\right) + 2Q'f = -2x_3 h(x) + 2Q'f(x) = 0.$$

■

Using the result above we are able to proof our main theorem.

THEOREM 12. *Let $f : \Omega \to \mathbb{H} \oplus \mathbb{H}$ be a hyperholomorphic function $\Omega \subset \mathbb{R}^4$ and $f \in C^3(\Omega)$. Then $\triangle f$ is monogenic in Ω. Conversely, if $g : \Omega \to \mathbb{H} \oplus \mathbb{H}$ is monogenic there exists locally a hyperholomorphic function f with values in $\mathbb{H} \oplus \mathbb{H}$ such that $\triangle(f) = g$.*

Proof. The function $h : \Omega \to \mathbb{H}$ defined in by Proposition 1 is harmonic and satisfies the equation $Df + 2h(x) = 0$ which implies the property $\triangle f = -2\overline{D}h$. Since h is harmonic the function $-2\overline{D}h$ is monogenic.

Conversely, assume that $g : \Omega \to \mathbb{H} \oplus \mathbb{H}$ is monogenic. Then there exists locally a harmonic function H with values in $\mathbb{H}$ satisfying $-2\overline{D}H = g$. The function H being quaternionic valued harmonic yields that there exists a hyperholomorphic function f such that $Q'f = x_3 H$. Hence we have $Df + 2\frac{Q'f}{x_3} = 0$, which implies $\triangle f = \overline{D}Df = -2\overline{D}H = g$. ■

We want to characterize the kernel of the mapping $\triangle$ from a set of hyperholomorphic functions to the set of monogenic functions. Naturally if $f : \Omega \to \mathbb{H} \oplus \mathbb{H}$ is a hyperholomorphic and $\triangle f = 0$, then f is also harmonic. We need the following Lemma.

LEMMA 13. *Let $f : \Omega \to \mathbb{H} \oplus \mathbb{H}$ be hyperholomorphic on an open subset Ω of $\mathbb{R}^4$. Then*

$$\triangle f = \frac{2}{x_3}\frac{\partial Pf}{\partial x_3} + \frac{2}{x_3}\left(\frac{\partial Qf}{\partial x_3} - \frac{Qf}{x_3}\right)e_3.$$

Proof. Using [10, Lemma 2] we infer

$$0 = \overline{M}Mf = M\overline{M}f = \triangle(Pf) - \frac{2}{x_3}\frac{\partial Pf}{\partial x_3}$$
$$+ \left(\triangle(Qf) - \frac{2}{x_3}\frac{\partial Qf}{\partial x_3} + \frac{2}{x_3^2}Qf\right)e_3.$$

Since $\triangle f = \triangle Pf + \triangle Qfe_3$, we obtain the assertion. ■

THEOREM 14. *A hyperholomorphic function* $f : \Omega \to \mathbb{H} \oplus \mathbb{H}$ *is harmonic if and only if there exist locally a harmonic function H and a monogenic function g depending only on x_0, x_1 and x_2 satisfying*

$$f(x) = g(x_0, x_1, x_2) - H'(x_0, x_1, x_2) + x_3 \overline{D}_2 H(x_0, x_1, x_2) e_3. \quad (12)$$

Proof. Assume first that a hyperholomorphic function $f : \Omega \to \mathbb{H} \oplus \mathbb{H}$ is harmonic. Applying the preceding result we obtain $\frac{\partial Pf}{\partial x_3} = 0$ and $\frac{\partial \frac{Qf}{x_3}}{\partial x_3} = 0$. Hence Pf is independent of x_3 and there exists a function w depending only on x_0, x_1 and x_2 such that $Qf(x) = w(x_0, x_1, x_2) x_3$. Using (8) we infer that $D_2 Qf = 0$ and so w is monogenic. By virtue of Theorem 5 we find locally a harmonic function H such that $w = \overline{D}_2 H$. Since f is hyperholomorphic the first equation of the system (8) implies that $D_2 Pf + w' = 0$. Hence $D_2(Pf + H') = 0$, which means that $Pf + H'$ is a monogenic function denoted by g. Consequently we have

$$
\begin{aligned}
f(x) &= Pf(x) + Qf(x) e_3 \\
&= g(x_0, x_1, x_2) - H'(x_0, x_1, x_2) + x_3 \left(\overline{D}_2 H(x_0, x_1, x_2) \right) e_3.
\end{aligned}
$$

Conversely, assume that a function f has locally the presentation (12), where g is monogenic and H harmonic. Then clearly f is harmonic. Moreover, applying (2) we obtain

$$Mf = -D_2 H'(x_0, x_1, x_2) + e_3 \overline{D}_2 H(x_0, x_1, x_2) e_3 + 2 D_2 H'(x_0, x_1, x_2) = 0,$$

completing the proof. ∎

The elementary monogenic polynomial P^ν is defined for a multi-index $\nu \in \mathbb{N}_0^3$ with $v_1 + v_2 + v_3 = m$ by

$$P^\nu(x) = \frac{1}{m!} \left(\sum_{(\sigma_1, \sigma_2, \ldots, \sigma_m) \in \sigma_\nu} (e_{\sigma_1} x_0 - x_{\sigma_1}) \ldots (e_{\sigma_m} x_0 - x_{\sigma_m}) \right),$$

where σ_ν is the set of all permutations of m elements from which ν_1 are equal to 1, ν_2 equal to 2 and ν_3 equal to 3. They satisfy the following differentiation rules ([8])

$$\frac{\partial}{\partial x_i} P^\nu(x) = -P^{\nu - \varepsilon_i} \quad \text{for} \quad i = 1, 2, 3,$$

$$\frac{\partial}{\partial x_0} P^\nu(x) = \sum_{i=1}^3 e_i P^{\nu - \varepsilon_i}.$$

THEOREM 15. *The polynomial P^ν has the representation*

$$P^\nu(x) = \frac{1}{m!} \sum_{\alpha + \beta = \nu} \binom{m}{\beta, \alpha} (-1)^{|\beta|} c(\alpha) x_0^{|\alpha|} \widetilde{x}^\beta,$$

where $x = (x_0, ..., x_3) = (x_0, \tilde{x})$ *and the coefficients* $c(\alpha)$ *with* $\alpha = (\alpha_1, \alpha_2, \alpha_3)$ *are given by the generalized binomial theorem of [6]*

$$c(\alpha) = \begin{cases} \dfrac{\binom{\frac{|\alpha|}{2}}{\frac{\alpha}{2}}}{\binom{|\alpha|}{\alpha}} (-1)^{\frac{|\alpha|}{2}}, & \text{if } \alpha \text{ even,} \\[3ex] \dfrac{\binom{\frac{|\alpha|-1}{2}}{\frac{\alpha-\varepsilon_i}{2}}}{\binom{|\alpha|}{\alpha}} (-1)^{\frac{|\alpha|-1}{2}} \varepsilon_i, & \text{if } \alpha - \varepsilon_i \text{ even,} \\[3ex] 0, & \text{otherwise.} \end{cases}$$

Proof. It is easy to check that the result is valid for $\nu = \varepsilon_i$ for $i = 1, 2, 3$. Assume that it holds for any multi-index v with $|\nu| = m$. Note first that

$$E^\alpha_{|\alpha|-1}(1) = L^\alpha_0 = \binom{|\alpha|}{\alpha} c(\alpha),$$

where the polynomial E^α_n and L^α_n are defined e.g. in [8, Section 2.]. Assume that ν is a multi-index with $|\nu| = m + 1$. Denote

$$K^\nu = \frac{1}{(m+1)!} \sum_{\alpha+\beta=\nu} \binom{m+1}{\beta, |\alpha|} (-1)^{|\beta|} E^\alpha_{|\alpha|-1}(1) x_0^{|\alpha|} \tilde{x}^\beta.$$

Then

$$\frac{\partial K^\nu}{\partial x_i} = \frac{1}{(m+1)!} \sum_{\alpha+\beta=\nu} \binom{m+1}{\beta, |\alpha|} \beta_i (-1)^{|\beta|} E^\alpha_{|\alpha|-1}(1) x_0^{|\alpha|} \tilde{x}^{\beta-\varepsilon_i}$$

$$= -\frac{1}{m!} \sum_{\alpha+\beta=\nu-\varepsilon_i} \binom{m}{\beta, |\alpha|} (-1)^{|\beta|} E^\alpha_{|\alpha|-1}(1) x_0^{|\alpha|} \tilde{x}^\beta$$

$$= -K^{\nu-\varepsilon_i} = -P^{\nu-\varepsilon_i} = \frac{\partial P^\nu}{\partial x_i}.$$

Using the recursion formula ([8, Lemma 10]) and $E^\alpha_{|\alpha|-2} = 0$ we infer

$$\frac{\partial K^\nu}{\partial x_0} = \frac{1}{(m+1)!} \sum_{\alpha+\beta=\nu} \binom{m+1}{\beta, |\alpha|} |\alpha| (-1)^{|\beta|} E^\alpha_{|\alpha|-1}(1) x_0^{|\alpha|-1} \tilde{x}^\beta$$

$$= \sum_{i=1}^3 \frac{1}{m!} \sum_{\alpha-\varepsilon_i+\beta=\nu-\varepsilon_i} \binom{m}{\beta, |\alpha|-1} (-1)^{|\beta|} \varepsilon_i E^{\alpha-\varepsilon_i}_{|\alpha|-2}(1) x_0^{|\alpha|-1} \tilde{x}^\beta$$

$$= \sum_{i=1}^3 \varepsilon_i K^{\nu-\varepsilon_i} = \sum_{i=1}^3 \varepsilon_i P^{\nu-\varepsilon_i} = \frac{\partial P^\nu}{\partial x_0}.$$

Since K^ν and P^ν are homogeneous polynomials of degree $m + 1$ depending only on $x_0, ..., x_3$, we obtain $K^\nu = P^\nu$, completing the proof. ∎

Acknowledgements

The author thanks the Mathematical Institute of the University of Erlangen-Nürnberg for the hospitality she enjoyed there while the work was being done, and the Academy of Finland for the financial support that made it possible.

References

1. F. Brackx, R. Delanghe and F. Sommen, *Clifford Analysis*, Pitman, Boston, London, Melbourne, 1982.
2. P. Cerejeiras, Decomposition of analytic hyperbolically harmonic functions. *Proceedings of Clifford Algebras and their Applications Aachen*, Kluwer, Dordrecht (1998), 45–51.
3. J. Cnops, *Hurwitz pairs and applications of Möbius transformation*, Ph.D. thesis, Univ. Gent, 1994.
4. J. Gilbert and M. Murray, *Clifford algebras and Dirac operators in harmonic analysis*, Cambridge studies in advanced mathematics **26**, Cambridge, New York (1991).
5. S.-L. Eriksson-Bique, Comparison of quaternionic analysis and its modification, To appear in the *Proceedings of Delaware*, 1–16.
6. S.-L. Eriksson-Bique, The binomial theorem for hypercomplex numbers, *Ann. Acad. Sci. Fenn.* **24** (1999), 225–229.
7. S.-L. Eriksson-Bique, Real analytic functions on modified Clifford analysis, Submitted for publication, 1–10.
8. S.-L. Eriksson-Bique, On modified Clifford analysis, To appear in *Complex Variables*,1–22.
9. S.-L. Eriksson-Bique and H. Leutwiler, On modified quaternionic analysis in $\mathbf{R}^3$, *Arch. Math.* **70** (1998), 228–234.
10. S.-L. Eriksson-Bique and H. Leutwiler, Hypermonogenic functions, In *Clifford Algenras and their Applications in Mathematical Physis*, Vol. 2, Birkhäuser, Boston, 2000, 287–302.
11. S.-L. Eriksson-Bique and H. Leutwiler, Hypermonogenic functions and Möbius transformations, Submitted for publication, 1–10.
12. Th. Hempfling, Beiträge zur Modifizierten Clifford-Analysis, *Dissertation, Univ. Erlangen-Nürnberg*, August, 1997.
13. Th. Hempfling, *Multinomials in modified Clifford analysis*, C. R. Math. Rep. Acad. Sci. Canada **18** (2,3) (1996), 99–102.
14. Th. Hempfling and H. Leutwiler, Modified quaternionic analysis in $\mathbf{R}^4$. *Proceedings of Clifford Algebras and their Applications Aachen*, Kluwer, Dordrecht (1998), 227–238.
15. W. Hengartner, H. Leutwiler, Hyperholomorphic functions in R^3. To appear.
16. H. Leutwiler, Modified Clifford analysis, *Complex Variables* **17** (1992), 153–171.
17. H. Leutwiler, Modified quaternionic analysis in $\mathbf{R}^3$. *Complex Variables* **20** (1992), 19–51.
18. H. Leutwiler, More on modified quaternionic analysis in R^3, *Forum Math.* **7** (1995), 279–305.

S-L Eriksson-Bique

19. H. Leutwiler, Rudiments of a function theory in $\mathbb{R}^3$, *Expo. Math* **14** (1996), 97-123.
20. A. Sudbery, Quaternionic analysis, *Math. Proc. Cambridge Philos. Soc.* **85** (1979), 199–225.

On Weighted Spaces of Monogenic Quaternion-valued Functions

Klaus Gürlebeck (guerlebe@fossi.uni-weimar.de)
Bauhaus-Universität Weimar

Abstract. We consider a scale of weighted spaces of quaternion-valued functions generalizing the idea of Q_p-spaces of the complex function theory. Different characterizations of these spaces will be described. Main goal of the paper is to prove that the inclusions of Q_p-spaces are strict.

Keywords: Clifford analysis, function spaces, monogenic functions

1991 Mathematics Subject Classification: 30G35

Dedicated to Richard Delanghe on the occasion of his 60th birthday.

1. Introduction

For $p > 0$ the spaces $\mathbf{Q_p}$ were introduced in [4] for functions f in the unit disc Δ by

$$\mathbf{Q_p} = \{f : f \text{ analytic in } \Delta \text{ and } \sup_{a \in \Delta} \int_\Delta |f'(z)|^2 g^p(z,a) dx dy < \infty\}. \quad (1)$$

Here, $g(z,a) = \ln|\frac{1-\bar{a}z}{a-z}|$ is the Green function of the real Laplacian in Δ. One idea of these $\mathbf{Q_p}$-spaces is to find a scale of spaces with the Dirichlet space

$$\mathbf{D} = \{f : f \text{ analytic in } \Delta \text{ and } \int_\Delta |f'(z)|^2 dx dy < \infty\}.$$

and the Bloch space

$$\mathbf{B} = \{f : f \text{ analytic in } \Delta \text{ and } B(f) = \sup_{z \in \Delta}(1 - |z|^2)|f'(z)| < \infty\}$$

respectively, "at the both end points" of the range. (see also [2], [3]). It should be mentioned that already in 1992 in [1] weighted spaces of Dirichlet type are studied and that it is observed there that these spaces form a scale between the Dirichlet space and the Hardy space H^2. In

F. Brackx et al. (eds.), Clifford Analysis and Its Applications, 81–89.
© 2001 *Kluwer Academic Publishers. Printed in the Netherlands.*

the sequel there was an intensive study of these spaces and properties
like

$$\mathbf{D} \subset \mathbf{Q_p} \subset \mathbf{Q_q} \subset BMOA \quad 0 < p < q < 1 \quad [4]$$

$$\mathbf{Q_p} = \mathbf{B} \quad \forall p > 1 \quad [2]$$

were proved. From [5] it is known that one of several equivalent definitions of BMOA is

$$f \in BMOA \leftrightarrow \{f \text{ analytic in } \Delta \text{ and } \sup_{a \in \Delta} \int_{\Delta} |f'(z)|^2 g(z,a) dx dy < \infty\}.$$

It was observed in [4] that this definition coincides with the definition
of $\mathbf{Q_1}$ and therefore the $\mathbf{Q_p}$-spaces may be seen also as generalizations
of BMOA.

There have been several attempts to generalize these ideas and the
corresponding approaches to higher dimensions. In ([19], [20], [7], [21])
the case of the unit ball in $\mathbb{C}^n$ was investigated but non-trivial $\mathbf{Q_p}$-
spaces could be defined only for $p \in (\frac{n-1}{n}, \frac{n}{n-1})$.

To overcome these difficulties and the restriction to even dimensions of
the real Euclidean space were reasons to look for other possibilities to
generalize the complex (one-dimensional) ideas.

In [10] hypercomplex generalizations of $\mathbf{Q_p}$-spaces were studied. Instead
of holomorphic functions in the unit disk monogenic quaternion-valued
functions $f : \mathbb{R}^3 \mapsto \mathbb{H}$ were taken as basis of the consideration.

Using the generalized Cauchy-Riemann operator D, its adjoint $\overline{D}$, the
hypercomplex Möbius transformation $\varphi_a(x) = (a - x)(1 - \bar{a}x)^{-1}$, and
a modified fundamental solution g of the real Laplacian, generalized
$\mathbf{Q_p}$-spaces were defined by

$$\mathbf{Q_p} = \{f \in \ker D : \sup_{a \in B_1(0)} \int_{B_1(0)} |\bar{D}f(x)|^2 (g(\varphi_a(x)))^p dx < \infty\}.$$

where $B_1(0)$ stands for the unit ball in $\mathbb{R}^3$. This definition was motivated by the fact that the function theory of monogenic functions
generalizes some aspects of the complex (one-variable) function theory
(see e.g. [6], [13], [14]) and that $\bar{D}$ can be understood as the derivative
of a monogenic function (see [22] and [18] for $f : \mathbb{R}^4 \mapsto \mathbb{H}$ and [11] for
arbitrary dimensions n).

In [8] another generalization of the complex case is considered. Instead
of the derivative of the function, a gradient norm is used, measured
by a more general weight function. The main goal in [8] was to construct generalizations of the complex $\mathbf{Q_p}$-spaces for all real dimensions,
conserving the Möbius invariance and the principal idea of a weight

function with some kind of mass concentration around the singularity. In [9] the two approaches are compared and it is proved that they are not equivalent.

A goal of this paper is to describe characterizations of Q_p-functions by properties of their Taylor coefficients and to prove that the inclusions of the scale are strict.

For simplicity we will restrict ourselves to the consideration of functions defined in $\mathbb{R}^3$ with values in the skew field $\mathbb{H}$ of quaternions. As usual we identify $\mathbb{H}$ with the Clifford algebra $Cl_{0,2}$ and write $\{1, i, j, k\}$ instead of $\{e_0, e_1, e_2, e_1 e_2\}$. The points of $\mathbb{R}^3$ are denoted by $x = (x_0, x_1, x_2)$, and as the analogue of the complex Cauchy-Riemann operator we use

$$D = \partial_0 + i\partial_1 + j\partial_2.$$

Then, $D\overline{D} = \overline{D}D = \Delta$, where $\overline{D} = \partial_0 - i\partial_1 - j\partial_2$ is the conjugate Cauchy-Riemann operator. An $\mathbb{H}$-valued function satisfying $Df = 0$ in a domain will be called monogenic, or left monogenic.

2. Definition of Q_p-spaces in $\mathbb{R}^3$

For $|a| < 1$ we will denote by

$$\varphi_a(x) = (a - x)(1 - \bar{a}x)^{-1}$$

the Möbius transform, which maps the unit ball onto itself. Let

$$g(x, a) = \frac{1}{4\pi}\left(\frac{1}{|\varphi_a(x)|} - 1\right)$$

be the modified fundamental solution of the Laplacian in $\mathbb{R}^3$ composed with the Möbius transform $\varphi_a(x)$ and let $f : B_1(0) \mapsto \mathbb{H}$ be a monogenic function. As in [10] we define the Bloch-space

$$\mathbf{B} = \{f \in \ker D : \sup_{x \in B_1(0)} (1 - |x|^2)^{3/2}|\overline{D}f(x)| < \infty\},$$

the Q_p-spaces

$$Q_p = \{f \in \ker D : \sup_{a \in B_1(0)} \int_{B_1(0)} |\overline{D}f(x)|^2 g^p(x, a) dB_x < \infty\}, \quad p > 0, \quad (2)$$

and the Dirichlet space

$$\mathbf{D} = \{f \in \ker D : \int_{B_1(0)} |\overline{D}f(x)|^2 dB_x < \infty.$$

Because of the order of the singularity of $g(x, a)$ the seminorms above defined make sense for $p < 3$ only. Obviously, these spaces are not Banach spaces. By addition of the L_1-norm of f over a small ball $U_\epsilon \subset B_1(0)$ to our seminorms, **B** as well as $\mathbf{Q_p}$ will become Banach spaces. Because this additional term is independent of p we will consider in the following only the spaces with the corresponding seminorm.

3. Basic properties of $\mathbf{Q_p}$-spaces

First one has to show that the $\mathbf{Q_p}$-spaces form a range of Banach IH-modules (with our additional term added to the seminorm), connecting the spatial Dirichlet space with the spatial Bloch space. This was done in [10] and we will repeat here only the results. Since $g(x, a)$ is non-negative in $B_1(0)$ we have

$$\mathbf{D} \subset \mathbf{Q_p}, \qquad 0 < p < 3.$$

LEMMA 1. *Let f be monogenic and $0 < p < 3$, then we have*

$$(1 - |a|^2)^3 |\overline{D}f(a)|^2 \le C_1 \int_{B_1(0)} |\overline{D}f(x)|^2 \left(\frac{1}{|\varphi_a(x)|} - 1 \right)^p dB_x, \qquad (3)$$

where the constant C_1 does not depend on a and f.

Considering on both sides of (3) the supremum we obtain the following corollary.

COROLLARY 1. *For $0 < p < 3$ we have $\mathbf{Q_p} \subset \mathbf{B}$.*

We recall that in the complex one-dimensional case all $\mathbf{Q_p}$-spaces with $p > 1$ are equal and coincide with the Bloch space. This leads to a corresponding question in the three-dimensional case which was answered also in [10].

THEOREM 1. *Let f be monogenic in the unit ball. Then the following conditions are equivalent:*

1. $f \in \mathbf{B}$.

2. $f \in \mathbf{Q_p}$ for all $2 < p < 3$.

3. $f \in \mathbf{Q_P}$ for some $p > 2$.

Theorem 1 means that all $\mathbf{Q_p}$-spaces for $p > 2$ coincide and are identical with the Bloch space.

4. Other characterisations of Q_p-spaces

The definition (2) was the first definition of Q_p-spaces in higher dimensions and it was directly influenced by the complex results. Because of the singularity of the Green function, difficulties arise in proving some properties of the scale. One of these properties is the inclusion property with respect to the index p. In this subsection we discuss other possibilities of characterizing Q_p-spaces, which are often easier to handle.

THEOREM 2. *Let f be monogenic in $B_1(0)$. Then, for $0 \leq p < 2.99$,*

$$f \in Q_p \Leftrightarrow \sup_{a \in B_1(0)} \int_{B_1(0)} |\overline{D}f(x)|^2 (1 - |\varphi_a(x)|^2)^p dB_x < \infty.$$

We have to keep in mind that theorem 1 means that all Q_p-spaces for $p > 2$ coincide, so in fact the condition $p < 2.99$ is only a technical requirement caused by the singularity of $g^p(x, a)$ for $p = 3$.

Using the result of theorem 2 it can be easily shown that the Q_p-spaces form a scale of Banach spaces. This is an obvious consequence of the weight function $(1 - |\varphi_a(x)|^2)^p$.

PROPOSITION 1. *For $0 < p < q$ we have: $Q_p \subset Q_q$.*

The equivalence result of theorem 2 opens a way to overcome the problems with the singularity of the Green function, which becomes stronger with increasing dimension. The norm from theorem 2 may serve to define the Q_p-spaces for all space dimensions.

We will now study some possibilities of characterizing Q_p-spaces with the help of the coefficients of their Taylor expansion, analogously to the complex one-dimensional case. A first attempt restricted to the case of monogenic functions, depending on only two real variables, was made in [9]. In the following we try to overcome this restriction.

One of the basic inequalities is a result on weighted L_p-norms of real analytic functions.

LEMMA 2. *Let $\alpha > 0$, $p > 0$, $n \geq 0$, $a_n \geq 0$, $I_n = \{k : 2^n \leq k < 2^{n+1}, k \in N\}$, $t_n = \sum_{k \in I_n} a_k$ and $f(x) = \sum_{n=1}^{\infty} a_n x^n$. Then there exists a constant K depending only on p and α such that*

$$\frac{1}{K} \sum_{n=0}^{\infty} 2^{-n\alpha} t_n^p \leq \int_0^1 (1 - x)^{\alpha-1} f(x)^p \, dx \leq K \sum_{n=0}^{\infty} 2^{-n\alpha} t_n^p. \qquad (4)$$

For the proof and a lot of more information on weighted $\mathbf{Q_p}$-spaces, Hardy spaces, Bergman spaces, and their connections we refer to [16]. In [9] the following estimate was established.

LEMMA 3.   *Let* $0 < p \le 2$, $|a| < 1$, $r \le 1$. *Then*

$$\iint\limits_{\partial B_1} \frac{1}{|1 - \bar{a}ry|^{2p}} \, d\Gamma_y \le C \frac{1}{(1 - |a|r)^p}$$

From [17] we will now use the idea of characterizing $\mathbf{Q_p}$-functions with the help of the coefficients of their Taylor series expansions. Referring to [15] we use the Taylor series expansion of a left monogenic function in terms of the totally regular variables z_k,

$$g(x) = \sum_{n=0}^{\infty} \left(\sum_{|\nu|=n} \underline{z}^{\nu} c_{\nu} \right),$$

where $\underline{z} = (z_1, z_2)$ is given by $z_1 = x_1 - ix_0$, $z_2 = x_2 - jx_0$, and $\nu = (\nu_1, \nu_2)$ is a multi-index with $|\nu| = (\nu_1 + \nu_2)$. If ν_1 elements of the set $a_1, \ldots, a_{|\nu|}$ are equal to z_1 and ν_2 elements are equal to z_2 then $\underline{z}^{\nu}$ is defined by

$$\underline{z}^{\nu} := \frac{1}{|\nu|!} \sum_{(i_1,\ldots,i_{|\nu|}) \in \pi(1,\ldots|\nu|)} a_{i_1} a_{i_2} \cdots a_{i_{|\nu|}} \tag{5}$$

where the sum runs over *all* permutations of $(1, \ldots, |\nu|)$. In the following we use the notation $H_n(x) := \sum_{|\nu|=n} \underline{z}^{\nu} c_{\nu}$ for such a homogeneous monogenic polynomial of degree n and consider monogenic functions of the form

$$f(x) = \sum_{n=0}^{\infty} H_n(x) b_n, \quad b_n \in \mathbb{H}.$$

Applying the triangle inequality we get for this type of functions

$$\left| \frac{1}{2} \overline{D} f(x) \right| \le \sum_{n=0}^{\infty} n \left(\sum_{|\nu|=n} |c_{\nu}| \right) |b_n| |x|^{n-1}. \tag{6}$$

Using the structure of (6) we abbreviate $a_n := \left(\sum_{|\nu|=n} |c_{\nu}| \right) |b_n|$, $(a_n \ge 0)$ and we get finally

$$\left| \frac{1}{2} \overline{D} f(x) \right| \le \sum_{n=0}^{\infty} n a_n |x|^{n-1}. \tag{7}$$

THEOREM 3. *Let* $I_n = \langle k : 2^n \leq k < 2^{n+1}, k \in N \rangle$, $f(x) = \sum_{n=0}^{\infty} H_n(x)b_n$, $b_n \in \mathbb{H}$, H_n *be a homogeneous monogenic polynomial of degree* n *of the aforementioned type, and* a_n *be defined as before,* $0 < p \leq 2$. *Then*

$$\sum_{n=0}^{\infty} 2^{n(1-p)} \left(\sum_{k \in I_n} a_k \right)^2 < \infty \Rightarrow f \in \mathbf{Q_p}.$$

The idea of the proof is the same as in the proof of the corresponding theorem in [9].

More interesting is to prove a converse theorem. This requires knowing some analogue of norm equalities related to $\bar{\partial} z^n = n z^{n-1}$ in the complex one-dimensional case.

LEMMA 4. *[12] Let* $\alpha = (\alpha_1, \alpha_2)$, $\alpha_i \in \mathbb{R}$, $i = 1, 2$, *with* $\alpha_1^2 + \alpha_2^2 \neq 0$ *and* $H_{n,\alpha}(x) = (z_1\alpha_1 + z_2\alpha_2)^n$. *Then*

$$\frac{\|(-\frac{1}{2}\overline{D}H_{n,\alpha})\|_{L_2(\partial B_1)}}{\|H_{n,\alpha}\|_{L_2(\partial B_1)}} = \sqrt{n(n + \frac{1}{2})}.$$

Notice that this fraction does not depend on the choice of α but only on the degree of the homogeneous monogenic polynomial.

THEOREM 4. *Let* $0 < p \leq 2$, $f(x) = (\sum_{k=0}^{\infty} \frac{H_{2^k,\alpha}}{\|H_{2^k,\alpha}\|_{L_2(\partial B_1)}} a_k) \in \mathbf{Q_p}$.

Then $\sum_{k=0}^{\infty} 2^{k(1-p)}|a_k|^2 < \infty$.

Proof.

$$\|f\|_{Q_p}^2 \geq \int\int_{B_1(0)} |\sum_{k=1}^{\infty} (-\frac{1}{2}\overline{D}) \left(\frac{H_{2^k,\alpha}}{\|H_{2^k,\alpha}\|_{L_2(\partial B_1)}} \right) a_k|^2 (1 - |x|^2)^p dG$$

$$= \int_0^1 \sum_{k=1}^{\infty} |a_k|^2 r^{2(2^k-1)} 2^k (2^k + \frac{1}{2}) r^2 (1 - r^2)^p dr$$

$$\geq \int_0^1 \sum_{k=1}^{\infty} |a_k|^2 r^{2(2^k-1)} (2^k - 1)^2 r^3 (1 - r^2)^p dr$$

$$= \frac{1}{2} \sum_{k=1}^{\infty} |a_k|^2 (2^k - 1)^2 \int_0^1 x^{2^k} (1 - x)^p dx$$

$$\geq \frac{1}{8} \sum_{k=1}^{\infty} |a_k|^2 2^{2k} \int_0^1 x^{2^k}(1-x)^p dx$$

$$\geq \frac{1}{96} \sum_{k=1}^{\infty} |a_k|^2 2^{2k} 2^{-k(p+1)} = \frac{1}{96} \sum_{k=1}^{\infty} |a_k|^2 2^{k(1-p)}$$

Hereby we used the quaternion-valued inner product $(f,g)_{L_2(\Gamma)} = \int_\Gamma \overline{f}(x)g(x)d\Gamma$ and the orthogonality of spherical mono-genics (c.f. [6]). Theorem 3 and Theorem 4 imply that

$$f = \sum_{k=0}^{\infty} H_{2k,\alpha} a_k \in Q_p \iff \sum_{k=0}^{\infty} 2^{-k(p+1)}|a_k|^2 < \infty \qquad (8)$$

for a wider class of functions than in [9].

THEOREM 5. *The inclusions* $\mathbf{Q_{p_1}} \subset \mathbf{Q_p}$ *are strict for* $0 < p_1 < p \leq 2$.

Proof. Let $f(x) = \sum_{k=0}^{\infty} H_{2^n,\alpha} a_n$,

$$H_{n,\alpha}(x) = (z_1\alpha_1 + z_2\alpha_2)^n, \quad |\alpha|^2 = \alpha_1^2 + \alpha_2^2 \neq 0, \quad a_n = \frac{1}{2^{n(1-p_1)/2}}.$$

Then,

$$\sum_{n=0}^{\infty} 2^{n(1-p)}|a_n|^2 = \sum_{n=0}^{\infty} \frac{1}{2^{n(p-p_1)}} < \infty \qquad \forall p > p_1$$

and

$$\sum_{n=0}^{\infty} 2^{n(1-p_1)}|a_n|^2 = \sum_{n=0}^{\infty} 1 = \infty.$$

By Theorem 3 and Theorem 4 it follows that $f \in \mathbf{Q_p}$ but $f \notin \mathbf{Q_{p_1}}$.

The idea of the proof is completely analogous to the ideas used in [16].

References

1. A. Aleman: 'Hilbert spaces of analytic functions between the Hardy and the Dirichlet spaces', *Proc. Amer. Math. Soc.* 115 (1992) 97-104.
2. R. Aulaskari and P. Lappan: 'Criteria for an Analytic Function to be Bloch and a Harmonic or Meromorphic Function to be Normal', in Chung-Chun Y. et al. (eds.): *Complex Analysis and its Applications (Hong Kong 1993)* Pitman Research Notes in Mathematics, 305, Longman 1994, pp.136-146.
3. R. Aulaskari and L.M. Tovar; 'On the function spaces B^p and $\mathbf{Q_p}$', Bull. Hong Kong Math. Soc. 1, No.2, 203-208 (1997)

4. R. Aulaskari, J. Xiao, and R. Zhao: 'On subspaces and subsets of BMOA and UBC', *Analysis* 15 (1995), 101–121
5. A. Baernstein: 'Analytic functions of bounded mean oscillation', Aspects of contemporary complex analysis, Proc. instr. Conf., Durham/Engl. 1979, 3-36 (1980)
6. F. Brackx, R. Delanghe and F. Sommen: *Clifford Analysis*, Research Notes in Mathematics **76**, Pitman Advanced Publishing Program, London, 1982.
7. J.S. Choa, H.O. Kim and Y.Y. Park: 'A Bergman-Carleson measure characterization of Bloch functions in the unit ball of $\mathbb{C}^n$', *Bull. Korean Math. Soc.*, 29 (1992), 285–293
8. J. Cnops and R. Delanghe: 'Möbius invariant spaces in the unit ball', *Appl. Analysis* (1-2) 73 (2000) 45-64.
9. K. Gürlebeck: 'On some weighted spaces of quaternion-valued functions', Proceedings of the Second ISAAC Congress, Kluwer Academic Publ. (2000).
10. K. Gürlebeck, U. Kähler, M. Shapiro and L. M. Tovar: 'On Q_p-Spaces of Quaternion-Valued Functions', *Complex Variables*, Vol. 39 (1999), pp. 115–135.
11. K. Gürlebeck and H.R. Malonek: 'A hypercomplex derivative of monogenic functions in $\mathbb{R}^{n+1}$ and its applications', *Complex Variables*, Vol. 39 (1999), pp.199–228.
12. K. Gürlebeck and H.R. Malonek: 'On strict inclusions of Q_p-spaces of quaternion-valued functions', Cadernos de matematica CM00/I-09, Dep. de Matematica, Universidade de Aveiro (2000)
13. K. Gürlebeck and W. Sprößig: *Quaternionic and Clifford calculus for Engineers and Physicists*, John Wiley &. Sons, Chichester, 1997.
14. V.V. Kravchenko and M.V. Shapiro: *Integral representations for spatial models of mathematical physics*, Pitman Research Notes in Math. Series 351, 1996.
15. H. Malonek: 'A new hypercomplex structure of the Euclidean space $\mathbb{R}^{m+1}$ and the concept of hypercomplex differentiability', *Complex Variables Theory Appl.*, 1990, Vol. **14**, 25-33
16. M. Mateljevic and M. Pavlovic: 'L_p-behavior of power series with positive coefficients and Hardy spaces', *Proc. Amer. Math. Soc.*, Vol. 87, No. 2, 1983, pp. 309–316
17. J. Miao: 'A property of analytic functions with Hadamard gaps', *Bull. Austral. Math. Soc.* 45 (1992), pp. 105–112
18. I.M. Mitelman and M.V. Shapiro: 'Differentiation of the Martinelli-Bochner Integrals and the Notion of Hyperderivability', *Math. Nachr.* 172 (1995), 211-238.
19. C. Ouyang, W. Yang and R. Zhao: 'Characterizations of Bergman spaces and Bloch space in the unit ball of $\mathbb{C}^n$', *Transactions of the American Math. Soc.*, Vol. 347, No. 11, Nov. 1995, pp. 4301–4313.
20. C. Ouyang, W. Yang and R. Zhao: 'Möbius invariant $\mathbf{Q_P}$ spaces associated with the Green's function on the unit ball of $\mathbb{C}^n$', *Pacific J. Math.*, to appear
21. K. Stroethoff: 'Besov-Type Characterisations For The Bloch Space', *Bull. Austral. Math. Soc.*, Vol. 39 (1989), 405 – 420.
22. A. Sudbery: 'Quaternionic analysis', *Math. Proc. Cambr. Phil. Soc.* 85, 199–225, 1979.

Plane Waves in Premetric Electrodynamics

Bernard Jancewicz (bjan@ift.uni.wroc.pl)
Institute of Theoretical Physics, University of Wrocław

Abstract. Classical electrodynamics can be divided into two parts. In the first one, a need of introducing a multiplicity of directed quantities occurs, namely multivectors and differential forms but no scalar product is necessary. We call it premetric electrodynamics. In this part, the principal equations of the theory can be tackled. The second part concerns solutions of the equations and requires the establishing of a scalar product and, consequently a metric. For anisotropic media two scalar products can be introduced depending on the electric permittivity and magnetic permeability tensors. We show which part of the description of plane electromagnetic waves is independent of scalar products.

Keywords: differential forms, multivectors, classical electrodynamics, plane waves

Mathematics Subject Classification: 15A75, 53B50, 78A40

Dedicated to Richard Delanghe on the occasion of his 60th birthday.

1. Introduction

There exist physical quantities which can be defined without using any scalar product; there are also physical laws which can be formulated without it. This is achieved for the price of introducing many more types of directed quantities. For the three-dimensional description, anti-symmetric contravariant tensors are needed of ranks from zero to three (they are known as multivectors) as well as antisymmetric covariant tensors of ranks from zero to three (called exterior forms or differential forms when they depend on position). This makes altogether eight types of directed quantities. They can be divided further by their behaviour under inversions (ordinary or even tensors on the one hand, and twisted or odd tensors on the other hand; pseudovectors are of the second kind) and this gives sixteen types of directed quantities.

In recent decades, a way of presenting electrodynamics has been proposed based on a broad use of differential forms; see Refs [1–9]. Most authors concentrate on algebraic definitions of the outer forms: nice exceptions are Refs. [2, 7, 10] where visualisations by geometric images are shown. Not all presentations put enough care to the use of odd forms. In Refs. [1–4], D is claimed to be a two-form. Only few authors applied odd forms in electrodynamics, under various names:

F. Brackx et al. (eds.), Clifford Analysis and Its Applications, 91–102.
© 2001 *Kluwer Academic Publishers. Printed in the Netherlands.*

covariant W-p-vectors [5], *twisted forms* [6, 7] or *odd forms* [8, 9]. In references [5–9] authors admit that D is an odd two-form.

When asked what directed quantity in three-dimensional space are the electric field strength E and electric induction D we usually answer: they are vectors. Similarly, when asked about the directed nature of the magnetic field strength H and the magnetic induction B we answer: they are pseudovectors or axial vectors. We do so because we do not realize that to exterior forms also attributes of magnitude and direction can be ascribed. There are arguments showing that E, D, H and B are differential forms and they also can be considered as directed quantities.

The even and odd forms are necessary to formulate electrodynamics in a scalar-product-independent way. We call it *premetric electrodynamics*. It turns out that only the principal equations of this theory can be tackled in this manner, namely Maxwell's equations, the potentials, the Lorentz force, and the continuity equation for charge. When one seeks their solutions, that is, specific electromagnetic fields as functions of position, a scalar product is needed for writing, among others, the constitutive equations involving electric permeability and magnetic permittivity. A special scalar product can be introduced in the case of an anisotropic dielectric, for which the <u>vectors</u> $\vec{E}$ and $\vec{D}$ become parallel. In this manner the medium can be treated analogously to the isotropic one. Then the counterpart of the Coulomb field and the fields for many electrostatic problems can be found in a very natural way [11]. Analogously in the case of anisotropic magnetic medium another scalar product can be introduced, for which <u>pseudovectors</u> $\vec{H}$ and $\vec{B}$ are parallel. Then the medium appears to be isotropic, so the counterpart of the Biot-Savart law can be found and solutions of many magnetostatic problems [11]. When plane electromagnetic waves are considered, electric and magnetic fields are present simultaneously, and hence both scalar products have to be taken into account.

The multivectors and exterior forms (called directed quantities) are reviewed and the possible products between them are recalled in the rest of this Section. The premetric electrodynamics in the three dimensional space is shortly presented in Section 2. In Section 3 plane electromagnetic waves are presented as far as possible without scalar products.

For the well known *vector*, depicted as a directed segment, the *direction* consists of a straight line (on which the vector lies), after Lounesto [13] called an *attitude*, and an arrow on that line which is called the *orientation*. Two vectors of the same attitude are parallel. The list of directed quantities in three-dimensional space consists of even and odd multivectors, even and odd forms.[1] For their more systematic introduction see Refs. [11, 12]. Each *directed quantity* has a separate *direction* which consists of *attitude* and *orientation*. Because of the lack of space I omit figures illustrating geometric images of the directed quantities. The reader is referred to refs. [11, 12]. Table 1 collects relevant features of eight quantities which, in the presence of a metric, can be replaced by vectors or pseudovectors. The upper part contains even and odd multivectors, the lower part contains their duals. Typical examples of geometric and electromagnetic quantities are added for all of them. The other eight quantities: even scalars, odd scalars, even zero-forms, odd zero-forms, even trivectors, odd trivectors, even three-forms and odd three-forms can be replaced by scalars and pseudoscalars in the presence of metric.

Table 1

features	even vector	odd vector	even bivector	odd bivector
attitude	straight line	straight line	plane	plane
orientation	arrow on	curved arrow around	curved arrow on	arrow piercing
magnitude	length 1	length 1^*	area S	area S^*

	even one-form	odd one-form	even two-form	odd two-form
attitude	plane	plane	straight line	straight line
orientation	arrow piercing	curved arrow on	curved arrow around	arrow on
magnitude	inverse length E	inverse length H	inverse area B	inverse area D

For each pair of even or odd multivectors including scalars and pseudoscalars the outer product can be defined, which is denoted by wedge sign "∧". The grades of factors add in such a product; but,

---

[1] Even vector is the ordinary or polar vector, odd vector is the pseudovector.

when their sum is greater than three, the product is zero. What is important for the geometric images is that the attitude of the product i spanned by the attitude of the factors . This outer product can be extended by the distributivity onto linear combinations of multivectors of various grades. Thus one gets an algebra which we could call the *extended Grassmann algebra* of even and odd multivectors. Linear combinations of even multivectors form a linear subspace and a subalgebra in this algebra, namely the traditional Grassmann algebra, whereas the combinations of odd multivectors form only a linear subspace.

The same remarks concern the set of all even and odd forms which also can be multiplied in the outer product "∧". The attitude of the product is now intersection of attitudes of the factors. The linear combinations of exterior forms of various grades constitute another algebra which I would call the *extended Grassmann algebra* of even and odd forms.

Another kind of product can be introduced, namely, contractions marked by symbols "⌊" or "⌋" in which the factors are of different nature: one is multivector, the other is exterior form. The edge of the broken line of the symbol is on the side of factor with larger grade. The result of contraction is parallel to both factors.

2. Premetric electrodynamics

We now present a list of physical quantities with their designation as directed quantities along with short justifications.

The most natural vectorial quantity is the *displacement vector* l which is of the same nature as the *radius vector* r of a point in space relative to a reference point. The electric dipole moment $d = q(r_+ - r_-)$, where r_+ is the position of positive charge and r_- the position of negative charge, is also a vector.

The best physical model of the bivector is a flat electric circuit. Its magnitude is just the area encompassed by the circuit; its attitude is the plane of the circuit and orientation is given by the sense of the current. This bivector could be called a *directed area* S of the circuit. A connected bivectorial quantity is then the *magnetic moment* $m = IS$ of the circuit, where I is the current.

A one-form occurs naturally in the description of the waves. The locus of points in space with the same phase of a plane wave is just a plane. The family of planes with phases differing by a natural number can be viewed as the geometric image of the physical quantity known as the *wave vector* k with magnitude $2\pi/\lambda$, where λ is the wavelength. The physical quantity in its true directed nature is the one-form, not a

vector, thus, in my opinion, it deserves another name. If I am allowed
to create an English word I would propose the name *wavity*.

Another one-form quantity is the *electric field strength* **E**, since we
consider it to be a linear map of the infinitesimal vector dr into the
infinitesimal potential difference: $-dV = \mathbf{E} \cdot dr$. The *magnetic induction*
B is an example of a two-form quantity, since it can be treated as a
linear map of the directed area bivector $d\mathbf{S}$ into the magnetic flux:
$d\Phi = \mathbf{B}(d\mathbf{S})$.

Now for some examples of odd quantities; the area **S*** of a surface,
through which a flow is measured, is the first one. The side of the
surface from which a substance (mass, energy, electric charge, etc.)
passes is important. Hence, the orientation of **S*** can be marked as an
arrow piercing the surface. We claim that the *area of a flow* is an odd
bivector quantity. Accordingly, the *current density* **j** has to be an odd
two-form quantity. It corresponds to the linear map $dI = \mathbf{j}(d\mathbf{S}^*)$ of the
area $d\mathbf{S}^*$ into the electric current dI.

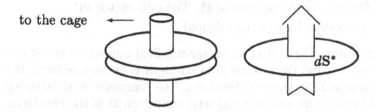

to the cage

Figure 1. Operational definition of **D**. The odd bivector $d\mathbf{S}^*$
corresponding to the disc.

The nature of the *electric induction* **D** can be deduced from the
following prescription of its measurement. Take two identical metal
discs, place one disc on top of the other, electrically discharge them
and then place them in the presence of a field. As you separate the
discs, the opposite sign charges induced on them are also separated.
Now measure one of them with the aid of a Faraday cage. It turns out
that for a small enough disc the charge is proportional to its area. One
will agree that the *disc area* $d\mathbf{S}^*$ is an odd bivector since its magnitude
is the area, its attitude is the plane and its orientation is given by an
arrow showing which disc is to be connected with the Faraday cage;
see Figure 1. Because of the proportionality relation $dQ = \mathbf{D}(d\mathbf{S}^*)$, we
ascertain that the *electric induction* is a linear map of the odd bivectors
into scalars, i.e. it is an odd two-form. Notice that **D** is of the same
directed nature as the electric current density.

The operational definition of the *magnetic field strength* **H** is as
follows: Take a very small wireless solenoid prepared from a super-

conducting material. Close the circuit in a region of space where the magnetic field vanishes. Afterwards, introduce the circuit into an arbitrary region in the field. A superconductor has the property that the magnetic flux enclosed by it is always the same; a current will be induced to compensate for this external field flux. Now measure the current dI flowing through the superconductor. It turns out to be proportional to the solenoid length: $dI = H(dl^*)$. The *solenoid length* dl^* in this experiment is apparently an odd vector, see Fig. 2, hence the *magnetic field strength* H is an odd one-form.

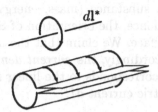

Figure 2. Operational definition of H. The odd vector dl^*
corresponding to the solenoid

As shown above, the four electromagnetic field quantities are of different directed nature: the electric field strength E is the one-form, the electric induction D is the odd two-form, the magnetic field strength H is the odd one-form, and the magnetic induction B is the two-form. With the aid of the outer derivative

$$d = f^1 \frac{\partial}{\partial x^1} + f^2 \frac{\partial}{\partial x^2} + f^3 \frac{\partial}{\partial x^3},$$

where f^i are basic one-forms, one can express the differential Maxwell's equations in terms of differential forms:

$$d \wedge E + \frac{\partial B}{\partial t} = 0, \tag{1}$$

$$d \wedge H - \frac{\partial D}{\partial t} = j, \tag{2}$$

$$d \wedge B = 0, \tag{3}$$

$$d \wedge D = \rho. \tag{4}$$

where ρ is the density of charge odd three-form and j is the electric current density odd two-form.

The energy density of the electromagnetic field can be expressed by the formula:

$$w = \frac{1}{2}(E \wedge D + B \wedge H). \tag{5}$$

The energy flux density of the electromagnetic field, as all flux densities, should be an odd two-form. Only the outer product $\mathbf{H} \wedge \mathbf{E}$ (or $\mathbf{E} \wedge \mathbf{H}$) gives such a quantity which replaces the traditional Poynting vector:

$$S = \mathbf{E} \wedge \mathbf{H}. \tag{6}$$

It may be called the *Poynting two-form*.

3. Plane waves

We are searching for solutions of the free Maxwell equations in shape of the plane waves:

$$\mathbf{E}(\mathbf{r}, t) = \psi(\mathbf{k}[\mathbf{r}] - \omega t)\mathbf{E}_0, \qquad \mathbf{B}(\mathbf{r}, t) = \psi(\mathbf{k}[\mathbf{r}] - \omega t)\mathbf{B}_0,$$

$$\mathbf{H}(\mathbf{r}, t) = \psi(\mathbf{k}[\mathbf{r}] - \omega t)\mathbf{H}_0, \qquad \mathbf{D}(\mathbf{r}, t) = \psi(\mathbf{k}[\mathbf{r}] - \omega t)\mathbf{D}_0, \tag{7}$$

where $\psi(\cdot)$ is a scalar function of a scalar argument, ω is a scalar constant, and $\mathbf{k}[\mathbf{r}]$ is the value of the linear form $\mathbf{k}$ of the wavity on the radius vector $\mathbf{r}$. In the expected solution (7) all fields maintain their attitudes for all times and positions; only magnitudes and orientations may change. This means that the wave is linearly polarized. Why these solutions are called plane? Because they are constant where the function $k(\mathbf{r}) = \mathbf{k}[\mathbf{r}]$ is constant, i.e. on planes. When the one-form $\mathbf{k}$ is used, the problem of propagation direction is open. This direction, understood as one-dimensional direction of a vector, can not be chosen as perpendicular to the planes of the constant fields, because any scalar product is not discriminated.

The argument $\phi = \mathbf{k}[\mathbf{r}] - \omega t$ of ψ, called *phase*, depends linearly both on position and time. The loci of points of constant phase are still planes, but these planes move when time flows. Can a phase velocity be introduced? If position $\mathbf{r}$ becomes a function of time, this means that we introduce the motion $\mathbf{r}(t)$ of a fictitious particle. Introduce thus a motion (of course uniform) such that, the fictitious particle is always on a plane of fixed phase:

$$\phi = \mathbf{k}[\mathbf{r}(t)] - \omega t = \text{const.}$$

Differentiate this equality with respect to time:

$$\frac{d\mathbf{k}[\mathbf{r}(t)]}{dt} - \omega = 0.$$

Since the mapping $\mathbf{r} \to \mathbf{k}[\mathbf{r}]$ is linear and continuous, the derivative may be put under the argument of $\mathbf{k}$:

$$\mathbf{k}\left[\frac{d\mathbf{r}(t)}{dt}\right] = \omega,$$

that is

$$\mathbf{k}[\mathbf{v}] = \omega,$$

where $\mathbf{v} = d\mathbf{r}/dt$ denotes velocity of the fictitious particle. We obtain:

$$\omega^{-1}\mathbf{k}[\mathbf{v}] = 1. \tag{8}$$

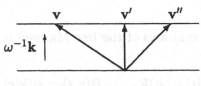

Figure 3

There are a lot of velocities $\mathbf{v}$ for which the one-form $\omega^{-1}\mathbf{k}$ gives value one, see Fig. 3. Which one can we admit as the phase velocity? As long as scalar product is not present, none of them. Once a scalar product is introduced, we choose velocity perpendicular to the planes of constant phase. For such velocity $\mathbf{v}$, relation (8) assumes the shape

$$\omega^{-1}\vec{\mathbf{k}} \cdot \mathbf{v} = \omega^{-1}kv = 1,$$

where $\vec{\mathbf{k}}$ is the vector also perpendicular to the planes of constant phase, $k = |\mathbf{k}|$, and hence $v = \frac{\omega}{k}$, which is the well known formula for the phase velocity. The equivalent formula $k/\omega = v^{-1}$ says that the quotient $\omega^{-1}k$ is the inverse of the phase velocity. If we want to designate $\omega^{-1}k$ by a separate word, the *slowness* is most suitable. Returning to (8), we may say that the linear form $\mathbf{u} = \omega^{-1}\mathbf{k}$ is *phase slowness*. This notion can be introduced when no scalar product is present, which is not the case for the phase velocity.

We proceed to consider the Maxwell equations. The equation $\mathbf{d} \wedge \mathbf{E} + \partial \mathbf{B}/\partial t = 0$ gives the condition

$$\psi' \, \mathbf{k} \wedge \mathbf{E}_0 - \psi'\omega\mathbf{B}_0 = 0$$

(where prime denotes the derivative with respect to the whole argument) and allows to write

$$\mathbf{B}_0 = \omega^{-1}\mathbf{k} \wedge \mathbf{E}_0 \quad \text{and} \quad \mathbf{B} = \omega^{-1}\mathbf{k} \wedge \mathbf{E} = \mathbf{u} \wedge \mathbf{E}. \tag{9}$$

This equality expressed in the traditional language has the shape $\vec{B} = \omega^{-1}\vec{k} \times \vec{E}$. We see also from (9) that the two forms are parallel: $\mathbf{B} \parallel \mathbf{E}$, which in terms of vectors is written as $\vec{B} \cdot \vec{E} = 0$. Moreover, $\mathbf{B} \parallel \mathbf{k}$ which corresponds to $\vec{B} \cdot \vec{k} = 0$. We illustrate all this observations on Fig. 4.

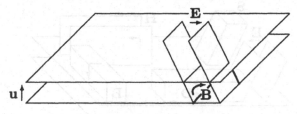

Figure 4

Next Maxwell equation, $\mathbf{d} \wedge \mathbf{B} = 0$, gives the condition

$$\psi' \, \mathbf{k} \wedge \mathbf{B_0} = 0.$$

This condition and this Maxwell equation are automatically satisfied by virtue of (9).

The third free Maxwell equation $\mathbf{d} \wedge \mathbf{H} - \partial\mathbf{D}/\partial t = 0$ reduces to the condition

$$\psi' \, \mathbf{k} \wedge \mathbf{H_0} + \psi'\omega\mathbf{D_0} = 0$$

and allows to write down

$$\mathbf{D_0} = -\omega^{-1}\mathbf{k} \wedge \mathbf{H_0} \quad \text{and} \quad \mathbf{D} = -\omega^{-1}\mathbf{k} \wedge \mathbf{H} = -\mathbf{u} \wedge \mathbf{H}. \qquad (10)$$

Therefore, the fourth free Maxwell equation $\mathbf{d} \wedge \mathbf{D} = 0$ is automatically satisfied. The constant forms $\mathbf{E_0}$, $\mathbf{H_0}$, present in relations (9) and (10), for the time being, are arbitrary. It can be seen from (10) that $\mathbf{D} \parallel \mathbf{H}$ and $\mathbf{D} \parallel \mathbf{k}$; we display this on Fig. 5. Condition (10) can be translated into the traditional language as $\vec{D} = -\omega^{-1}\vec{k} \times \vec{H}$, and the parallelity conditions as $\vec{D} \cdot \vec{H} = 0$ and $\vec{D} \cdot \vec{k} = 0$.

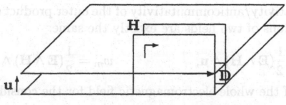

Figure 5

Summarizing, we write now the plane-wave solutions of the Maxwell equations:

$$\mathbf{E}(\mathbf{r},t) = \psi(\mathbf{k}[\mathbf{r}] - \omega t)\mathbf{E_0}, \qquad \mathbf{B}(\mathbf{r},t) = \psi(\mathbf{k}[\mathbf{r}] - \omega t)\mathbf{u} \wedge \mathbf{E_0},$$

$$\mathbf{H}(\mathbf{r}, t) = \psi(\mathbf{k}[\mathbf{r}] - \omega t)\mathbf{H}_0, \qquad \mathbf{D}(\mathbf{r}, t) = -\psi(\mathbf{k}[\mathbf{r}] - \omega t)\mathbf{u} \wedge \mathbf{H}_0, \quad (11)$$

where $\mathbf{E}_0$, $\mathbf{H}_0$ are arbitrary constant one-forms. This could be considered as a premetric form of the plane waves.

The configuration of four field quantities is depicted in Fig. 6.

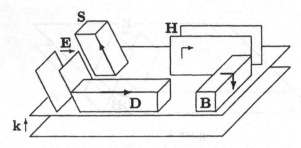

Figure 6

As is known from refs. [11, 12], the one-dimensional attitude of the Poynting odd two-form as the outer product $\mathbf{S} = \mathbf{E} \wedge \mathbf{H}$ of one-forms is the intersection of the planes of the two factors. If $\mathbf{E}$ and $\mathbf{H}$ are as in Fig. 6, the attitude of $\mathbf{S}$ is oblique with respect to the planes of $\mathbf{k}$ as is already shown in Fig. 6.

The direction of $\mathbf{S}$ must be regarded as the direction of wave propagation. We have now settled the dilemma considered immediately after Fig. 3 – a phase velocity is not given uniquely, hence it should be abandoned in the anisotropic medium. On the other hand, the propagation direction of the plane wave is determined by the energy flux density i.e. by the Poynting odd two-form.

The energy densities of the electric and magnetic field $w_e = \frac{1}{2}\mathbf{E} \wedge \mathbf{D}$, $w_m = \frac{1}{2}\mathbf{H} \wedge \mathbf{B}$ after a use of (9) and (10) are

$$w_e = \frac{1}{2}\mathbf{E} \wedge (\mathbf{H} \wedge \mathbf{u}), \qquad w_m = \frac{1}{2}\mathbf{H} \wedge (\mathbf{u} \wedge \mathbf{E}).$$

The commutativity/anticommutativity of the outer product shows that the contributions of two fields are exactly the same:

$$w_e = \frac{1}{2}(\mathbf{E} \wedge \mathbf{H}) \wedge \mathbf{u}, \qquad w_m = \frac{1}{2}(\mathbf{E} \wedge \mathbf{H}) \wedge \mathbf{u}.$$

The energy of the whole electromagnetic field for the considered plane waves is thus

$$w = w_e + w_m = (\mathbf{E} \wedge \mathbf{H}) \wedge \mathbf{u} = \mathbf{S} \wedge \mathbf{u}. \qquad (12)$$

We obtain an interesting identity relating the energy density with the energy flux density.

One may also introduce the velocity of the energy transport by the electromagnetic wave. In the traditional approach (solely in terms of vectors and scalars) the following equality is written

$$\vec{S} = w\vec{v}, \tag{13}$$

which defines $\vec{v}$ as the energy transport velocity. This relation is analogous to $\vec{j} = \rho\vec{v}$ linking the current density $\vec{j}$ with the density ρ of charges and their velocity $\vec{v}$.

Because of another type of the directed quantity we should write (13) rather as the contraction

$$\mathbf{S} = w\lfloor\mathbf{v}, \tag{14}$$

It is an inverse relation to (12). Can one calculate $\mathbf{v}$ from this formula? For this purpose it is worth introducing an odd trivector T inverse to the odd three-form w in the sense that the condition

$$w[T] = 1$$

is satisfied. If w is the energy density, T may be called *volume of unit energy*, since it has the physical dimension $\frac{\text{volume}}{\text{energy}}$. We may look at (14) as a mapping of a vector into an odd two-form The inverse mapping is the contraction of the odd two-form with the inverse odd trivector, hence we have

$$\mathbf{v} = -T\lfloor\mathbf{S}. \tag{15}$$

Since the odd trivector T is positive, the direction of $\mathbf{v}$ is the same as that of $\mathbf{S}$. In this way (15) is the sought formula for the *velocity of the energy transport by the electromagnetic field*:

In this sense we may claim that the phase slowness and the velocity of the energy transport by the plane electromagnetic wave are mutually inverse quantities. The energy transport velocity can be one of vectors depicted on Fig. 3.

References

1. H. Grauert and I. Lieb: *Differential und Integralrechnung*, vol 3, Springer Verlag, Berlin 1968.
2. Charles Misner, Kip Thorne and John Archibald Wheeler: *Gravitation*, Freeman and Co., San Francisco 1973, Sec. 2.5.
3. Walther Thirring: *Course in Mathematical Physics*, vol. 2: *Classical Field Theory*, Springer Verlag, New York 1979.
4. G.A. Deschamps: "Electromagnetics and differential forms", *Proc. IEEE* **69**(1981)676.

5.   Jan Arnoldus Schouten: *Tensor Analysis for Physicists*, Dover Publ., New York 1989 (first edition: Clarendon Press, Oxford 1951).

6.   Theodore Frankel: *Gravitational Curvature. An Introduction to Einstein's Theory*, Freeman and Co., San Francisco 1979.

7.   William L. Burke: *Applied Differential Geometry*, Cambridge University Press, Cambridge 1985.

8.   Roman Ingarden and Andrzej Jamiołkowski: *Classical Electrodynamics*, Elsevier, Amsterdam 1985.

9.   Kurt Meetz, Walter L. Engel: *Elektromagnetische Felder*, Springer Verlag, Berlin 1980.

10.   William L. Burke: *Spacetime, Geometry, Cosmology*, University Science Books, Mill Valley 1980.

11.   Bernard Jancewicz: "A variable metric electrodynamics. The Coulomb and Biot-Savart laws in anisotropic media", *Ann. Phys* **245** (1996)227.

12.   Bernard Jancewicz: "The extended Grassmann algebra of R^3", in *Clifford (Geometric) Algebras*, W.E. Baylis, editor, Birkhäuser, Boston 1996.

13.   Pertti Lounesto, Risto Mikkola and Vesa Vierros: *J. Comp. Math. Sci. Teach.*, **9**(1989)93.

Contact Symplectic Geometry in Parabolic Invariant Theory and Symplectic Dirac Operator

Lenka Kadlčáková (kadlcak@karlin.mff.cuni.cz)
Mathematical Institute of Charles University

Abstract. The authors of (Stein, Weiss) introduced the general method of construction of first order differential operators based on covariant derivatives composed with projections onto irreducible components of the target space. Following this general construction, we introduce, inside parabolic geometry (parabolic invariant theory), the symplectic Dirac operator first defined via analytical methods by K.Haberman. The role of the spinor bundle in orthogonal case is played here (in the symplectic case) by Segal-Shale-Weil representation. This representation is infinite-dimensional, so the Harish-Chandra category of $K = U(n)$-finite modules must be introduced.

Keywords: Clifford Analysis, Dirac Operators, Parabolic Geometries, Segal-Shale-Weil Representation, Symplectic Contact Geometry

Mathematics Subject Classification: 34L40, 35A02, 81R05

Dedicated to Richard Delanghe on the occasion of his 60th birthday.

1. Metaplectic Group and Segal-Shale-Weil Representation

For the reader's convenience, we submit here a short summary concerning the metaplectic Lie group, metaplectic Lie algebra and their representation theory. General references to this section are (Kashiwara, Vergne), (Kostant), (Haberman95).

The symplectic group $Sp(n, \mathbb{R})$ is the subgroup of $GL(2n, \mathbb{R})$ preserving symplectic form ω. According to this fact, we introduce the notation

$$e_i = (0, \ldots, 1_i, \ldots, 0, 0, \ldots, 0)^T \in \mathbb{R}^{2n}, \ i = 1, \ldots, n;$$
$$e_{n+i} = (0, \ldots, 0, 0, \ldots, 1_{(n+i)}, \ldots, 0)^T \in \mathbb{R}^{2n}, \ i = 1, \ldots, n \quad (1)$$

where the suffix means the number of components of the vector, for canonical symplectic basis of $\mathbb{R}^{2n}$.

Definition 1.1
The symplectic Clifford algebra of the symplectic space $(\mathbb{R}^{2n}, \omega)$ is the algebra over $\mathbb{R}$ multiplicatively generated by elements

F. Brackx et al. (eds.), *Clifford Analysis and Its Applications*, 103–111.
© 2001 *Kluwer Academic Publishers. Printed in the Netherlands.*

$\{e_1, \ldots, e_n, e_{n+1}, \ldots, e_{2n}\}$ given in (1), together with the relations

$$e_i e_j = e_j e_i, \quad i, j = 1, \ldots n$$

$$e_{n+i} e_{n+j} = e_{n+j} e_{n+i}, \quad i, j = 1, \ldots n$$

$$e_i e_{n+j} - e_{n+j} e_i = -\omega(e_i, e_{n+j}) = -\delta_{i,j} \quad i, j = 1, \ldots n$$

These relations give us, in contrast to the standard Clifford algebra in the orthogonal case, the structure of the Lie bracket, and so the symplectic Clifford algebra becomes a (real) Lie algebra. If we take the polynomials of degree ≤ 1 in $\{e_1, \ldots, e_n, e_{n+1}, \ldots, e_{2n}\}$, we generate the subalgebra $\mathbb{R} \oplus \mathbb{R}^{2n}$ in the symplectic Clifford algebra, which is isomorphic to the Heisenberg algebra. The elements $\{e_i e_j, e_{n+i} e_{n+j}\}_{1 \leq i \leq j \leq n}$ and $\{e_i e_{n+j} + e_{n+j} e_i\}_{1 \leq i,j \leq n}$ form a basis of the Lie subalgebra m of the symplectic Clifford algebra.

The algebra m is isomorphic to the symplectic Lie algebra $sp(n, \mathbb{R})$, thus, when $\mathfrak{M}$ is the simply connected Lie group corresponding to the algebra m, it is isomorphic to the simply connected covering group of $Sp(n, \mathbb{R})$. Looking for more details about the covering groups of $Sp(n, \mathbb{R})$, it is necessary to compute the first homotopy group of $Sp(n, \mathbb{R})$, or, because all topological information about any group can be elicited from its maximal compact subgroup, it is necessary to know the first homotopy group of $U(n)$. As is well known,

$$\pi_1(Sp(n, \mathbb{R})) \simeq \pi_1(U(n)) = \mathbb{Z}.$$

Hence, there exists, up to isomorphism, a uniquely determined two-fold covering group of $Sp(n, \mathbb{R})$, the metaplectic group $Mp(n, \mathbb{R})$, and its algebra m is isomorphic to the metaplectic Lie algebra $mp(n, \mathbb{R})$.

Now, to define the Segal-Shale-Weil (or the metaplectic) representation of the group $Mp(n, \mathbb{R})$, we consider the following generators of $Sp(n, \mathbb{R})$:

$$g(a) = \begin{pmatrix} a & 0 \\ 0 & (a^t)^{-1} \end{pmatrix} \quad \text{for } a \in Gl(n, \mathbb{R}),$$

$$t(b) = \begin{pmatrix} 1 & b \\ 0 & 1 \end{pmatrix} \quad \text{for } b \in S(n),$$

$$\sigma = \begin{pmatrix} 0 & -1 \\ 1 & 0 \end{pmatrix}$$

Since $\{t(b); b \in S(n)\}$ is simply connected, $\widetilde{t(b)} = (t(b), \pm 1)$ can be viewed as an element in $Mp(n, \mathbb{R})$ so that $\tilde{t}(0)$ is the identity of

$Mp(n, \mathbb{R})$. For each $a \in GL(n, \mathbb{R})$ we take $(\det a)^{\frac{1}{2}}$ and identify $\widetilde{g(a)} = (g(a), (\det a)^{\frac{1}{2}})$ and $\tilde{\sigma} = (\sigma, i^{\frac{1}{2}})$ as the generators of $Mp(n, \mathbb{R})$.

Now, the metaplectic representation L on the space $L^2(\mathbb{R}^n)$ of complex valued functions is given as follows:

$$
\begin{aligned}
(L(\widetilde{g(a)})f)(x) &= (deta)^{\frac{1}{2}} f(a^t x), \\
(L(\widetilde{t(b)})f)(x) &= e^{-\frac{i}{2}(bx,x)} f(x), \\
(L(\tilde{\sigma})f)(x) &= (\frac{i}{2\pi})^{n/2} \int_{\mathbb{R}^n} e^{i(x,y)} f(y) dy.
\end{aligned}
\tag{2}
$$

For completeness, let us also mention the definitions of representations of metaplectic Lie algebra $mp(n, \mathbb{R})$ and Heisenberg algebra $\mathbb{R}^{2n} \oplus \mathbb{R}$.

Remark 1.2 The representation of Heisenberg algebra $\mathbb{R}^{2n} \oplus \mathbb{R}$ on $L^2(\mathbb{R}^n)$ is given by

$$
1 \in \mathbb{R} \to i ; \quad e_j \in \mathbb{R}^{2n} \to i x_j ; \quad e_{n+j} \in \mathbb{R}^{2n} \to \frac{\partial}{\partial x_j}
\tag{3}
$$

Remark 1.3 The differential L_* of the Segal-Shale-Weil representation is given by

$$
\begin{aligned}
L_*(e_i e_j) &= i x_i x_j \\
L_*(e_{n+i} e_{n+j}) &= -i \frac{\partial^2}{\partial x_{n+i} \partial x_{n+j}} \\
L_*(e_i e_{n+j} + e_{n+j} e_i) &= x_i \frac{\partial}{\partial x_{n+j}} + \frac{\partial}{\partial x_{n+j}} x_i \quad i, j = 1, \ldots, n
\end{aligned}
\tag{4}
$$

1.1. METAPLECTIC STRUCTURE

Let (M, ω) be a $2n$-dimensional symplectic manifold, let $\mathcal{R}$ be the $Sp(n, \mathbb{R})$-principal fiber bundle over M, and let $\rho : Mp(n, \mathbb{R}) \to Sp(n, \mathbb{R})$ be a two-fold covering map.

Definition 1.4
A pair $(\mathcal{P}, f)$ consisting of the $Mp(n, \mathbb{R})$-principal fiber bundle $\mathcal{P}$ over M and a continuous surjective map $f : \mathcal{R} \to \mathcal{P}$ is called metaplectic structure, if following diagram commutes:

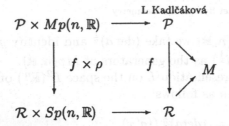

The metaplectic structure on (M, ω) is then a lift of the symplectic structure with respect to the two-fold covering ρ.

Definition 1.5
the metaplectic manifold is a symplectic manifold admitting a metaplectic structure.

Answering the question, whether the symplectic manifold admits a metaplectic structure, suggests a strong analogy with the spinor case. The manifold M admits a metaplectic structure if the second Stiefel-Whitney cohomology class in $H^2(M, \mathbb{Z}_2)$ vanishes. The set of metaplectic structures is then an affine space modelled on $H^1(M, \mathbb{Z}_2)$.

2. Decomposition

Let us denote our group $Sp(n, \mathbb{R})$ by G_0 and its maximal compact subgroup $U(n)$ by K.

As described in (Moshinsky, Quesne), the metaplectic representation L on the space $L^2(\mathbb{R}^n)$ decomposes into two irreducible infinite-dimensional representations of the group $Sp(n, \mathbb{R})$, whose leading weights[1] are $(\frac{1}{2}, \frac{1}{2}, \ldots, \frac{1}{2})_K$ and $(\frac{3}{2}, \frac{1}{2}, \ldots, \frac{1}{2})_K$.

We denote these representations L_+ and L_- and will regard them as the harmonic series representations of finite K-modules.

The group $U(n)$ is reductive, and we can write it as

$$U(n) = SU(n) \times U(1).$$

So, what is to be done is to decompose the metaplectic representation L into irreducible $SU(n)$-representations or, in accordance with the Weyl unitary trick, into irreducible $SL(n, \mathbb{C})$-representations?

We will not derive the branching rules here. The sympathetic reader is asked to believe that the metaplectic representation decomposes as

$$L_+ \rightarrow \varepsilon^{\frac{1}{2}}(\{0\} + \{2\} + \{4\} + \ldots),$$

---

[1] By the expression 'leading weights' we mean the weights closest to the origin of the weight lattice corresponding to the irreducible K-representation.

$$L_- \to \varepsilon^{\frac{1}{2}}(\{1\} + \{3\} + \{5\} + \ldots),$$

where by symbol $\{\mu\}$ we denote the $SU(n)$-dominant weight and by $\varepsilon^{\frac{1}{2}}$ the square root of the determinant representation of the group $U(n)$ (see also (King, Wybourne)).

Remark 2.6 We can regard $SU(n)$-dominant weight $(k, 0, 0, \ldots, 0)$ as a polynomial of degree k. These polynomials are embedded into the space $L^2(\mathbb{R}^n)$ with the norm $e^{-|x|^2}$ and their images form the orthogonal system in $L^2(\mathbb{R}^n)$ with respect to this measure (Jarník). So, if it will be convenient, we will take $SU(n)$-dominant weights $(k, 0, 0, \ldots, 0)$ as symmetric polynomials, and write them in the form

$$\sum_{i_1 \ldots i_k} a_{i_1 i_2 \ldots i_k} x_{i_1} x_{i_2} \ldots x_{i_k}$$

Note, that this remark changes nothing in our definition of the metaplectic representation given in equation (2).

2.1. SYMPLECTIC CLIFFORD MULTIPLICATION

As usually let $\{e_1, \ldots, e_{2n}\}$ be the canonical symplectic basis of $\mathbb{R}^{2n}$. To define symplectic Clifford multiplication, we use the representation of the Heisenberg algebra $\mathbb{R}^{2n} \oplus \mathbb{R}$ on the space $L^2(\mathbb{R}^n)$ defined in equation (3). Let σ be the restriction of this representation to the symplectic vector space $\mathbb{R}^{2n}$.

The operators ix_j and $\frac{\partial}{\partial x_j}$ are continuous unbounded operators on $L^2(\mathbb{R}^n)$ with dense domain $S(\mathbb{R}^n) \subset L^2(\mathbb{R}^n)$ (Schwartz space).

Definition 2.7
The symplectic Clifford multiplication is the map

$$\mu : \mathbb{R}^{2n} \times L^2(\mathbb{R}^n) \to L^2(\mathbb{R}^n)$$

given by $\mu(v, f) = \sigma(v)f$.

To simplify the notation, we will also write $v.f$ instead of $\mu(v, f)$.

Remark 2.8 Note that L_*, defined in equation (4), can be expressed by symplectic Clifford multiplication; for all $u, v \in \mathbb{R}^{2n}$,

$$L_*(u, v) = -i\sigma(u)\sigma(v).$$

Let $\{\varepsilon^1, \varepsilon^2, \ldots, \varepsilon^{2n}\}$ be the dual basis of $\{e_1, e_2, \ldots, e_{2n}\}$, i.e. $\varepsilon^i(e_j) = \delta_{ij}$. We decompose the space $(L^2(\mathbb{R}^n) \otimes \mathbb{R}^{2n})$ on the subspaces given by the kernel and image subspaces of Clifford multiplication. In coordinates, the decomposition looks as follows:

$$V_1 = \{\sum_{i=1}^{2n} e_i s_i \otimes \varepsilon^i; \ s_i \in L_\pm\} \subset L_\mp \otimes \mathbb{R}^{2n}$$

$$\tag{5}$$

$$V_2 = \{\sum_{i=0}^{2n} s_i \otimes \varepsilon^i; \ s_i \in L_\pm; \sum_{i=1}^{2n} e_i s_i = 0\} \subset L_\mp \otimes \mathbb{R}^{2n}$$

Note a strong analogy with the spinor representation of the orthogonal group. The space V_1 corresponds to the image of Clifford multiplication, the space V_2 to its kernel.

It seems that this decomposition requires a brief comment, because we have slightly simplified the notation:

Lemma 2.9 The symplectic Clifford multiplication is a map $L_\pm \to L_\mp$, which means $\mu : \mathbb{R}^{2n} \times L_\pm \to L_\mp$.

Proof

Fix the $SU(n)$-type $s_k = (k, 0, \dots, 0) \in L_\pm$. For every $i = 1, \dots n$,

$$e_i s_k = a_{i_1 \dots i_k} e_i x_{i_1} x_{i_2} \dots x_{i_k} + a_{i_1 \dots i_k} e_{n+i} x_{i_1} x_{i_2} \dots x_{i_k}. \tag{6}$$

Passing to the coordinate basis $(x_i, \frac{\partial}{\partial x_i})$ of $\mathbb{R}^{2n}$, gives the representation

$$a_{i_1 \dots i_k} x_i x_{i_1} x_{i_2} \dots x_{i_k} + a_{i_1 \dots i_k} (\frac{\partial}{\partial x_i})(x_{i_1} x_{i_2} \dots x_{i_k}),$$

and we really end up in $L_\mp$. $\square$

Theorem 2.10

Let $\rho : Sp(n, \mathbb{R}) \to Aut(\mathbb{R}^{2n})$ denote the standard representation of the group $Sp(n, \mathbb{R})$, let ρ^* denote its dual representation and let L denote the metaplectic representation. The spaces V_1 and V_2 defined above are $Sp(n, \mathbb{R})$-invariant subspaces as the representation spaces of $(\rho^* \otimes L)$. Moreover, the space V_1 is isomorphic as the $Sp(n, \mathbb{R})$-module to the $L_\mp$ part in the decomposition of the product $(L_\pm \otimes \mathbb{R}^{2n})$.

Proof

Let $m \in Mp(n, \mathbb{R}), e_i \in \mathbb{R}^{2n}$. Then the metaplectic representation and the standard representation of the group $Sp(n, \mathbb{R})$ are related by the formula

$$m^{-1} \tau(e_i) = \tau(\sum_j \rho_{ij}(m^{-1}) e_j),$$

described by commutative diagram

$$
\begin{array}{ccccc}
Sp(n,\mathbb{R}) & \xrightarrow{(*,0)} & Sp(n,\mathbb{R}) \times \mathbb{R}^{2n} & \xrightarrow{\rho} & \mathbb{R}^{2n} \\
\uparrow{\scriptstyle\pi} & & \downarrow{\scriptstyle\tau} & & \searrow{\scriptstyle i} \quad \mathbb{R}^{2n} \times \mathbb{R} \\
Mp(n,\mathbb{R}) & \xrightarrow{(*,0)} & Mp(n,\mathbb{R}) \times L^2(\mathbb{R}^n) & \xrightarrow{L} & L^2(\mathbb{R}^n)
\end{array}
$$

Note that in what follows, we omit the projection π and write $\rho(m^{-1})$ instead of $\rho(\pi(m^{-1}))$. It is also evident that we should write $\tau(e_i)s$ instead of $e_i s$ in the coordinate decomposition (5).

The dual representation ρ^* acting on $\Lambda^1(\mathbb{R}^{2n}) = (\mathbb{R}^{2n})^*$ is defined by

$$\rho^*(m)(\varepsilon^i) = \sum_j \rho_{ij}(m^{-1})\varepsilon^j.$$

Invariance of V_1 follows from the computation:

$$(L \otimes \rho^*)(\sum_{j=1}^{2n} e_j s_j \otimes \varepsilon^j) = \sum_{j=1}^{2n}\sum_{k=1}^{2n} m(e_j s_j) \otimes \rho_{kj}(m^{-1})\varepsilon^k =$$

$$= \sum_k m(\sum_j \rho_{kj}(m^{-1})e_j)s_j \otimes \varepsilon^k = \sum_k m(m^{-1}e_k m)s_k \otimes \varepsilon^k =$$

$$= \sum_k e_k m s_k \otimes \varepsilon^k \in V_1.$$

And for V_2:

$$(L \otimes \rho^*)(\sum_j s_j \otimes \varepsilon^j) = \sum_j \sum_k m s_j \otimes \rho_{kj}(m^{-1})\varepsilon^k =$$

$$= \sum_k (\sum_j \rho_{kj}(m^{-1})m s_j) \otimes \varepsilon^k = \sum_k s'_k \otimes \varepsilon^k.$$

To complete the proof we need to show that $\sum_k e_k s'_k = 0$.

$$\sum_k e_k s'_k = \sum_k e_k \sum_j \rho_{kj}(m^{-1})m s_j = \sum_j (\sum_k (e_k \rho_{kj}(m^{-1})))m s_j =$$

$$= \sum_j (\sum_k \rho_{jk}(m)e_k)m s_j = \sum_j (m e_j m^{-1})\dot{m} s_j =$$

$$= \sum_j \dot{m} e_j s_j = 0.$$

Note, that we proved also the surjectivity of the Clifford multiplication.
□.

We can now write down, how the projections onto V_1 and V_2 appear.
Take $\sum_k s_k \otimes \varepsilon^k \in L^2(\mathbb{R}^n) \otimes (\mathbb{R}^{2n})^*$, then

$$P_{V_1} : L_{\pm} \otimes \mathbb{R}^{2n} \to V_1$$

$$P_{V_1}(\sum_k s_k \otimes \varepsilon^k) = \frac{1}{2n} \sum_k [e_k(\sum_j e_j s_j)] \otimes \varepsilon^k \qquad (7)$$

$$P_{V_2} : L_{\pm} \otimes \mathbb{R}^{2n} \to V_2$$

$$P_{V_2}(\sum_k s_k \otimes \varepsilon^k) = \sum_k [s_k - \frac{1}{2n} e_k(\sum_j e_j s_j)] \otimes \varepsilon^k \qquad (8)$$

3. Symplectic Dirac Operator

Let (M, ω) be a $2n$–dimensional metaplectic manifold with a fixed
metaplectic structure P.

Definition 3.11
*The vector bundle associated to the metaplectic structure via the Segal-
Shale-Weil representation L, i.e. the bundle*

$$Q = P \times_L L^2(\mathbb{R}^n),$$

*is called the symplectic spinor bundle of (M, ω). A symplectic vector
field φ on (M, ω) is a section $\varphi \in \Gamma(Q)$ of Q.*

Definition 3.12
Let (M, ω) be a symplectic manifold. A covariant derivative

$$\nabla : \Gamma(TM) \to \Gamma(T^*M \otimes TM)$$

is called symplectic if and only if $\nabla \omega = 0$.

By analogy with Riemannian geometry, there is a bijection between
symplectic covariant derivatives on the associated tangent bundle TM
and symplectic connections Z in the symplectic frame bundle, i.e. $Z :
T\mathcal{R} \to sp(n, \mathbb{R})$.

In the symplectic case, we have no uniqueness property for connections as in the case of the Levi-Civita connection in Riemannian
geometry.

Definition 3.13

Let (M, ω) be the metaplectic manifold with symplectic spinor bundle Q. The symplectic Dirac operator of (M, ω) attached to the symplectic covariant derivative ∇ is the differential operator defined by

$$D = \mu \circ \nabla : \Gamma(Q) \to \Gamma(T^*M \otimes Q) \overset{\omega}{\cong} \Gamma(TM \otimes Q) \to \Gamma(Q) \qquad (9)$$

Rewriting the general definition given in (Haberman95), we get

$$D = P_{V_1} \circ \nabla : \Gamma(L_\pm) \to \Gamma(L_\pm \otimes \mathbb{R}^{2n}) \to \Gamma(L_\mp).$$

References

Baston, R.J., Eastwood, M.G.; The Penrose Transform, Its Interaction with Representation Theory; Oxford University Press; 1989

Black, G.R.E., King, R.C., Wybourne, B.G.; Kronecker Products for Compact Semisimple Lie Groups; Journal of Physics - A 16, 1555-1589; 1983

Bureš, J., Souček, V.; Regular Spinor Valued Mappings; Seminari di Geometria, Universita Degli Studi di Bologna; 1984

Haberman, K.; Basic Properties of Symplectic Dirac Operators; Communications in Mathematical Physics 184, 629-652; 1997

Haberman, K.; Harmonic symplectic Spinors on Riemann Surfaces; Manuscripta Math. 94, 465-484; 1997

Haberman, K.; The Dirac Operator on Symplectic Spinors; Annals Of Global Analysis and Geometry 13, 155-168; 1995

Jarník, V.; Integrálnípočet II.; Academia Praha; 1984

Kashiwara, M., Vergne, M.; On The Segal-Shale-Weil Representation and Harmonic Polynomials; Inventiones Mathematicae 44, 1-47; 1978

King, R.C., Wybourne, B.G.; Holomorphic Discrete Series and Harmonic Series Unitary Irreducible Representations of Non-Compact Lie Groups: $Sp(2n, \mathbb{R}), U(p, q)$ and $SO^*(2n)$; Journal of Physics - A 18, 3113-3139; 1985

Kostant, B.; Symplectic Spinors; Symposia Mathematica XIV, 139-152; 1974

Moshinsky, M., Quesne, C.; Linear Canonical Transformations and Their Unitary Representations; Journal of Mathematical Physics, Vol. 12, No. 8, 1772-1780; 1971

Slovák, J.; Parabolic Geometries; Masaryk University Brno; 1997

Slovák, J., Souček, V.; Invariant Operators of the First Order on Manifolds with a Given Parabolic Structure; to appear in the Proceedings of the conference "Harmonic Analysis"; Luminy; 1999

Stein, E., Weiss, G.; Generalization of the Cauchy-Riemann Equations and Representations of the Rotation Group; Amer. J. Math. 90, 163-196; 1968

Communication via Holomorphic Green Functions [*]

Gerald Kaiser (kaiser@wavelets.com)
Virginia Center for Signals and Waves, www.wavelets.com

Abstract. Let $G(x_r - x_e)$ be the causal Green function for the wave equation in four spacetime dimensions, representing the signal received at the spacetime point x_r due to an impulse emitted at the spacetime point x_e. Such emission and reception processes are highly idealized, since no signal can be emitted or received at a single (mathematical) point in space and time. We present a simple model for *extended* emitters and receivers by continuing G analytically to a function $\tilde{G}(z_r - z_e)$, where $z_e = x_e + iy_e$ is a complex spacetime point representing a circular *pulsed-beam emitting antenna dish* centered at x_e and oriented by y_e, and $z_r = x_r - iy_r$ represents a circular *pulsed-beam receiving antenna dish* centered at x_r and oriented by y_r. The holomorphic Green function $\tilde{G}(z_r - z_e)$ represents the *communication* between the emission from z_e and the reception at z_r. To preserve causality and give nonsingular coupling, the orientation vectors y_e and y_r must belong to the *future cone* V_+ in spacetime. Equivalently, z_e and z_r belong to the *future and past tubes* in complex spacetime, respectively. The space coordinates of y_e and y_r give the spatial orientations and radii of the dishes, while their time coordinates determine the *duration and focus* of the emission and reception processes. The *directivity $D(y)$* of the communication process is a convex function on V_+, i.e., $D(y_r + y_e) \leq D(y_r) + D(y_e)$. This shows that the efficiency of the communication can be no better than the sum of its emission and reception components.

Keywords: Wave equation, Green function, holomorphic, pulsed beams, wavelets, communication.

Mathematics Subject Classification: 35L20,65T60

Dedicated to Richard Delanghe on the occasion of his 60th birthday.

1. Introduction

I begin by summarizing earlier work. *Physical wavelets* were defined in (Kaiser, 1994) as wavelet-like bases for spaces of solutions of the homogeneous wave equation *(acoustic wavelets)* or Maxwell's equations *(electromagnetic wavelets)*. This was motivated by the observation that information is often communicated by acoustic or electromagnetic waves, and this fact should be taken into account when "processing" the resulting signals. All such wavelets can be obtained

---

[*] Supported by US AFOSR contract #F49620-98-C-0013.

113

from a single "mother" wavelet by translations, scaling, rotations and Lorentz transformations.

The construction of physical wavelets was based on a holomorphic extension $\tilde{F}(x + iy)$ of solutions $F(x)$ to complex spacetime, with the imaginary spacetime variables y interpreted as singling out approximate directions and frequencies of propagation. Thus $\tilde{F}(x + iy)$ is a description of the wave intermediate between the spacetime domain (where exact positions and times are known but no directional or frequency information is given) and the Fourier domain (where exact directional and frequency information is known but no local spacetime information is given). This is an extension to spacetime of continuous wavelet analysis of one-dimensional *time signals*, whose wavelet transform is intermediate between the time domain and the frequency domain representations.

The physical wavelets of the homogeneous wave- and Maxwell equations were then shown to split into a sum of *causal* and *anticausal* wavelets. Essentially, the causal wavelets are holomorphic extensions, in the sense of *positive-frequency analytic signals*, of the causal (retarded) Green function, and the anticausal ones are similar extensions of the anticausal (advanced) Green function for the appropriate equation. The causal wavelets are *pulsed-beam solutions emitted by disk-like sources*. That is, that they represent well-directed acoustic or electromagnetic beams that are *pulsed* in time rather than going on forever. The direction, pulse width, and duration of these beams are determined by the imaginary spacetime variables y. Such objects have appeared previously in the engineering literature under the name *complex-source pulsed beams* (see Heyman and Felsen, 1989, and the references therein).

In this paper we further develop the above analysis by showing that the holomorphic extension of the causal Green function describes not only the *emission* but also the *reception* of a pulsed beam, and so represents a *communication* between the emitting and receiving antenna dishes.

2. Holomorphic Green Functions

For simplicity, we concentrate on the wave equation in four-dimensional spacetime $\mathbf{R}^4$. The causal Green function is a fundamental solution of the wave equation,

$$(\partial_t^2 - \Delta)G(\mathbf{x}, t) = \delta(\mathbf{x}, t), \qquad \Delta \equiv \Delta_{\mathbf{x}}, \qquad (1)$$

given by

$$G(\mathbf{x}, t) = \frac{\delta(t - |\mathbf{x}|)}{4\pi|\mathbf{x}|}. \qquad (2)$$

Its analytic extension to complex spacetime is obtained as follows. First we extend the delta function to the lower-half time plane by taking its positive-frequency (analytic signal) part. This gives the Cauchy kernel:

$$\delta(t) = \frac{1}{2\pi} \int_{-\infty}^{\infty} e^{-i\omega t} d\omega \;\to\; \tilde{\delta}(\tau) = \frac{1}{2\pi} \int_{0}^{\infty} e^{-i\omega\tau} d\omega = \frac{1}{2i\pi\tau}, \tag{3}$$

where

$$\tau = t - is \quad \text{with} \quad s > 0 \tag{4}$$

is necessary for convergence. Next, we extend the Euclidean distance $r \equiv |\mathbf{x}|$ to complex space:

$$r = \sqrt{\mathbf{x} \cdot \mathbf{x}} \;\to\; \tilde{r} \equiv \sqrt{\mathbf{z} \cdot \mathbf{z}}, \qquad \mathbf{z} = \mathbf{x} - i\mathbf{y} \in \mathbf{C}^3. \tag{5}$$

Writing

$$|\mathbf{x}| = r \quad \text{and} \quad |\mathbf{y}| = a, \tag{6}$$

we see that

$$\tilde{r} = \sqrt{r^2 - a^2 - 2iar\cos\theta}, \tag{7}$$

where θ is the angle between $\mathbf{x}$ and $\mathbf{y}$. The complex root has branch points when $r = a$ and $\theta = 0$. For fixed $\mathbf{y}$, these form a circle of radius a in the plane orthogonal to $\mathbf{y}$. In order to make $\tilde{r}$ a single-valued function, we choose the branch defined by

$$\Re\tilde{r} \geq 0, \text{ so that } \mathbf{y} \to \mathbf{0} \;\Rightarrow\; \tilde{r} \to r. \tag{8}$$

The branch cut (again, for fixed $\mathbf{y}$) is then the *disk*

$$S(\mathbf{y}) = \{\mathbf{x} : \tilde{r} = 0\} = \{\mathbf{x} : r \leq a, \; \theta = 0\}. \tag{9}$$

As with ordinary branch cuts in the complex plane, the disk $S(\mathbf{y})$ can be *deformed* continuously to a *membrane*, as long as its boundary ($r = a$, $\theta = 0$) remains invariant. The extended Coulomb potential

$$\tilde{\phi}(\mathbf{z}) \equiv -\frac{1}{4\pi\tilde{r}(\mathbf{z})} \tag{10}$$

is a holomorphic extension of the fundamental solution for the Laplacian in $\mathbf{R}^3$. The distribution defined by

$$\tilde{\delta}(\mathbf{z}) \equiv \Delta\tilde{\phi}(\mathbf{z}), \tag{11}$$

where Δ is the distributional Laplacian with respect to $\mathbf{x}$, is an *extended source distribution* which contracts to the delta function as $\mathbf{y} \to \mathbf{0}$ (Kaiser, 2000):

$$\mathbf{y} \to \mathbf{0} \;\Rightarrow\; \tilde{\delta}(\mathbf{x} - i\mathbf{y}) \to \delta(\mathbf{x}). \tag{12}$$

Since the Coulomb potential $\phi(\mathbf{x})$ is harmonic outside the origin, it follows that $\tilde{\phi}(\mathbf{z})$ is harmonic outside the branch disk $S(\mathbf{y})$, and so the distribution $\tilde{\delta}$ is supported on $S(\mathbf{y})$. Thus $S(\mathbf{y})$ acts as an *extended source* generalizing the usual point source of the Coulomb potential, and this source has been constructed simply by analytic continuation.

We now have all the ingredients for extending the causal Green function $G(\mathbf{x}, t)$ in (2) to complex spacetime. To simplify the notation, denote real spacetime points by

$$x = (\mathbf{x}, t) \in \mathbf{R}^4, \quad y = (\mathbf{y}, s) \in \mathbf{R}^4 \tag{13}$$

and complex spacetime points by

$$z = (\mathbf{z}, \tau) \in \mathbf{C}^4, \quad \mathbf{z} = \mathbf{x} - i\mathbf{y} \in \mathbf{C}^3, \quad \tau = t - is, \ s > 0, \tag{14}$$

so that

$$z = x - iy. \tag{15}$$

The *holomorphic Green function* for the wave equation is now defined by

$$\tilde{G}(z) = \tilde{G}(\mathbf{z}, \tau) = \frac{\tilde{\delta}(\tau - \tilde{r}(\mathbf{z}))}{4\pi\tilde{r}(\mathbf{z})} = \frac{1}{8i\pi^2\tilde{r}(\tau - \tilde{r})}. \tag{16}$$

Bur recall from (4) that the imaginary part of the argument of the numerator had to be negative. We must therefore require that

$$-\Im(\tau - \tilde{r}) = s + \Im\tilde{r} > 0. \tag{17}$$

It can be shown (Kaiser, 2000) that this is equivalent to requiring the imaginary spacetime coordinates $y = (\mathbf{y}, s)$ to satisfy

$$|\mathbf{y}| < s, \quad \text{or} \quad y \in V_+, \tag{18}$$

where V_+ is the *future cone* in spacetime. This means that the argument z of $\tilde{G}$ belongs to the *past tube* $\mathcal{T}_-$ in complex spacetime (Kaiser, 1994).

3. Pulsed-Beam Wavelets

We now show that $\tilde{G}(x - iy)$ describes the emission of a pulsed beam by an elementary "antenna dish" that can be identified with the imaginary spacetime variable $y \in V_+$, as observed at the spacetime point x. Note that when $y \to 0$, this reduces to the usual interpretation of $G(x)$ as the signal observed at x due to an idealized *impulse* emitted at the origin.

To simplify the analysis, we suppose that the observer is far from the source disk $S(\mathbf{y})$ of (9). By (7),

$$r \gg a \implies \tilde{r} \approx r - ia\cos\theta, \tag{19}$$

where the choice of branch $\Re\, \tilde{r} \geq 0$ has been enforced. Substituting this into (16) gives the *far-zone approximation*

$$\tilde{G}(\mathbf{x} - i\mathbf{y}, t - is) \approx \frac{1}{8i\pi^2 r} \cdot \frac{1}{t - r - iT(\theta)}, \tag{20}$$

where

$$T(\theta) = s - a\cos\theta > 0 \quad \text{since} \quad (\mathbf{y}, s) \in V_+. \tag{21}$$

At a fixed position $\mathbf{x}$, (20) is easily seen to be a *pulse* passing the observer at time $t = r$, in accordance with causality and *Huygens' principle*. The *duration* of this pulse is given by $T(\theta)$. The pulse is *shortest* and *strongest* when the observer is on the front axis of the disk $S(\mathbf{y})$ ($\mathbf{x}$ parallel to $\mathbf{y}$), and *longest and weakest* on the rear axis ($\mathbf{x}$ antiparallel to $\mathbf{y}$). By making $s - a$ small, we obtain a *well-focused pulsed beam* concentrated around the front axis. The smaller $s - a$, the better the focus.

Thus $y = (\mathbf{y}, s) \in V_+$ controls the shape of the pulsed beam $\tilde{G}(x - iy)$ observed at x. Namely, $\mathbf{y}$ determines the radius $a = |\mathbf{y}|$ and orientation $\mathbf{y}/a$ of the source disk $S(\mathbf{y})$, while $s - a$ controls the focus of the emitted pulsed beam and its duration along the beam axis. We will label these emission parameters by a subscript e:

$$y \to y_e \equiv (\mathbf{y}_e, s_e) \in V_+. \tag{22}$$

The above pulsed beam is emitted near the origin $\mathbf{x} = \mathbf{0}$ around the time $t = 0$. To emit a pulsed beam from any point $\mathbf{x}_e$ at any time t_e, we need only perform a spacetime translation:

$$\tilde{G}(x - y_e) \to \tilde{G}(x - x_e - iy_e) = \tilde{G}(x - z_e), \tag{23}$$

where

$$z_e = (\mathbf{x}_e, t_e) + i(\mathbf{y}_e, s_e) = x_e + iy_e \tag{24}$$

belongs to the *future tube* $\mathcal{T}_+$ in complex spacetime since $y_e \in V_+$.

4. Reception and Communication of Pulsed Beams

The holomorphic Green function $\tilde{G}(x - z_e)$, defined in the past tube $\mathcal{T}_-$, represents a wave emitted by an extended source described by $z_e = x_e + $

iy_e, with x_e giving the spacetime coordinates of the *center* of the source
and y_e giving the *spacetime extension* about this center (the radius and
orientation of the emitting disk, as well as the duration of the emitted
pulse). By contrast, the original Green function $G(x - x_e)$ describes an
idealized *spherical impulse* emitted from the single spacetime point x_e.
By making the coordinates x_e complex, we have thus obtained a more
realistic and physically interesting model for emission.

However, our model for *reception* is still highly idealized since the
observer is supposed to measure the pulsed beam at the single spacetime point x. We now remedy this by making the observation point
complex as well:

$$x \to z_r \equiv (\mathbf{x}_r, t_r) - i(\mathbf{y}_r, s_r) = x_r - iy_r, \qquad (25)$$

where we have labeled the complex *reception point* z_r with a subscript
r. The change in sign as compared with (24) will be explained below.

With the formal substitution (25), we have

$$\tilde{G}(x - z_e) \to \tilde{G}(z_r - z_e) = \tilde{G}(x_r - x_e - i(y_r + y_e)). \qquad (26)$$

Since the argument of $\tilde{G}$ must belong to the past tube $\mathcal{T}_-$, we have to
require that

$$y_r + y_e \in V_+ \quad \text{for all} \quad y_e \in V_+. \qquad (27)$$

This implies that $y_r \in V_+$, which explains our choice of sign in (25).

> The emission point z_e must belong to the future tube $\mathcal{T}_+$,
> and the reception point z_r must belong to the past tube $\mathcal{T}_-$.
> $\tilde{G}(z_r - z_e)$ represents the coupling between z_e and z_r, giving
> the strength of the overall communication process.

These requirements also make intuitive sense, since emission creates a
signal in the future while reception measures a signal from the past.
¿From now on we identify $z_e \in \mathcal{T}_+$ with the *emitting dish* and $z_r \in \mathcal{T}_-$
with the *receiving dish*. Note that this includes the *durations* of the
emission and the reception processes. (In reception, "duration" is interpreted as the *integration time*.) Our use of the term "dish" therefore
stretches the usual meaning, being a *spacetime* concept rather than
merely spatial.

The condition $z_r \in \mathcal{T}_-$ was derived from the *mathematical* requirement that $z_r - z_e \in \mathcal{T}_-$ for all $z_e \in \mathcal{T}_+$. We now confirm that our model
also makes *physical* sense by studying the communication $\tilde{G}(z_r - z_e)$
in the far-zone approximation. Writing

$$r = |\mathbf{x}_r - \mathbf{x}_e|, \quad t = t_r - t_e, \quad a = |\mathbf{y}_r + \mathbf{y}_e|, \quad s = s_e + s_e, \qquad (28)$$

(20) gives

$$r \gg a \implies \tilde{G}(z_r - z_e) \approx \frac{1}{8i\pi^2 r} \cdot \frac{1}{t - r - iT(\theta)}, \qquad (29)$$

where θ is the angle between $\mathbf{x}_r - \mathbf{x}_e$ and $\mathbf{y}_r + \mathbf{y}_e$ and

$$T(\theta) = s - a \cos \theta \qquad (30)$$

now denotes the *duration* of the overall communication process. Let us fix the distance r between the centers of the emitting and receiving disks $S(\mathbf{y}_e)$ and $S(\mathbf{y}_r)$, as well as their radii $a_e = |\mathbf{y}_e|$ and $a_r = |\mathbf{y}_r|$ and the duration parameters s_e and s_r.

To maximize the communication (29), we need to minimize the duration function $T(\theta)$. By Schwarz's inequality,

$$a \le a_r + a_e, \quad \text{with} \quad a = a_r + a_e \text{ iff } \mathbf{y}_r \text{ is } \textit{parallel} \text{ to } \mathbf{y}_e. \qquad (31)$$

Thus (30) shows that the communication is maximal when

1. the emitting and receiving dishes are synchronized for causal communication, so that $t = r$;

2. the spatial direction vectors $\mathbf{y}_r$ and $\mathbf{y}_e$ are parallel to one another (to maximize a) and also parallel to $\mathbf{x}_r - \mathbf{x}_e$ (to make $\theta = 0$).

We have seen that $\mathbf{y}_e$ gives the direction of propagation of the emitted pulsed-beam wavelet. If we similarly assume that $\mathbf{y}_r$ gives the direction in which the receiving disk is pointed, the above result clearly runs against common sense since it states that reception is greatest when the receiver points directly *away* from the transmitter. Rather, we must interpret $\mathbf{y}_r$ as a vector pointing *into* the receiver, so that the dish $z_r = x_r - iy_r$ is configured to receive receive signals coming *from* the direction of $\mathbf{y}_r$. With this interpretation, the above results are in complete harmony with intuition.

The communication between an emitting dish $z_e = x_e - iy_e$ and a receiving dish $z_r = x_r - iy_r$ is greatest when the two dishes are synchronized for causal commnication and each is pointed towards the center of the other.

5. The Convex Directivity Function

According to the above, the *peak value* of the pulsed beam emitted by z_e and received by z_r is obtained when $\mathbf{y}_r, \mathbf{y}_e$ and $\mathbf{x}_r - \mathbf{x}_e$, are all parallel, so that

$$r = t, \quad a = a_e + a_r, \quad \theta = 0 \qquad (32)$$

and

$$\tilde{G}(z_r - z_e) \approx \frac{1}{8\pi^2 r} \cdot \frac{1}{s - a}. \qquad (33)$$

A dimensionless measure of the *directivity* of the communication, independent of r, may be given as

$$D \equiv \frac{a}{s-a} = \frac{a_r + a_e}{s_r + s_e - a_r - a_e}. \tag{34}$$

Since this expression depends only on $y_r + y_e \in V_+$, it defines a function $D(y)$ on V_+. Note that

$$0 \leq D(y) < \infty, \quad \text{with } D(y) = 0 \text{ iff } a = 0. \tag{35}$$

But under the above assumptions, $a = 0$ implies $\mathbf{y}_r = \mathbf{y}_e = \mathbf{0}$. The directivity $D(y_r + y_e)$ therefore vanishes if and only if the emitting and receiving disks both shrink to points, making the communication process entirely direction-free. (But note that we still have $s_e > 0$ and $s_r > 0$, so that the communicated signal remains a *pulse* rather than an *impulse*.) This helps justify the term "directivity."

But D has another attractive property that goes deeper than the above. For all $y_r, y_e \in V_+$ we have $s_r - a_r > 0$ and $s_e - a_e > 0$, hence

$$D(y_r + y_e) = \frac{a_r + a_e}{(s_r - a_r) + (s_e - a_e)} \leq \frac{a_r}{s_r - a_r} + \frac{a_e}{s_e - a_e}, \tag{36}$$

thus

$$D(y_r + y_e) \leq D(y_r) + D(y_s). \tag{37}$$

Now V_+ is a *convex cone* in $\mathbf{R}^4$, and (37) shows that D is a *convex function* on V_+. This is an important property with an immediate physical interpretation. $D(y_e)$ measures the directivity of the communication when the receiver is a *spacetime point* ($y_r = 0$), so that $\tilde{G}(x_r - z_e)$ represents a pure emission. Similarly, $D(y_r)$ measures the directivity of the communication when the emitter is a spacetime point, so that $\tilde{G}(z_r - x_e)$ represents a pure reception. Then (37) states that *the efficiency of the overall communication can be no better than the sum of its separate emission and reception components.* Further developments of these ideas will appear in (Kaiser, 2001).

References

Heyman, E. and Felsen, L., Complex source pulsed beam fields, *J. Optical Soc. America*, **6**: 806–817, 1989.

Kaiser, G., *A Friendly Guide to Wavelets*. Birkhäuser, Boston, 1994.

Kaiser, G., *Radar Analysis with Causal Pulsed-Beam Wavelets*. Lecture notes for a short course given at *EuroElectromagnetics 2000*. Edinburgh, Scotland, May 29, 2000.

Kaiser, G., Complex-distance potential theory and hyperbolic equations. In *Clifford Analysis*, J. Ryan and W. Sprössig, eds. Birkhäuser, Boston, 2000.

Kaiser, G., *Physical Wavelets and Wave Equations*. Birkhäuser, Boston, to be published, 2001.

Hyper-holomorphic Cells and Riemann-Hilbert Problems

Georg Khimshiashvili

Department of Theoretical Physics, University of Lodz

Abstract. We consider images of cells under mappings satisfying certain linear elliptic systems of the first order. It is shown that in some cases families of such cells attached to a given submanifold may be described by Fredholm operators in appropriate function spaces. Using recent results of I.Stern and of the present author on existence of elliptic Riemann-Hilbert problems for generalized Cauchy-Riemann systems, we indicate some classes of systems which give rise to non-linear Fredholm operators of such type.

Keywords: generalized Cauchy-Riemann system, Clifford algebra, elliptic cell, hyper-holomorphic mapping, Dirac operator, Riemann-Hilbert problem, Fredholm operator

Mathematics Subject Classification: 30G35, 58B15

Dedicated to Richard Delanghe on the occasion of his 60th birthday.

1. Introduction

We deal with collections of differentiable functions on affine domains satisfying certain elliptic system of linear partial differential equations of first order with constant coefficients. Such systems are sometimes called "canonical first order systems" (CFOS) and they gained a lot of attention [3], [6]. An especially important class of such systems is provided by so-called "generalized Cauchy-Riemann systems" (GCRS) [13]. Their solutions are often called hyper-holomorphic (or regular, or monogenic) mappings [3], [6] and in many problems it is necessary to consider images of some standard domains (e.g., balls) under such mappings. Standard examples of such systems in low dimensions are the classical Cauchy-Riemann system on the plane, the Moisil-Theodorescu system in $\mathbf{R}^3$, and Fueter system for functions of one quaternionic variable [3].

We will usually fix a smooth bounded domain homeomorphic to a ball of appropriate dimension and consider its images under solutions of a fixed GCRS. Such images will be called elliptic cells and they will be our main concern in this paper. In this paper we work in a natural setting suggested by non-linear boundary value problems of Riemann-Hilbert type for holomorphic functions of one complex variable [15].

123

F. Brackx et al. (eds.), Clifford Analysis and Its Applications, 123–133.
© 2001 *Kluwer Academic Publishers. Printed in the Netherlands.*

In other words, we consider elliptic cells with boundaries in a fixed submanifold M of the target space of the system. They are called elliptic cells attached to M. In such situation, following terminology of [15], M is called the target manifold. Actually, we will only consider hyper-holomorphic cells, i.e., those which are defined by solutions of a fixed GCRS. A classical example of this type is given by analytic (holomorphic) discs attached to a fixed totally real submanifold [4]. There also exist important generalizations of this classical example in the framework of symplectic geometry [8].

Recently, similar objects appeared in mathematical physics under the name of D-branes and found useful applications in topological field theory and string theory [2], [5]. It is worthy of noting that in this physical context there also appear elliptic cells attached to certain submanifolds. This confirms our belief that such objects deserve some attention on their own.

With this in mind, we undertake an attempt to investigate situations where families of attached hyper-holomorphic cells may be locally described as kernels of some (non-linear) Fredholm operator. Such situations are very important in the classical case of analytic discs [8]. We try to do the same for general elliptic cells and study properties of emerging non-linear operator (we call it Gromov's operator) using a comprehensive theory of linear Riemann-Hilbert problems (RHPs) for GCRS developed in [14], [9].

In particular, we show that, for certain values of dimensions n, k, m, there exist non-compact k-submanifolds in affine m-space such that families of hyper-holomorphic n-cells attached to such submanifolds are locally described by Fredholm operators. Such submanifolds are called admissible targets (for a given GCRS). Existence of admissible targets and fredholmness of the corresponding Gromov's operator are derived from results on existence of elliptic boundary value problems for GCRS [14], [9] (cf. also [12]).

Up to our knowledge, this topic has not been investigated earlier and our goal only was to make first few steps. There remain many points to be clarified, some of them are mentioned in appropriate places of the text. In conclusion, we briefly discuss some related results and further perspectives.

2. Generalized Cauchy-Riemann systems

We present here some necessary notions from [13] and [16].

DEFINITION 1. *([13], cf. also [6]) An elliptic system of first order with constant coefficients is called a generalized Cauchy-Riemann sys-*

*tem if it is invariant under a natural action of the orthogonal group on
the source space and all components of its twice differentiable solutions
are harmonic functions. Solutions of such systems are called hyperholomorphic (hh) mappings. For a given GCRS S, its solutions will be
also called S-mappings.*

It is known that, without loss of generality, one may always assume
that such a system in $\mathbf{R}^{n+1}$ may be written in a canonical form:

$$(1) \qquad E\frac{\partial w}{\partial x_0} + A_1\frac{\partial w}{\partial x_1} + \ldots + A_n\frac{\partial w}{\partial x_n} + Dw = f,$$

where A_j, D are constant complex $(m \times m)$ matrices, $E = A_0$ is the
identity matrix, and for all $i, j = 1, \ldots, n$, one has:

$$(2) \qquad A_iA_j + A_jA_i = -2\delta_{ij}E.$$

We will consider such system in a smooth domain $U \in \mathbf{R}^{n+1}$ and assume that the unknown vector-function w belongs to the class $C^1(U, \mathbf{C}^m)$.

It is well known that system (1) is elliptic, in the usual sense [16],
i.e.,

$$det(t_0E + t_1A_1 + \ldots + t_nA_n) \neq 0,$$

for all $(t_0, \ldots, t_n) \in \mathbf{R}^{n+1} - \{0\}$.

It is also clear that such a system defines a representation of Clifford
algebra Cl_n on $\mathbf{C}^m$ [6]. So the (complex) target dimension m, being the
sum of dimensions of irreducible representations of Cl_n, is an integer
multiple of $2^{[n/2]}$ [3]. If for a given system S this dimension is the
minimal possible $m(n) = 2^{[n/2]}$, we will say that system S is irreducible.
In many problems it is sufficient to consider only irreducible GCRSs
and we will also simplify things by doing so.

For the sake of visuality, we explicitly write down some examples of
such systems in low dimensions. For $n = 1$, one has $m(1) = 1$ and the
corresponding irreducible system is just the classical Cauchy-Riemann
system.

For $n = 2$, one has $m(2) = 2$, and the corresponding irreducible
system for four real functions s, u, v, w of three real variables is the
well-known Moisil-Theodorescu system which may be written using
standard operators on vector-functions in $\mathbf{R}^3$ [3]:

$$div\,(u, v, w) = 0,$$
$$grad\,s + rot\,(u, v, w) = 0.$$

For $n = 3$, one has $m(3) = 2$, and the corresponding irreducible
system becomes so-called Fueter system for four real functions f_i of

four real variables x_j [3]:

$$\frac{\partial f_0}{\partial x_0} - \frac{\partial f_1}{\partial x_1} - \frac{\partial f_2}{\partial x_2} - \frac{\partial f_3}{\partial x_3} = 0,$$

$$\frac{\partial f_0}{\partial x_1} + \frac{\partial f_1}{\partial x_0} - \frac{\partial f_2}{\partial x_3} + \frac{\partial f_3}{\partial x_2} = 0,$$

$$\frac{\partial f_0}{\partial x_2} + \frac{\partial f_1}{\partial x_3} + \frac{\partial f_2}{\partial x_0} - \frac{\partial f_3}{\partial x_1} = 0,$$

$$\frac{\partial f_0}{\partial x_3} - \frac{\partial f_1}{\partial x_2} + \frac{\partial f_2}{\partial x_1} + \frac{\partial f_3}{\partial x_0} = 0.$$

As is well known, this system is a direct analogue of Cauchy-Riemann system for a function of one quaternionic variable. Its solutions, called quaternonic-regular functions, have many interesting properties similar to those of usual holomorphic functions of one complex variable [3].

As suggests the general theory of PDE, for such systems one can formulate various reasonable boundary value problems (BVPs) in bounded domains [16]. For our purposes the most relevant are classical local boundary value problems of Riemann-Hilbert type defined as follows. One searches for solutions of (1) satisfying a boundary condition of the form:

(3) $$(B_1 B_2) \cdot w = g,$$

where B_1, B_2 are continuous complex $(\frac{m}{2} \times \frac{m}{2})$-matrix-functions on bU such that the rows of $(\frac{m}{2} \times m)$-matrix-function (B_1, B_2) are linearly independent at every boundary point, and g is a continuous vector-function with values in $\mathbf{C}^m$.

For our purposes especially important are those RHPs which are elliptic in the usual sense (i.e., satisfy Shapiro-Lopatinski condition [16]) because then the problem (1), (3) is described by a Fredholm operator in appropriate function spaces [16]. It is well-known that not all systems of the type (1) admit elliptic boundary conditions (3) [16], so the first natural question is to establish which GCRS possess elliptic RHPs. We will give an answer to this question in Section 4 and this will enable us to indicate cases in which elliptic cells are described by Fredholm operators.

For notational convenience, in the sequel we denote by V the target space $\mathbf{C}^m$ of system (1).

3. Hyper-holomorphic cells

We fix a GCRS of the form (1) and denote by B a $(n+1)$-ball in its source space. We also take some submanifold M in V and refer to it as a target.

DEFINITION 2. *A hyper-holomorphic (hh) cell attached to M is defined as (the image of) a hh mapping $u : \overline{B} \to V$ such that $u(bB) \subset M$. For a fixed GCRS S, we will speak of S-cells attached to M.*

The usual way of dealing with hh cells attached to a given submanifold is to consider families of cells passing through a given point. Such families may be described by certain non-linear operators in appropriate function spaces and if these operators appear to be Fredholm, then one may obtain a reasonable structural theory of such cells, as it happens, for example, for pseudo-holomorphic discs and curves [8]. So it is natural to begin with looking for such situations where hh cells may be related to Fredholm operators. In order to make this idea precise, we need some constructions and definitions.

To this end, consider an irreducible GCRS S in $\mathbf{R}^{n+1}$ with values in V. Consider also some smooth (C^∞) submanifold M of V of real dimension equal to complex dimension of V (in such case we will speak of submanifold of middle dimension). Let B denote an $(n+1)$-ball centered at the origin of the source space of S and let q be some fixed point at its boundary n-sphere bB. Furthermore, we fix a point $p \in M$ and a non-integer $r > 1$, and let H^r denote the usual Hölder class.

Let F be the space of H^{r+1} maps $f : (B, bB, q) \to (V, M, p)$ which are homotopic to the constant map $f_p = p$ in $\pi_{n+1}(V, M, p)$. In a standard way one checks that F is a smooth complex-Banach manifold (cf. [8]). Let G be the complex Banach space of all H^r maps $g : B \to V$. Define also a submanifold in $F \times G$ by putting

$$(4) \qquad H = \{(f, g) \in F \times G : D(f) = g\},$$

where by D is denoted the matricial partial differential operator defined by the left-hand-side of (1).

Then it is easy to see that H is a connected submanifold of $F \times G$ and one may define the projection map $L_p : H \to G$ given by $L_p(f, g) = g$. It is also easy to check that L_p is a smooth map of H into Banach space G.

DEFINITION 3. *Mapping L_p is called Gromov's operator of the pair (S, M) at point p. Manifold M is called an S-admissible target if Gromov's operator $L_p(S, M)$ is a Fredholm operator (mapping) for every $p \in M$.*

Similar operators were introduced by M.Gromov for analytic discs [8]. General principles of non-linear functional analysis (Sard-Smale theorem, theory of Fredholm structures) guarantee that if these operators appear to be Fredholm, one obtains finite-dimensional families of attached elliptic cells. Actually, this is a quite natural way of formulating a sort of Fredholm theory for elliptic cells and now we present a typical result of this type available in our context. More precisely, we will show that admissible targets exist for certain GCRS.

Consider an irreducible GCRS S in $\mathbf{R}^{n+1}$ with values in V. We first construct another GCRS $D(S)$ with values in $W = V \times V$ which is a sort of "double" of S. If n is even, than $D(S)$ simply consists of two identical copies of S. If n is odd, then one adds to S the canonical GCRS corresponding to the second irreducible representation of Cl_n. Obviously, complex dimension of the target space of $D(S)$ is $2m(n)$.

Consider now a submanifold M in W of real dimension $2m(n)$ and let B denote an $(n+1)$-ball centered at the origin of the source space of $D(S)$.

Granted all this, the first main result of this paper may be stated as follows.

THEOREM 1. *There exists an open set of embeddings* $f : \mathbf{R}^{2m(n)} \to W$ *such that, for every point* $p \in M$, *Gromov's operator of the pair* $(D(S), f(\mathbf{R}^{2m(n)}))$ *at point* p *is a (non-linear) Fredholm operator.*

In other words, non-compact admissible targets exist for systems of the form $D(S)$. For such targets, a standard application of implicit function theorem for Banach spaces implies the collection of attached elliptic cells is locally finite-dimensional.

COROLLARY 1. *The family of* $D(S)$-*hyper-holomorphic cells attached to* M *at* p *is finite-dimensional.*

This result can be considered as a description of the subset of all hh cells attached to M which are close to the "degenerate" cell $f_p = p$. This situation can be generalized by considering the subset of all hh cells attached to M which are sufficiently close to an arbitrary given cell g attached to M. One need not even assume vanishing of the class of g in $\pi_{n+1}(V, M)$.

COROLLARY 2. *For a given* S-*cell* g *attached to an* S-*admissible target* M, *the set of all* S-*cells attached to* M *near* g *is finite-dimensional.*

Of course in both these cases there arises a natural problem of computing this "virtual dimension of nearby attached hh cells" in terms of S, M, and of the given cell g. Such formulas are available for (pseudo-)analytic discs (or Cauchy-Riemann cells, in our terminology) and in terms of notion of the Maslov index of a curve along a totally real submanifold M [4]. In the general case this requires obtaining explicit index formulas for Riemann-Hilbert problems, which seems to be an unsolved problem.

The proof of Theorem 1 is presented in the next section. Using a natural linearization procedure it may be derived from general results on existence of elliptic boundary value problems for GCRSs which are also presented in the next section.

REMARK 1. *Of course, it is natural to ask why have we formulated the theorem only for such special GCRSs and what can be said for an arbitrary GCRS. These points will be clarified in the next section.*

4. Elliptic Riemann-Hilbert problems

Here we will give a comprehensive description of those GCRS which possess elliptic local boundary value problems. By definition with every GCRS S there are associated integers m and n. Since the target $\mathbf{C}^m$ is a representation space for Clifford algebra Cl_n, its dimension m is an integer multiple l of the dimension of irreducible representations $m(n) = 2^{[\frac{n}{2}]}$ [14]. Thus there appears the third integer l (of course it is completely determined by m and n). For odd n, there also appear integers l_1, l_2 ($l_1 + l_2 = l$) showing how many irreducible representations of every of the two possible non-isomorphic types do participate in the direct sum decomposition of the representation defined by system S [14]. It turns out that these parameters completely determine existence of elliptic BVPs.

THEOREM 2. *([9]) Suppose that n is odd and $n \geq 3$. If l is also odd, then there exist no elliptic RHPs for the given GCRS. If l is even, then elliptic RHPs exist if and only if $l_1 = l_2$. In the latter case, the set of elliptic boundary conditions is open and dense in the space of all local boundary conditions of the form (3).*

THEOREM 3. *([14], [9]) If $n \geq 2$ is even, $n \neq 4, 6$, then there always exist elliptic RHPs for GCRS as above and the set of elliptic boundary conditions is open and dense in the space of all local boundary conditions of the form (3).*

Proofs of these results are based on K-homology theory. More precisely, we use the so-called Baum-Douglas criterion of existence of elliptic boundary problems which was recently proved in [7].

REMARK 2. *The restriction that $n \neq 4, 6$ results from the method of the proof of Theorem 3 used in [9]. It is related to some delicate questions of K-theory and it appeared for the first time in the paper [7] which is substantially used in our proof. At the moment it still remains unclear for us if this restriction is really necessary.*

Taking into account these results we may now prove Theorem 1 and in course of proving we will also see the way of generalizing it to arbitrary GCRSs on odd-dimensional spaces, which is the second main result of this paper.

Proof of Theorem 1. Let us first determine the derivative (differential) of L_p at some point (f_0, g_0) and show that it may be interpreted as a boundary value problem (1), (3) for system $D(S)$, i.e., that it is an RHP for the GCRS $D(S)$.

Using the usual description of the tangent space to a manifold of mappings in terms of vector fields along a given mapping [8], it is easy to see that $T_{(f_0, g_0)}H$ may be identified with the space

$$Z = \{f : B \to W : f \in H^{r+1}(B, W), f(q) = p, f(x) \in T_{f_0(x)}M, \forall x \in bB\}.$$

Granted that, it becomes clear that the derivative of L_p at (f_0, g_0) may be identified with the map $\delta : Z \to G$ given by $\delta(f) = Df$.

Let N_M denote the (geometric) normal bundle of M. This is a real vector bundle with fibre dimension $k = 2m(n)$. Consider its pull-back $E_0 = (f_0 \mid bB)^*(N_M)$. From the homotopy condition in the definition of F it follows that E_0 is a trivial bundle over bB, generated by k global sections, say, $p_1, \ldots, p_k$. Using P_j as rows we may form the $(k \times 2k)$-matrix-function $p \in H^{r+1}(bB)$. By the very construction of P, $T_{f_0(x)}M = \{w \in W : P(x)w = 0\}$ and one immediately observes that matrix P has exactly the same form as the matrix of boundary condition (3) for system $D(S)$.

Let us set $X = H^{r+1}(B, W), Y = H^r(B, W) \times H^{r+1}(bB, \mathbf{C}^{m(n)})$ and define a map $R : X \to Y$ by $R(f) = (D(f), (Pf) \mid bB)$. It is obvious that R is exactly the operator of a Riemann-Hilbert problem (1),(3) for system $D(S)$.

Our next goal is to understand under which conditions one may guarantee that R is a Fredholm operator. Notice that if the corresponding RHP is elliptic (i.e., satisfies the Shapiro-Lopatinski condition), then R is a Fredholm operator in virtue of the general theory of elliptic linear boundary value problems [16]. So we should only arrange that matrix P defines an elliptic boundary condition for $D(S)$.

Now the desired conclusion follows from the above two theorems. Notice that one of them is always applicable in the case of a system of the form $D(S)$. If n is even this is evident. If n is odd (this is a more delicate case), from the definition of $D(S)$ it follows that each of the two irreducible representations of Cl_n enters with multiplicity one in the decomposition of the target space $\mathbf{C}^{2m(n)}$ of $D(S)$.

Hence in all cases one can conclude that system $D(S)$ has a plenty of elliptic boundary conditions. In particular, there exist constant matrices P_0 which define elliptic RHPs (1), (3). Let us embed $\mathbf{R}^{2m(n)}$ in W in such a way that the normal space of the image M is orthogonal to the subspace spanned by rows of such a P_0. For such a target M, R is obviously Fredholm, so L_p is also Fredholm at any $p \in M$. Taking into account stability of Fredholm property under small perturbations, we see that all sufficiently small perturbations of M will be admissible targets, which finishes the proof.

REMARK 3. *We used the fact that systems of the form $D(S)$ possess elliptic boundary conditions (3) defined by constant matrix-functions B_1, B_2. This fact is by no means self-evident but it may be derived from results of [9]. The "raison d'être" of this result is the fact (see [9]) that RHPs for systems of the form $D(S)$ are equivalent to so-called transmission problems (also called linear conjugation problems) for system S [10]. Existence of constant elliptic coefficients for general transmission problems was established in [10]. For $n = 1$, this is just a trivial consequence of the classical theory of linear conjugation problems for holomorphic functions (actually every non-degenerate constant matrix generates an elliptic transmission problem because its partial indices obviously vanish). For irreducible systems with $n = 2$ (Moisil-Theodorescu system) and $n = 3$ (Fueter system), existence of constant elliptic transmission conditions also follows from the criteria of fredholmness for such problems obtained in [12].*

The situation with existence of compact admissible targets remains unclear. It is well known that there might be topological obstructions to existence of such targets, which happens already for the classical Cauchy-Riemann system [4]. In order to clarify this issue it is necessary to achieve better understanding of geometric conditions on admissible targets, which can be hopefully done in terms of transversality to certain subspace of the Grassmannian $G(2m(n), 4m(n))$. For $n = 3$, such conditions may be derived from results of [12] but we cannot see how to extend that to the general case.

Analyzing the proof of Theorem 1 and taking into account the previous remark, one sees that, for certain irreducible systems, the same conclusion may be achieved even without taking their doubles.

THEOREM 4. *For any even $n \neq 2$ except $n = 4$ and $n = 6$, there exists an open set of embeddings of $\mathbf{R}^m(n)$ into $\mathbf{C}^{m(n)}$ such that their images are admissible targets for attached hh cells.*

For "exceptional" values $n = 4$ and $n = 6$ the situation remains unclear, but we feel that the result should remain valid. This suggests that one should try to invent an explicit construction of constant elliptic boundary conditions (3) for irreducible systems in $\mathbf{R}^5$ and R^7.

We would like to point our that despite certain apparent analogies with analytic discs, the situation with hh cells is much more subtle. In particular, our restriction to systems of the $D(S)$ type cannot be just removed.

Indeed, the most straightforward generalization of analytic discs attached to totally real surfaces [4] would be to consider the Fueter system in $\mathbf{R}^4 = \mathbf{H}$ (quaternionic regular functions [3]) and try to construct Fueter cells attached to 4-dimensional submanifolds in $\mathbf{R}^8$. However, it turns out that in this way one cannot obtain a reasonable Fredholm theory for such cells, since in this situation Gromov's operator is never Fredholm. Indeed, this follows directly from Theorem 2 because the resulting system has $2 = l_1 \neq l_2 = 0$.

Other related problems and perspectives are briefly discussed in the last section.

5. Concluding remarks

As was already mentioned, for $n = 2, 3$, one can indicate explicit geometric conditions on $T_p M$ for target manifold M to be admissible. This follows from explicit criteria of fredholmness for RHPs for Moisil-Theodorescu and Fueter systems obtained in [12]. It would be interesting to obtain similar results for general GCRSs.

In the classical case of pseudo-holomorphic discs admissible targets are exactly the totally real submanifolds of $\mathbf{C}^n$ and for them it is known that Gromov's operators are Fredholm of index zero [8]. In many cases the same may be proven for Gromov's operators in our setting but these aspects require a separate discussion. For Moisil-Theodorescu and Fueter systems this may be derived from results of [12].

Here naturally enters the general problem of computing the index of an elliptic RHP for GCRS. In principle this is possible using general results of Atiyah and Bott, which should lead to explicit formulas of Dynin-Fedosov type, but it does not seem that somebody have ever written down those explicit formulas in terms of the characteristic matrix and boundary condition. Thus it would be illuminating to obtain an exact recipe, or even an algorithm applicable in concrete situations. In low dimensions, some results in this direction were obtained in [12].

Another promising perspective is related with the problem of characterization of tangent spaces to elliptic cells as subsets of appropriate grassmanians. For the classical Cauchy-Riemann system it is quite simple: tangent spaces of analytic discs constitute the subset of totally real subspaces [4]. In fact, many properties of families of analytic discs may be derived from this interpretation [11], so one may hope that finding such characterizations will be also useful in other cases.

A closer look at this problem reveals that it may be naturally put in the framework of certain general problems of linear algebra over polynomial rings. In particular, there seems to exist a connection with the algebraic analysis of first order systems in the spirit of [1]. These aspects will be considered in next publications of the author.

References

1. W.Adams, P.Loustaunau, V.Palamodov, D.Struppa. Hartog's phenomenon for polyregular functions and projective dimension of related modules, Ann. Inst. Fourier 47(1997), 623–640.
2. M.Bershadsky, V.Sadov, C.Vafa. D-branes and topological field theory, Nucl. Phys. B463(1996), 420–434.
3. F.Brackx, R.Delanghe, F.Sommen. Clifford analysis. Pitman, 1982.
4. F.Forstneric. Complex tangents of real surfaces in complex surfaces. Duke Math. J. 67(1992), 353–376.
5. D.Freed, E.Witten. Anomalies in string theory with D-branes, Asian J. Math. 3(1999), 819–852.
6. J.Gilbert, M.Murray. Clifford algebras and Dirac operators in harmonic analysis, Cambridge Univ. Press, 1990.
7. G.Gong. Relative K-cycles and elliptic boundary conditions, Bull. Amer. Math. Soc. 28(1993), 104–108 .
8. M.Gromov. Pseudo-holomorphic curves in symplectic manifolds, Invent. Math. 82(1985), 307–347.
9. G.Khimshiashvili. Elliptic boundary values problems for Cauchy-Riemann systems. Proc. Conf. Quaternionic and Clifford Analysis, Seiffen, 1996, 77–83.
10. G.Khimshiashvili. Lie groups and transmission problems on Riemann surfaces. Contemp. Math. 131(1992), 121–134.
11. G.Khimshiashvili, E.Wegert. Analytic discs and totally real surfaces. Bull. Acad. Sci. Georgia 162(2000), 41–44.
12. M.Shapiro, N.Vasilevski. Quaternionic hyper-holomorphic functions, singular integral operators and boundary value problems. I, II. Complex Variables 27(1995), 17–44, 67–96.
13. E.Stein, G.Weiss. Generalization of Cauchy-Riemann equations and representations of the rotation group, Amer. J. Math. 90(1968), 163–196.
14. I.Stern. On the existence of Fredholm boundary value problems for generalized Cauchy-Riemann systems, Complex Variables 21(1993), 19–38.
15. E.Wegert. Non-linear boundary value problems for holomorphic functions and singular integral equations. Akademie-Verlag, 1992.
16. J.Wloka. Partielle Differentialgleichungen, Teubner-Verlag, 1982.

Nilpotent Lie Groups in Clifford Analysis and Mathematical Physics

Five Directions for Research

Vladimir V. Kisil (kisilv@e-math.ams.org)
School of Mathematics, University of Leeds

Abstract. The aim of the paper is to popularise nilpotent Lie groups (notably the Heisenberg group and alike) in the context of Clifford analysis and related models of mathematical physics. It is argued that these groups are underinvestigated in comparison with other classical branches of analysis. We list five general directions which seem to be promising for further research.

Keywords: Clifford analysis, Heisenberg group, nilpotent Lie group, Segal-Bargmann space, Toeplitz operators, singular integral operators, pseudodifferential operators, functional calculus, joint spectrum, quantum mechanics, spinor field

Mathematics Subject Classification 2000: 30G35; 22E27, 43A85, 47A60, 47G, 81R30, 81S10

Dedicated to Richard Delanghe on the occasion of his 60th birthday.

1. Introduction

> *One simpleton can ask more questions*
> *than a hundred of wise men find answers.*

The purpose of this short note is to advertise nilpotent Lie groups among researchers in Clifford analysis. It is the author's feeling that this interesting subject is a fertile area still waiting for an appreciation.

The rôle of symmetries in mathematics and physics is widely acknowledged: the Erlangen program for geometries of F. Klein and the theory of relativity of A. Einstein are probably the most famous examples became common places already. The field of Clifford analysis is not an exception in this sense: the rôle of symmetries was appreciated from the very beginning, see e.g. [4, 30] (to mention only very few papers). Moreover rich groups of symmetries are among corner stones which distinguish analytic function theories from the rest of real analysis [19, 20, 21, 24]. This seems to be widely accepted today: in the proceedings of the recent conference in Ixtapa [31] seven out of total seventeen contributions explicitly investigate symmetries in Clifford analysis. But

135

F. Brackx et al. (eds.), Clifford Analysis and Its Applications, 135–141.
© 2001 *Kluwer Academic Publishers. Printed in the Netherlands.*

all of these paper concerned with the semisimple group of *Möbius (conformal) transformations* of Euclidean spaces. This group generalises the $SL_2(\mathbb{R})$—the group of overwhelming importance in mathematics in general [13, 25] and complex analysis in particular [20, 24].

On the other hand the Möbius group is not the only group which may be interesting in Clifford analysis. It was argued [11, 12] that the *Heisenberg group* is relevant in many diverse areas of mathematic and physics. The simplest confirmations are that for example

- Differentiation $\frac{\partial}{\partial x}$ and multiplication by x—two basic operation of not only analysis but also of *umbral calculus* in combinatorics [3, 29] , for example—generate a representation of the Lie algebra of the Heisenberg group.

- any quantum mechanical model gives a representation of the Heisenberg group.

An impressive continuation of the list can be found in [11, 12].

Clifford analysis is not only a subfield of analysis but also has rich and fertile connections with other branches: real harmonic analysis [26], several complex variables [10, 27], operator theory [14, 17], quantum theory [8], etc. Therefore it is naturally to ask about (cf. [11])

the rôle of the Heisenberg group in Clifford analysis.

Unfortunately not enough has been written on the subject. The early paper [16] just initiated the topic and the recent joint paper [5] only hinted at the richness of a possible theory. Thus the whole field seems to be unexplored till now. In order to bring researchers' attention to the above question we list in the next Section five (rather wide) directions for future advances.

2. Five Directions for Research

In the joint paper [5] with Jan Cnops we constructed two examples of spaces of monogenic functions generated by nilpotent Lie groups. The first example is based on the Heisenberg group and gives a monogenic space which is isometrically isomorphic to the classic *Segal-Bargmann space* $F_2(\mathbb{C}^n, e^{-|z|^2}dz)$ of holomorphic functions in $\mathbb{C}^n$, square integrable with respect to the Gaussian measure $e^{-|z|^2}dz$. The second example gives monogenic space of Segal-Bargmann type generated by a Heisenberg-like group with an n dimensional centre. The following propositions are motivated by these examples.

1. **Representation Theory**
   Representation theory of nilpotent Lie group by unitary operators in linear spaces over the field of complex numbers is completely described by the Kirillov theory of *induced representations* [15]. In particular all irreducible unitary representations are induced by a character (one dimensional representation) of the centre of the group. Therefore the image of the centre is always one-dimensional. Representations in linear spaces with Clifford coefficients open a new possibility: unitary irreducible (in an appropriate sense) representations can have a multidimensional image of the groups centre [5]. Various aspects of such representations should be investigated.

2. **Harmonic Analysis**
   Clifford analysis technique is useful [26] for investigation of classic real harmonic analysis questions about *singular integral operators* [32]. Classical real analytical tools are closely related to the harmonic analysis on the Heisenberg and other nilpotent Lie groups [9, 12, 32]. A combination of both—the Heisenberg group and Clifford algebras—could combine the power of two approaches in a single device.

3. **Operator Theory**
   The Segal-Bargmann space F_2 and associated orthogonal projection $P : F_2 \to F_2$ produce an important class of *Toeplitz operators* $T_a = Pa(z, \bar{z})I$ [6] which is a base for the *Berezin quantisation* [2]. Connections between properties of an operator P_a and its symbol $a(z, \bar{z})$ are the subject of important theory [7]. Moreover translations of Toeplitz operators to the Schrödinger representation gives interesting information on *pseudodifferential operators* and their *symbolic calculus* [12]. Monogenic space of the Segal-Bargmann type [5] could provide additional insights in these important relations.

4. **Functional Calculus and Spectrum**
   Functions of several operators can be defined by means of the *Weyl calculus* [1], i.e. essentially using the Fourier transform and its connections with representations with the Heisenberg group [22]. Another opportunity of a *functional calculus* from Segal-Bargmann type spaces is also based on this group [22]. On the other hand a functional calculus can be defined by the *Cauchy formula* for monogenic functions [14, 17]. Simultaneous usage of monogenic functions and nilpotent groups could give a fuller picture for functional calculus of operators and associated notions of *joint spectrums*.

5. Quantum Mechanics

Observables of coordinates and momenta satisfy the Heisenberg commutation relations $[p, q] = i\hbar I$, thus generated the *algebra of observables* representing the Heisenberg group. Even better: the representation theory of Heisenberg and other nilpotent Lie groups provides us with both non-relativistic *classic and quantum* descriptions of the world and a correspondence between them [18, 23, 28]. On the other hand Clifford modules provide a natural description for *spinor degrees of freedom* of particles or fields. Therefore a mixture of these two objects provides a natural model for quantum particles with spin. Such models and their advantages should be worked out.

It should not be difficult to extend the list of open problems (cf. the epigraph).

ACKNOWLEDGEMENTS

This paper was prepared during authors research visit to University of Aveiro, Portugal (August-September 2000) supported by the IN-TAS grant 93–0322–Ext. I am grateful to the Prof. Helmuth Malonek for his hospitality and many useful discussions. Dr. B.Veytsman and O. Pilipenko gave me useful advice.

REFERENCES

Any bibliography for a paper with a small size and a wide scope is necessarily grossly incomplete. I ask for the understanding of readers may be disappointed that their relevant papers are not listed among (almost random) references as well as an excuse for the extensive self-citing.

1. Anderson, R. F.: 1969, 'The Weyl Functional Calculus'. *J. Funct. Anal.* 4, 240–267.

2. Berezin, F. A.: 1988, *Method of Second Quantization*. Moscow: "Nauka".

3. Cigler, J.: 1978, 'Some Remarks on Rota's Umbral Calculus'. *Nederl. Akad. Wetensch. Proc. Ser. A* 81, 27–42.

4. Cnops, J.: 1994, 'Hurwitz Pairs and Applications of Möbius Transformations'.

Habilitation dissertation, Universiteit Gent, Faculteit van de Wetenschappen.
Eprint ftp://cage.rug.ac.be/pub/clifford/jc9401.tex.

5. Cnops, J. and V. V. Kisil: 1998, 'Monogenic Functions and Representations of Nilpotent Lie Groups in Quantum Mechanics'.
   Mathematical Methods in the Applied Sciences **22**(4), 353–373.
   Eprint http://arXiv.org/abs/math/9806150/math/9806150.

6. Coburn, L. A.: 1994, 'Berezin-Toeplitz Quantization'.
   In: *Algebraic Methods in Operator Theory*.
   New York: Birkhäuser Verlag, pp. 101–108.

7. Coburn, L. A. and J. Xia: 1995, 'Toeplitz Algebras and Rieffel Deformation'.
   Comm. Math. Phys. **168**(1), 23–38.

8. Dirac, P.: 1967, *Lectures on Quantum Field Theory*.
   New York: Yeshiva University.

9. Folland, G. and E. Stein: 1982, *Hardy Spaces on Homogeneous Group*.
   Princeton, New Jersey: Princeton University Press.

10. Gürlebeck, K. and H. R. Malonek: 1999, 'A hypercomplex derivative of monogenic functions in $\mathbf{R}^{n+1}$ and its applications'.
    Complex Variables Theory Appl. **39**(3), 199–228.

11. Howe, R.: 1980a, 'On the Role of the Heisenberg Group in Harmonic Analysis'.
    Bull. Amer. Math. Soc. (N.S.) **3**(2), 821–843.

12. Howe, R.: 1980b, 'Quantum Mechanics and Partial Differential Equations'.
    J. Funct. Anal. **38**, 188–254.

13. Howe, R. and E. C. Tan: 1992, *Non-Abelian Harmonic Analysis: Applications of $SL(2, \mathbb{R})$*, Universitext.
    New York: Springer-Verlag.

14. Jefferies, B. and A. McIntosh: 1998, 'The Weyl Calculus and Clifford Analysis'.
    Bull. Austral. Math. Soc. **57**(2), 329–341.

15. Kirillov, A. A.: 1962, 'Unitary Representations of Nilpotent Lie Groups'.
    Russian Math. Surveys **17**, 53–104.

16. Kisil, V. V.: 1993, 'Clifford Valued Convolution Operator Algebras on the Heisenberg Group. A Quantum Field Theory Model'.
In: F. Brackx, R. Delanghe, and H. Serras (eds.): *Proceedings of the Third International Conference held in Deinze, 1993*, Vol. 55 of *Fundamental Theories of Physics*.
Dordrecht: Kluwer Academic Publishers Group, pp. 287–294.
MR 1266878.

17. Kisil, V. V.: 1996a, 'Möbius Transformations and Monogenic Functional Calculus'.
http://www.ams.org/era/Electron. Res. Announc. Amer. Math. Soc. **2**(1)
(electronic) MR 98a:47018.

18. Kisil, V. V.: 1996b, 'Plain Mechanics: Classical and Quantum'.
J. Natur. Geom. **9**(1), 1–14.
MR 96m:81112 Eprint
http://arXiv.org/abs/funct-an/9405002/funct-an/9405002.

19. Kisil, V. V.: 1998, 'How Many Essentially Different Function Theories Exist?'.
In: V. Dietrich, K. Habetha, and G. Jank (eds.): *Clifford Algebras and Applications in Mathematical Physics. Aachen 1996*.
Netherlands: Kluwer Academic Publishers, pp. 175–184.
Eprint http://www.clifford.org/clf-alg/kisi9602.

20. Kisil, V. V.: 1999a, 'Analysis in $\mathbb{R}^{1,1}$ or the Principal Function Theory'.
Complex Variables Theory Appl. **40**(2), 93–118.
Eprint http://arXiv.org/abs/funct-an/9712003/funct-an/9712003.

21. Kisil, V. V.: 1999b, 'Two Approaches to Non-Commutative Geometry'.
In: H. Begehr, O. Celebi, and W. Tutschke (eds.): *Complex Methods for Partial Differential Equations*.
Netherlands: Kluwer Academic Publishers, Chapt. 14, pp. 219–248.
Eprint http://arXiv.org/abs/funct-an/9703001/funct-an/9703001.

22. Kisil, V. V.: 1999c, 'Wavelets in Banach Spaces'.
Acta Appl. Math. **59**(1), 79–109.
Eprint http://arXiv.org/abs/math/9807141/math/9807141.

23. Kisil, V. V.: 2000, 'Quantum and Classic Brackets'. p. 19.
(Submitted) Eprint http://arXiv.org/abs/math-ph/0007030

24. Kisil, V. V.: 2000–2001, 'Groups, Wavelets, and Spaces of Analytic Functions'. p. 89.
    lecture notes, preliminary draft available on
    Eprint http://www.amsta.leeds.ac.uk/~kisilv/coimbraf.html.

25. Lang, S.: 1985, $SL_2(\mathbf{R})$, Vol. 105 of *Graduate Text in Mathematics*.
    New York: Springer-Verlag.

26. McIntosh, A.: 1995, 'Clifford Algebras, Fourier Theory, Singular Integral Operators, and Partial Differential Equations on Lipschitz Domains'.
    In: J. Ryan (ed.): *Clifford Algebras in Analysis and Related Topics*.
    Boca Raton: CRC Press, pp. 33–88.

27. Mitelman, I. M. and M. V. Shapiro: 1995, 'Differentiation of the Martinelli-Bochner integrals and the notion of hyperderivability'.
    Math. Nachr. **172**, 211–238.

28. Prezhdo, O. V. and V. V. Kisil: 1997, 'Mixing Quantum and Classical Mechanics'.
    Phys. Rev. A (3) **56**(1), 162–175.
    MR 99j:81010 Eprint http://arXiv.org/abs/quant-ph/9610016/.

29. Roman, S. and G.-C. Rota: 1978, 'The Umbral Calculus'.
    Adv. in Math. **27**, 95–188.

30. Ryan, J.: 1995, 'Conformally covariant operators in Clifford analysis'.
    Z. Anal. Anwendungen **14**(4), 677–704.
    MR 97a:30062.

31. Ryan, J. and W. Sprößig (eds.): 2000, *Clifford algebras and their applications in mathematical physics. Proceedings of the 5th conference, Ixtapa-Zihuatanejo, Mexico, June 27–July 4, 1999. vol. 2: Clifford Analysis*, Vol. 19 of *Progress in Physics*.
    Boston, MA: Birkhaüser.
    xxv, 461 p. $ 69.95 (2000). [ISBN 0-8176-4183-1].

32. Stein, E. M.: 1993, *Harmonic Analysis: Real-Variable Methods, Orthogonality and Oscillatory Integrals*.
    Princeton, New Jersey: Princeton University Press.

A Quaternionic Generalization of the Riccati Differential Equation

Viktor Kravchenko
Departamento de Matemática, Universidade do Algarve

Vladislav Kravchenko (vkravche@maya.esimez.ipn.mx)*
Esc. Sup. de Ing. Mec. y Eléc. del Inst. Polit. Nac., Dept. de Telecom.

Benjamin Williams[†]
Esc. Sup. de Fís. Mat. del Inst. Polit. Nac.

Abstract. A quaternionic partial differential equation is shown to be a generalisation of the traditional Riccati equation and its relationship with the Schrödinger equation is established. Various approaches to the problem of finding particular solutions to this equation are explored, and the generalisations of two theorems of Euler on the Riccati equation, which correspond to this partial differential equation, are stated and proved.

Keywords: quaternionic Riccati equation

Mathematics Subject Classification: 30G35, 35J60

Dedicated to Richard Delanghe on the occasion of his 60th birthday.

1. Introduction

The Riccati equation

$$\partial u = pu^2 + qu + r,$$

where p, q and r are functions, has received a great deal of attention since a particular version was first studied by Count Riccati in 1724, owing to both its peculiar properties and the wide range of applications in which it appears. For a survey of the history and classical results on this equation, see for example [13], [5] and [12]. This equation can be reduced to its canonical form [4],

$$\partial y + y^2 = -v, \tag{1}$$

and this is the form that we will consider.

* supported by CONACYT project 32424-E
† supported by a scholarship from the Instituto Mexicano de Cooperación Internacional

F. Brackx et al. (eds.), Clifford Analysis and Its Applications, 143–154.
© 2001 *Kluwer Academic Publishers. Printed in the Netherlands.*

One of the reasons for which the Riccati equation has so many applications is that it is related to the general second order homogeneous differential equation. In particular, the one-dimensional Schrödinger equation

$$-\partial^2 u - vu = 0 \qquad (2)$$

where v is a function, is related to the (1) by the easily inverted substitution $y = \frac{\partial u}{u}$. This substitution, which, as its most spectacular application, reduces Burger's equation to the standard one-dimensional heat equation, is the basis of the well-developed theory of logarithmic derivatives for the integration of nonlinear differential equations [11]. A generalisation of this substitution will be used in this work.

A second relation between the one-dimensional Schrödinger equation and the Riccati equation is as follows. The one-dimensional Schrödinger operator can be factorised in the form

$$-\partial^2 - v(x) = -(\partial + y(x))(\partial - y(x))$$

if and only if (1) holds.

Among the peculiar properties of the Riccati equation two theorems of Euler, dating from 1760, stand out The first of these states that if a particular solution y_0 of the Riccati equation is known, the substitution $y = y_0 + z$ reduces (1) to a Bernoulli equation which in turn is reduced by the substitution $z = \frac{1}{u}$ to a first order linear equation. Thus, given a particular solution of the Riccati equation, the general solution can be found in two integrations. The second of these theorems states that given two particular solutions y_0, y_1 of the Riccati equation, the general solution can be found in the form

$$y = \frac{ky_0 \exp(\int y_0 - y_1) - y_1}{k \exp(\int y_0 - y_1) - 1} \qquad (3)$$

where k is a constant. That is, given two particular solutions of (1), the general solution can be found in one integration.

Other interesting properties are those discovered by Picard and Weyr ([13],[5]). The first is that given a third particular solution y_3, the general solution can be found without integrating. That is, an explicit combination of three particular solutions gives the general solution. The second is that given a fourth particular solutions y_4, the cross ratio

$$\frac{(y_1 - y_2)(y_3 - y_4)}{(y_1 - y_4)(y_3 - y_2)}$$

is a constant.

This article is concerned with a quaternionic generalisation of the Riccati equation and versions of the above-mentioned theorems of Euler corresponding to this generalisation. Some necessary notation will be introduced in Section 2. In Section 3 we propose the quaternionic generalisation of the Riccati equation, which is shown to be a good generalisation for various reasons, including the fact that it is related to the three-dimensional Schrödinger equation

$$\Delta u + vu = 0 \qquad (4)$$

(here Δ is the three-dimensional Laplacian) in the same way as the Riccati equation is related to (2). In Section 4, we turn our attention to cases in which particular solutions of (9) can be found, some of which differ considerably from the one-dimensional case. Finally, in Section 5, generalisations of Euler's theorems will be stated and proved.

2. Preliminaries

The complex numbers and complex quaterions are denoted by $\mathbb{C}$, $\mathbb{H}(\mathbb{C})$ respectively. The latter consists of elements of the form

$$a = \sum_{k=0}^{3} a_k i_k$$

where the $a_k \in \mathbb{C}$ and the base units i_k satisfy the following rules of multiplication

$$i_0^2 = i_0 = -i_k^2, \ i_0 i_k = i_0 i_k = i_k, \ k = 1,2,3,$$

$$i_1 i_2 = -i_2 i_1 = i_3, \ i_2 i_3 = -i_3 i_2 = i_1, \ i_3 i_1 = -i_1 i_3 = i_2.$$

The complex unit i commutes with the i_k. Frequently it is useful to consider a quaternion a as being the sum of a scalar and a vector part, denoted respectively

$$a_0 := \mathrm{Sc}(a), \ \vec{a} := \mathrm{Vec}(a) = \sum_{k=1}^{3} a_k i_k.$$

Conjugation is defined as follows,

$$\bar{a} := a_0 - \vec{a}$$

and the modulus is

$$|a|^2 = a \cdot \bar{a} = a_0^2 + a_1^2 + a_2^2 + a_3^2$$

so that in particular $\vec{a}^2 = -|\vec{a}|^2$.

Note that in terms of scalars and vectors, the quaternionic product can be written

$$ab = (a_0 + \vec{a})(b_0 + \vec{b}) = a_0 b_0 + a_0 \vec{b} + b_0 \vec{a} - \left\langle \vec{a}, \vec{b} \right\rangle + \left[\vec{a} \times \vec{b} \right]$$

where $\langle a, b \rangle$ is the standard inner product and $[a \times b]$ the standard cross product in $\mathbb{R}^3$. In particular

$$\{\vec{a}, \vec{b}\} = -2 \left\langle \vec{a}, \vec{b} \right\rangle \tag{5}$$

where $\{a, b\} = ab + ba$ is the standard anticommutator.

In what follows functions $g : \Omega \to \mathbb{H}(\mathbb{C})$ will be considered, where Ω is some domain in $\mathbb{R}^3$. The Moisil-Theodorescu operator D is defined on differentiable functions g as follows:

$$Dg = \sum_{k=1}^{3} i_k \partial_k g$$

where $\partial_k = \frac{\partial}{\partial x_k}$. Due to properties of the quaternionic product, this can be written

$$Dg = -\mathrm{div}\,\vec{g} + \mathrm{grad}\,g_0 + \mathrm{rot}\,\vec{g}. \tag{6}$$

Thus it follows that

$$\mathrm{Sc}(D\vec{g}) = -\mathrm{div}\,\vec{g},$$

$$\mathrm{Vec}(D\vec{g}) = \mathrm{rot}\,\vec{g},$$

and for scalar functions u

$$Du = \mathrm{grad}\,u.$$

The theorem of Leibnitz for this operator is the following: given a differentiable scalar function u and a differentiable quaternionic function g,

$$D(ug) = D(u)g + uD(g). \tag{7}$$

The logarithmic derivative (in the sense of Marchenko) of a scalar function u such that $u \neq 0$ in Ω, is defined as

$$\breve{\partial} u = u^{-1} Du.$$

The function $\breve{\partial} u$ is a vector. The derivative is logarithmic in the following sense: given two scalar functions u_1, u_2 that do not vanish in Ω, formula (7) implies that

$$\breve{\partial}(u_1 u_2) = \breve{\partial} u_1 + \breve{\partial} u_2.$$

3. A quaternionic generalisation of the Riccati equation

The substitution of the Jackiw-Nohl-Rebbi-'t Hooft ansatz [8] in the self-duality equation can be written [7] in the following quaternionic form

$$\partial_t g + Dg + |g|^2 = 0 \tag{8}$$

where the subscript t denotes differentiation with respect to time. The solutions of this equation, which has obvious formal similarities with equation (1), are called instantons. In [9] the relation of (8) to the Fueter operator

$$\partial_t + D$$

was shown. In particular, it was shown that for any $f \in \ker(\partial_t + D)$, the function

$$\frac{2(\operatorname{grad} f_0 - \operatorname{div} \vec{f})}{f_0}$$

is a solution of (8), that is, a class of instantons was obtained. In what follows we will concentrate on the case of time independent, purely vectorial quaternionic functions, but the nonhomogeneous equation will be considered.

The following result generalises to three dimensions the relation, mentioned in the introduction, between the one-dimensional Schrödinger operator and the Riccati differential equation via the logarithmic derivative.

PROPOSITION 1. φ is a solution of (4) if and only if $\vec{f} := \check{\partial}\varphi$ is a solution of

$$D\vec{f} + \vec{f}^2 = v \tag{9}$$

Proof. Suppose that there exists a function φ such that $\vec{f} = \check{\partial}\varphi$. Applying (7) gives

$$D\vec{f} = \frac{1}{\varphi^2}\langle \nabla\varphi, \nabla\varphi\rangle - \frac{1}{\varphi}\Delta\varphi,$$

$$\vec{f}^2 = -\frac{1}{\varphi^2}\langle \nabla\varphi, \nabla\varphi\rangle,$$

so that $-\frac{1}{\varphi}\Delta\varphi = v$, or equivalently $\Delta\varphi + v\varphi = 0$. Conversely, given a solution φ of (4), $\vec{f} = \check{\partial}\varphi$ is a solution of (9).

In [2] and [3] it was shown that the three-dimensional Schrödinger operator can be factorised in the following way

$$-\Delta - vI = (D + M^{\vec{f}})(D - M^{\vec{f}})$$

where I is the identity operator and $M^{\vec{f}} q := q\vec{f}$, if and only if equation (9) holds.

Thus the two relations between the Riccati equation (1) and the one-dimensional Schrödinger equation (2) mentioned in the introduction have natural counterparts relating (9) and the three-dimensional Schrödinger equation (4). These relationships suggest that (9) can be considered a good generalisation of (1). It should also be noted that if $\vec{f} = f_k(x_k)i_k$, then (9) is reduced to

$$\partial_k f_k + f_k^2 = -v.$$

That is, a one-dimensional solution of (9) is a solution of (1). Equation (9) will be referred to as the Riccati PDE.

Note that the scalar and vector components of equation (9) are respectively

$$-\mathrm{div}\,\vec{f} + \vec{f}^2 = v,$$

$$\mathrm{rot}\,\vec{f} = 0.$$

The second equation implies that for a simply-connected domain Ω, there exists a scalar function φ such that $\vec{f} = \mathrm{grad}\varphi$. Substituting this in the first equation gives

$$\Delta\varphi + \langle \nabla\varphi, \nabla\varphi \rangle = -v. \tag{10}$$

This equivalence of the Riccati PDE with a scalar elliptic partial differential equation will be used frequently in what follows.

It should also be noted that if the function v in (9) is zero, the substitution $\vec{f} = \breve{\partial}\varphi$ reduces the equation to

$$\Delta\varphi = 0.$$

Thus the homogeneous Riccati equation can be solved explicitly, and its solutions are of the form $\breve{\partial}\varphi$, where $\varphi \in \ker \Delta$. This generalises the highly restricted class of solutions of the homogeneous equation (1), which are precisely functions of the form

$$\frac{1}{x + c},$$

where c is a constant.

4. Particular solutions of the equation

In the next section it will be shown that given one solution of the Riccati PDE it can be linearised. Thus in this section we discuss some possibilities for obtaining particular solutions of (9). First we note that if the function v is of the form

$$v(x) = v_1(x_1) + v_2(x_2) + v_3(x_3),$$

then assuming that $\vec{f} = f_1(x_1)i_1 + f_2(x_2)i_2 + f_3(x_3)i_3$, equation (9) reduces to the system of Riccati ordinary differential equations

$$\partial_k f_k + f_k^2 = -v_k, \quad k = 1, 2, 3.$$

Thus in this case, a particular solution of (9) can be found if and only if each of the above equations can be solved. Obviously, if any of the v_k's are zero, this task is greatly simplified. This situation corresponds to the Schrödinger equation (4) with potential v of the form given above, in which case the variables can be separated.

The existence of a large class of solutions of the homogeneous equation, as described in Section 3, motivates the following procedure, which reduces (9) to various scalar differential equations. Substituting the sum of two functions $\vec{f}_1, \vec{f}_2 \in C^1(\Omega)$ into equation (9) gives the following possible decomposition:

$$D\vec{f}_1 + \vec{f}_1^2 = v_1,$$

$$D\vec{f}_2 + \vec{f}_2^2 = v_2,$$

$$\{\vec{f}_1, \vec{f}_2\} = v_3,$$

$$v = v_1 + v_2 + v_3.$$

In particular, if $\vec{f}_1, \vec{f}_2$ are solutions of the homogeneous equation, this is reduced to

$$\{\vec{f}_1, \vec{f}_2\} = -2\langle \partial \varphi_1, \partial \varphi_2 \rangle = v, \tag{11}$$

where φ_1, φ_2 are harmonic functions.

If $\varphi_1 = x_1$, $\vec{f}_1 = \frac{i_1}{x_1}$, and this becomes

$$\partial_1 \varphi_2 = -\frac{1}{2}v x_1 \varphi_2$$

which has solution

$$\varphi_2 = A(x_2, x_3)\exp(-\frac{1}{2}\int v x_1 dx_1)$$

where $A(x_2, x_3)$ is an arbitrary function. Thus if $\varphi_2 \in \ker \triangle$ and φ_2 is of the above form, the sum

$$\frac{i_1}{x_1} + \mathring{\partial}\varphi_2$$

is a particular solution of the Riccati PDE.

If instead of choosing φ_1 as above, $\varphi_1 = \varphi_2$ is substituted in (11), the eikonal equation

$$2(\mathring{\partial}\varphi)^2 = -v \tag{12}$$

results. Thus for a scalar function $\varphi \in \ker \triangle$, which is also a solution of the above eikonal equation, $\overrightarrow{f} = 2\mathring{\partial}\varphi$ is a solution of (9).

Example 1. Let φ be the standard fundamental solution of the laplacian $\triangle$,

$$\varphi = \frac{1}{4\pi |x|}.$$

This function is harmonic and positive in any domain Ω which does not include the origin. Furthermore it satisfies equation (12) with

$$v = \frac{1}{|x|^2}.$$

Thus

$$\overrightarrow{f} = \frac{-2\overrightarrow{x}}{|x|^2}$$

is a solution of (9).

5. Generalisations of Euler's theorems on the Riccati equation

We now state and prove the generalisations to the Riccati PDE of Euler's theorems on the Riccati equation that were mentioned in the introduction. The first of these theorems states that, given a particular solution of the Riccati differential equation, the equation can be linearised.

PROPOSITION 2. *(Generalization of Euler's first theorem)*
Let $\overrightarrow{h} = grad\xi$ be an arbitrary particular solution of (9). Then

$$\overrightarrow{f} = \overrightarrow{g} + \overrightarrow{h} \tag{13}$$

is also a solution of (9), where $\vec{g} = \check{\partial}\Psi$ *and* Ψ *is a solution of the equation*

$$\Delta\Psi + 2\langle\nabla\xi, \nabla\Psi\rangle = 0, \tag{14}$$

or equivalently of

$$div(e^{2\xi}\nabla\Psi) = 0. \tag{15}$$

Proof. Substituting (13) in (9) gives

$$D\vec{g} + \{\vec{h}, \vec{g}\} + \vec{g}^2 = 0$$

or alternatively, using (5)

$$D\vec{g} - 2\langle\vec{h}, \vec{g}\rangle + \vec{g}^2 = 0. \tag{16}$$

Note that, as in (9), the vector part of (16) is $\mathrm{rot}\,\vec{g} = 0$, so that

$$\vec{g} = \mathrm{grad}\,\Phi$$

for some function Φ. If $\Psi = e^{\Phi}$, this is equivalent to

$$\vec{g} = \check{\partial}\Psi.$$

Equation (16), written in terms of Ψ, is

$$-\frac{1}{\Psi^2}(\nabla\Psi)^2 - \frac{1}{\Psi}\Delta\Psi - \frac{2}{\Psi}\langle\nabla\xi, \nabla\Psi\rangle + \frac{1}{\Psi^2}(\nabla\Psi)^2 = 0,$$

so that (16) is equivalent to

$$\Delta\Psi + 2\langle\nabla\xi, \nabla\Psi\rangle = 0.$$

Noting that

$$div(e^{2\xi}\nabla\Psi) = 2e^{2\xi}\langle\nabla\xi, \nabla\Psi\rangle + e^{2\xi}\Delta\Psi = e^{2\xi}(\Delta\Psi + 2\langle\nabla\xi, \nabla\Psi\rangle),$$

this equation can be rewritten in the form

$$div(e^{2\xi}\nabla\Psi) = 0.$$

Equation (14) is the well-known transport equation, which appears for example in the Ray Method of approximations of solutions to the wave equation, coupled with the eikonal equation [1]. Equation (15) appears in various applications, for example in electrostatics [10], where $e^{2\xi}$ is the dielectric permeability and Ψ is the electric field potential, and as the continuity equation of hydromechanics in the case of a steady flow, where $e^{2\xi}$ is the density of the medium [6].

Remark 1. From (15) follows that

$$e^{2\xi}\nabla\Psi = \operatorname{rot}\vec{s}$$

for some vector-valued function $\vec{s}$, or

$$\nabla\Psi = e^{-2\xi}\operatorname{rot}\vec{s},$$

where $\vec{s}$ must satisfy the condition

$$\operatorname{rot}(e^{-2\xi}\operatorname{rot}\vec{s}) = 0,$$

since $e^{-2\xi}\operatorname{rot}\vec{s}$ must be the gradient of some function.

As mentioned in the introduction, given two particular solutions y_1, y_2 of the Riccati ordinary differential equation the general solution can be found in one integration. A natural question is whether this property extends to the Riccati PDE. The form (3) of the general solution found in this case suggests a similar substitution in (9). This line of reasoning gives the following result.

PROPOSITION 3. *(Generalisation of Euler's second theorem)*
Let $\vec{h}_1 = grad\xi_1$, $\vec{h}_2 = grad\xi_2$ be two particular solutions of (9). Then there exists a scalar function φ such that

$$\vec{f} = grad\varphi = \frac{\vec{h}_1 w - \vec{h}_2}{w - 1}$$

is also a solution of (9), where $w = Ae^{\xi_1-\xi_2}$, $A \in \mathbb{C}$.
    Proof. Equation (9) is equivalent to

$$\Delta\varphi + \langle\nabla\varphi, \nabla\varphi\rangle = -v \tag{17}$$

where $\vec{f} = grad\varphi$. The scalar functions ξ_1 and ξ_2 are two solutions of (17). Substituting the expression

$$\frac{\nabla\xi_1 w - \nabla\xi_2}{w - 1} \tag{18}$$

for $\nabla\varphi$ into this equation, where w is a scalar function, gives

$$\Delta\varphi = \operatorname{div}\left(\frac{\nabla\xi_1 w - \nabla\xi_2}{w - 1}\right)$$

$$= \frac{(w-1)(w\Delta\xi_1 + \nabla w \cdot \nabla\xi_1 - \Delta\xi_2) - \nabla w \cdot (w\nabla\xi_1 - \nabla\xi_2)}{(w-1)^2}$$

and

$$\langle\nabla\varphi, \nabla\varphi\rangle = \frac{(w\nabla\xi_1)^2 - 2w\nabla\xi_1 \cdot \nabla\xi_2 + (\nabla\xi_2)^2}{(w-1)^2}.$$

Simplifying and using the fact that ξ_1 and ξ_2 are solutions of (17), this same equation is reduced to

$$\nabla \log w = \frac{\nabla w}{w} = \nabla(\xi_1 - \xi_2),$$

so that

$$w = A e^{\xi_1 - \xi_2}$$

for an arbitrary constant $A \in \mathbb{C}$. It remains to show that the expression (18) is the gradient of some scalar function φ. This is the case if the rotational part of (18) disappears. This is shown as follows, where the identities $\operatorname{rot} \nabla \varphi = 0$, $\vec{f} \times \vec{f} = 0$ are used.

$$
\begin{aligned}
\operatorname{rot}\left(\frac{\nabla \xi_1 w - \nabla \xi_2}{w - 1}\right) &= \operatorname{rot}\left(\frac{\nabla \xi_1 w}{w-1}\right) - \operatorname{rot}\left(\frac{\nabla \xi_2}{w-1}\right) \\
&= \operatorname{grad}\frac{w}{w-1} \times \nabla \xi_1 + \frac{w}{w-1}\operatorname{rot}\nabla\xi_1 \\
&\quad -\operatorname{grad}\frac{1}{w-1} \times \nabla \xi_2 - \frac{1}{w-1}\operatorname{rot}\nabla\xi_2 \\
&= \operatorname{grad}\frac{w}{w-1} \times \nabla \xi_1 - \operatorname{grad}\frac{1}{w-1} \times \nabla \xi_2 \\
&= \frac{(w-1)\nabla w - w\nabla w}{(w-1)^2} \times \nabla \xi_1 + \frac{\nabla w}{(w-1)^2} \times \nabla \xi_2 \\
&= \frac{\nabla w}{(w-1)^2} \times \nabla(\xi_2 - \xi_1) \\
&= -\frac{A e^{\xi_1 - \xi_2}}{(w-1)^2} \nabla(\xi_2 - \xi_1) \times \nabla(\xi_2 - \xi_1) \\
&= 0.
\end{aligned}
$$

It must be noted that the new solution gained is not necessarily the general solution, but it consists of a larger class of solutions. The following example illustrates this.

Example 2. Consider (9) with $v = -1$,

$$D\vec{f} + \vec{f}^2 = -1.$$

Two solutions of this equation are $h_1 = i_1 = \operatorname{grad}x_1$, $h_2 = i_2 = \operatorname{grad}x_2$. Applying the above result gives the class of solutions

$$\frac{i_1 A e^{x_1 - x_2} - i_2}{A e^{x_1 - x_2} - 1}, \quad A \in \mathbb{C}.$$

However a third solution $h_3 = i_3$ is *not* included in the above expression as a special case.

References

1. Babic, V.M. and V.S. Buldyrev: *Short-wavelength diffraction theory*, (Series on wave phenomena: Vol.4), Berlin, Springer-Verlag, 1991.
2. Bernstein, S.: *Factorization of solutions of the Schrödinger equation*, In: Proceedings of the symposium *Analytical and numerical methods in quaternionic and Clifford analysis*, Seiffen (Germany) 1996 (eds. K. Gürlebeck and W. Sprössig). Freiberg (Germany): Techn. Univ. Bergakad. 1996, p.207-216.
3. Bernstein, S. and K. Gürlebeck: *On a higher dimensional Miura transform*, Complex Variables, 1999, v.38, p.307-319.
4. Bogdanov, Y., S. Mazanik and Y. Syroid: *Course of differential equations*, Minsk, Universitetskae, 1996 (in Russian).
5. Davis, H.: *Introduction to nonlinear differential and integral equations*, New York, Dover Publications, 1962.
6. Fox, R. and A. McDonald, *Introduction to fluid mechanics*, New York, John Wiley & Sons, 1978.
7. Gürsey, F. and C. Tze: *Complex and quaternionic analyticity in chiral and gauge theories, Part I*, Ann. Phys, 1980, v.12, p.29-130.
8. Jackiw, R. and C. Nohl, C. Rebbi: *Conformal properties of pseudoparticle configurations*, Phys Rev D, v.15, 1977, p.1642-1647.
9. Kravchenko, V.G. and V.V. Kravchenko: *On some nonlinear equations generated by Fueter type operators*, Zeit. für Anal. und ihre Andwend, v.13, 1994, p.599-602.
10. Landau, L. and E. Lifschitz: *Electrodynamics of continuous media*, Oxford, Pergamon Press, 1968.
11. Marchenko, V.: *Nonlinear equations and operator algebras*, Dordrecht, Kluwer Academic Publishers, 1988.
12. Reid, W.:*Riccati differential equations*, New York, Academic Press, 1972.
13. Watson, G.: *A treatise on the Bessel functions*, Cambridge, Cambridge University Press, 1922.

Invariant Operators for Quaternionic Structures

Lukáš Krump (krumpkarlin.mff.cuni.cz) *
Mathematical Institute of Charles University Prague

Abstract. This paper is an application of the method of weight graphs to the case of the quaternionic structure. The result is the BGG sequence of operators (regular and singular versions) containing all standard invariant operators.

Keywords: quaternionic structure, BGG sequences, Fueter operator

Mathematics Subject Classification: 53A30, 53A40, 53A55

Dedicated to Richard Delanghe on the occasion of his 60th birthday.

1. Introduction

This paper deals with quaternionic structures (on manifolds) and regards them as an example of so-called $|1|$-graded parabolic structures. It is then a part of a general problem to describe all possible invariant differential operators on a space with such a given structure. A method has been recently developed to describe and to construct a broad class of such operators, called standard operators, see (Čap et al., 2000). It turns out that the resulting operators are described by representation-theoretical properties of the Lie algebras that define the structure, but also they are uniquely determined (up to a multiple) by the source and target spaces of sections. This means, if M is a manifold with a $|1|$-graded (or specifically, quaternionic) structure, and λ, μ are two weights of the Lie algebra $\mathfrak{g} = \mathfrak{sl}_{n+2}$ dominant for $\mathfrak{g}_0$, and V_λ, V_μ appropriate irreducible representations, then there exists one (up to a multiple) standard invariant operator

$$D : C^\infty(M, V_\lambda) \longrightarrow C^\infty(M, V_\mu).$$

The notion of Bernstein-Gelfand-Gelfand (BGG) sequences means that operators are composed (identifying the target space of one with the source space of another) into a connected graph. It turns out that in the regular case, all such graphs are isomorphic to a common pattern, called the Hasse diagram, and in the singular case the resulting sequences are isomorphic to subgraphs of the Hasse diagram. Singular

---

* Supported by GAČR, Grant No. 201/00/P068

F. Brackx et al. (eds.), Clifford Analysis and Its Applications, 155–162.
© 2001 *Kluwer Academic Publishers. Printed in the Netherlands.*

sequences are less in number, but most operators that appear naturally in mathematical theories, can be found just in the singular sequences. In (Struppa, 2000) can be found complexes that correspond to singular BGG sequences for quaternionic structure.

This gives a systematic classification of invariant operators and allows us to read out of the BGG sequence information about possible operators and some of their properties.

2. Basic settings

Let $\mathfrak{g} = A_{n+1} = \mathfrak{sl}_{n+2}\mathbb{C}$ be the special linear complex Lie algebra, i.e. the algebra of complex $(n+2) \times (n+2)$ matrices with zero trace. Consider the parabolic structure on $\mathfrak{g}$ given by the crossed Dynkin diagram notation:

$$\overset{\alpha_1}{\bullet}\!\!-\!\!\overset{\alpha_2}{\bullet} \cdots \overset{\alpha_n}{\bullet}\!\!-\!\!\overset{\alpha_n}{\times}\!\!-\!\!\overset{\alpha_{n+1}}{\bullet}$$

This structure on $\mathfrak{g}$, called the quaternionic structure, is an example of $|1|$-graded Lie algebra. For the description of the algebras of this type and for the notation see (Krump, 2000). We have

$$\mathfrak{g} = \mathfrak{g}_{-1} \oplus \mathfrak{g}_0 \oplus \mathfrak{g}_1$$

where

$$\mathfrak{g}_0 = \mathfrak{sl}_n\mathbb{C} \oplus \mathfrak{sl}_2\mathbb{C} \oplus \mathbb{C}$$

and

$$\mathfrak{g}_{-1} = \mathbb{C}^{n*} \otimes \mathbb{C}^2, \mathfrak{g}_1 = \mathbb{C}^n \otimes \mathbb{C}^{2*}.$$

In matrix notation,

$$\mathfrak{g} = \begin{pmatrix} \mathfrak{sl}_n\mathbb{C} & \mathbb{C}^n \otimes \mathbb{C}^{2*} \\ \mathbb{C}^{n*} \otimes \mathbb{C}^2 & \mathfrak{sl}_2\mathbb{C} \end{pmatrix}$$

We also denote by $\mathfrak{g}_0^s = \mathfrak{sl}_n\mathbb{C} \oplus \mathfrak{sl}_2\mathbb{C}$ the semisimple part of $\mathfrak{g}_0$, whereas $\mathbb{C}$ is a one-dimensional commutative part of $\mathfrak{g}_0$. We also denote by $\mathfrak{p} = \mathfrak{g}_0 \oplus \mathfrak{g}_1$ the parabolic subalgebra of $\mathfrak{g}$.

As for every $|1|$-graded Lie algebra, we have

$$[\mathfrak{g}_i, \mathfrak{g}_j] \subset \mathfrak{g}_{i+j}$$

for every i, j (with $\mathfrak{g}_k = 0, |k| \geq 2$). Consequently, $\mathfrak{g}_{-1}$ and $\mathfrak{g}_1$ can be considered as representations of $\mathfrak{g}_0$ (the action is just the restriction of the adjoint action) and also of $\mathfrak{g}_0^s$.

3. Hasse diagram and regular BGG sequences

Denote by $\{e_1, \ldots, e_{n+2}\}$ the standard orthonormal basis of the $\mathbb{R}^{n+2}$. The weight lattice is $\mathbb{R}^{n+2}/(\sum e_j = 0)$. The simple roots of $\mathfrak{g}$ are $\alpha_1, \ldots, \alpha_{n+1}$, where $\alpha_j = e_j - e_{j+1}$. The crossed root is α_n. The fundamental weights are $\Lambda_1, \ldots, \Lambda_{n+1}$ with $\Lambda_j = e_1 + \ldots + e_j$.

As explained in (Krump, 2000), the representation $\mathfrak{g}_1$ of $\mathfrak{g}_0^s$ has a special importance for determining the BGG sequence. We consider $\mathfrak{g}_1$ as a subrepresentation of the adjoint representation of $\mathfrak{g}$; its weights are then exactly those positive roots of $\mathfrak{g}$ which are spanned by the crossed root α_n (together with other roots), namely

$$e_i - e_j, \quad i = 1, \ldots, n, \quad j = n+1, n+2.$$

The highest weight of $\mathfrak{g}_1$ is $e_1 - e_{n+2}$ and the weight graph is

Recall that if β_1, β_2 are weights of $\mathfrak{g}_1$, we assign to the arrow $\beta_1 \longrightarrow \beta_2$ a label α if $\beta_2 = \beta_1 - \alpha$, where α is a simple root. The weight graph is just the set of all weights of the representation $\mathfrak{g}_1$ with the ordering generated by the relation $\beta_1 \geq \beta_2 \iff \beta_1 \longrightarrow \beta_2$.

Remark: in this section, all facts will be stated for the general case $\mathfrak{g} = A_{n+1}$, but the pictures will be drawn for the case $n = 5$, in order to visualise the results more clearly. For shortness, use the notation $ij = (e_i - e_j)$.

The weight graph for $n = 5$ is (without labels on the arrows):

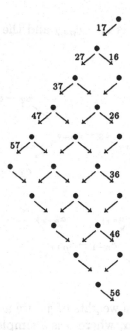

The Hasse diagram H is defined as a graph whose vertices are so-called acceptable subgraphs of the weight graph of $\mathfrak{g}_1$ (see (Krump, 2000)) Every arrow in H has a label which is a weight of $\mathfrak{g}_1$. Perpendicular lines have the same labels.

The Hasse diagram H looks like:

Now we are ready to draw a BGG sequence. We draw, by convention, an arrow $\lambda \longrightarrow \mu$ instead of an operator $D : C^\infty(M, V_\lambda) \longrightarrow C^\infty(M, V_\mu)$.

Every BGG sequence is determined by its initial weight – i.e. the weight λ that corresponds to the highest vertex of H.

Every regular BGG sequence is an isomorphic image of the Hasse diagram H in the sense that it is an isomorphic graph with determined directions: if $h_1 \longrightarrow h_2$ is an arrow labelled β in H, then the corresponding arrow in the BGG sequence is $\lambda \longrightarrow \mu$ with $\mu = \lambda + k_\beta \beta$, where k_β is the order of the arrow, determined by the order formula (see (Krump, 2001)).

We only mention a result of the order formula. It turns out that the orders of the arrows in a BGG sequence can only take values $k_j = \lambda_j + 1$, where λ_j are the coordinates of the initial weight $\lambda = (\lambda_1, \ldots, \lambda_{n+1})$ in the basis of fundamental weights $\Lambda_1 \ldots, \Lambda_{n+1}$.

These values are distributed in the weight graph in the following way:

$$
\begin{array}{c}
k_5 \\
\swarrow \quad \searrow \\
k_1 \qquad k_6 \\
\swarrow \quad \searrow \quad \swarrow \\
k_2 \qquad k_5 \\
\swarrow \quad \searrow \quad \swarrow \\
k_3 \qquad k_1 \\
\swarrow \quad \searrow \quad \swarrow \\
k_4 \qquad k_2 \\
\searrow \quad \swarrow \\
k_3
\end{array}
$$

The BGG sequence is called regular if all $\lambda_i \geq 0$. A regular BGG sequence looks like:

$$
\begin{array}{c}
\lambda^{01} \\
k_5 . 17 \nearrow \swarrow \\
\lambda^{11} \\
k_1 . 27 \nearrow \searrow k_6 . 16 \\
\lambda^{21} \qquad \lambda^{22} \\
k_2 . 37 \nearrow \searrow \qquad \nearrow \\
\lambda^{31} \qquad \lambda^{32} \\
k_3 . 47 \nearrow \searrow \qquad \nearrow \searrow k_5 . 26 \\
\lambda^{41} \qquad \lambda^{42} \qquad \lambda^{43} \\
k_4 . 57 \nearrow \searrow \qquad \nearrow \searrow \qquad \nearrow \\
\lambda^{51} \qquad \lambda^{52} \qquad \lambda^{53} \\
\searrow \qquad \nearrow \searrow \qquad \nearrow \searrow k_1 . 36 \\
\lambda^{61} \qquad \lambda^{62} \qquad \lambda^{63} \\
\searrow \qquad \nearrow \searrow \qquad \nearrow \\
\lambda^{71} \qquad \lambda^{72} \\
\searrow \qquad \nearrow \searrow k_2 . 46 \\
\lambda^{81} \qquad \lambda^{82} \\
\searrow \qquad \nearrow \\
\lambda^{91} \\
\searrow k_3 . 56 \\
\lambda^{(10)1}
\end{array}
$$

where

$$\lambda^{01} = \lambda$$
$$\lambda^{11} = \lambda^{01} + k_5.(e_1 - e_7)$$
$$\lambda^{21} = \lambda^{11} + k_1.(e_2 - e_7)$$
$$\lambda^{22} = \lambda^{11} + k_6.(e_1 - e_6)$$
$$\lambda^{31} = \lambda^{21} + k_2.(e_3 - e_7)$$
$$\lambda^{22} = \lambda^{21} + k_6.(e_1 - e_6) = \lambda^{22} + k_1.(e_2 - e_7)$$

etc.

4. Singular BGG sequences

The singular BGG sequences appear if one of the coordinates λ_j is equal to -1. This causes some of weights in the BGG sequence to be only "virtual" – these weights are not dominant. (They are represented by a cross in the pictures.) Moreover, the operators which correspond to $\lambda_j = -1$ have order 0 and by uniqueness they are (a multiple of) the identity.

Case $\lambda_6 = -1$. (Dual picture, i.e. symmetric by the horizontal axis, is obtained for $\lambda_4 = -1$.)

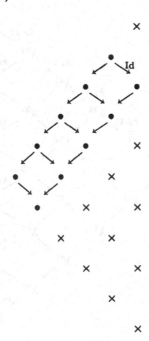

Case $\lambda_5 = -1$. (Dual picture is obtained for $\lambda_3 = -1$.)

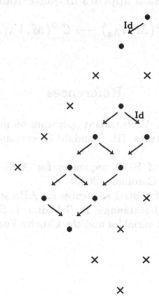

Case $\lambda_1 = -1$. (Dual picture is obtained for $\lambda_2 = -1$.)

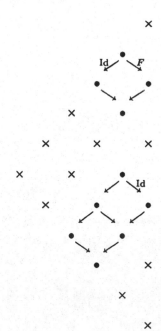

If n is even, the operator F in this sequence is the Fueter operator whose complexified version appears in quaternionic analysis:

$$F : C^\infty(M, V_{\Lambda_6}) \longrightarrow C^\infty(M, V_{\Lambda_1 + \Lambda_5}).$$

References

Čap, A., Slovák, J., Souček, V., Invariant operators on manifolds with almost Hermitian symmetric structures, III. Standard Operators, Jour. Diff. Geom. Appl. 12 (2000), 51–84

Krump, L., Construction of BGG sequences for AHS structures, to appear in Comment. Math. Univ. Carolinae (2000)

Krump, L., Order formula for BGG sequences for AHS structures, to be published

Adams, W., Bernstein, C., Loustaunau, P., Sabadini, I., Struppa, D., Regular functions of several complex variables and the Cauchy-Fueter complex, to appear in Jour. Geom. Anal.

On Generalized Clifford Algebras
- a Survey of Applications

A. Krzysztof Kwaśniewski (kwandr@noc.uwb.edu.pl)

Institute of Computer Science, Białystok University
Higher School of Mathematics and Applied Computer Science, Białystok

Abstract. The generalized Clifford numbers are defined [1] via a k-ubic form Q_k replacing a quadratic one for ordinary construction of an appropriate ideal of tensor algebra. One of the epimorphic image of universal algebras $k - C_{p,q} \cong T(V)/I(Q_k)$ is the algebra $Cl_{p,q}$ with $p + q = \dim V = n$ generators [2].

One enumerates and quotes here some applications of the $Cl_{p,q}$ algebras - mostly those to which the present author has had contributed. Generalized Clifford algebras $Cl_{p,q}$ possess inherent $Z_k \oplus Z_k \oplus \ldots \oplus Z_k$ grading. This makes generalized Clifford numbers an efficient apparatus to deal with spin lattice systems [3]. For the same reason $Cl_{p,q}$ algebras served in providing the explicit cohomological classification and construction of ε-Lie Γ graded algebras [1,4,5,6] and to prove the PBW theorem [7].

Another application [8] of generalized Clifford numbers originates from Herman Weyl's example [9] of finite dimensional quantum mechanics. There, one degree of freedom is represented by a toroidal grid $Z_k \times Z_k$ i.e. classical phase space.

The seeds of the "k-th order idea" may be traced back to Weierstrass [10] who considered possible commutative extensions of complex numbers to the case of arbitrary number of real dimensions. This possibility was afterwards realized [11-13] yielding quasi-number systems which form commutative subalgebras of generalized Clifford algebras $Cl_{p,q}$. These "special" Clifford numbers are perfectly suited for the development of generalized "hyperbolon & ellipton" trigonometries as exposed in [11] and [13]. These quasinumber algebras [13] enable explicit construction of specific generalizations of Tchebyshev polynomials [14].

The generators of the generalized Pauli algebra and the corresponding group of automorphisms preserving k-ubic form are strictly related to the Last Fermat Theorem [15].

Recently the non-commuting matrix elements of matrices from the quantum group $GL_q(2; C)$ with q being the n-th root of unity were given a representation as operators in Hilbert space with help of $C_4^{(n)}$ generalized Clifford algebra generators [16]. This is also described here.

Also recently [17] the quantum Torii Lie algebra and quantum universal enveloping algebra $U_q(sl(2))$ where $q = \omega = \exp\left\{\frac{2\pi i}{n}\right\}$ - were embedded in Generalized Clifford algebra $C_r^{(n)}$.

Keywords: generalized Clifford algebras, Potts models, quantum mechanics

Mathematics Subject Classification: 15A63, 15A66, 15A81

Dedicated to Richard Delanghe on the occasion of his 60th birthday.

F. Brackx et al. (eds.), Clifford Analysis and Its Applications, 163–171.
© *2001 Kluwer Academic Publishers. Printed in the Netherlands.*

1. General information on generalized Clifford and Grassmann algebras

Universal $k - C_n$ Clifford algebras are defined via the following commutative diagram [1]

$$\mathcal{V} \xrightarrow{\alpha_0} n - C_k$$
$$\alpha \searrow \quad \swarrow \sigma$$
$$\mathcal{A}$$

where $\mathcal{V}$ ia a complex vector space of dim $\mathcal{V} = k$, $n - C_k$ and $\mathcal{A}$ are associative algebras, $\sigma \in Hom\,(n - C_k; \mathcal{A})$ while α_0, α are monomorphisms with the property:
$[\alpha_0\,(x)]^n = Q_n\,(x)\,\mathbf{1}$, $[\alpha\,(x)]^n = Q_n\,(x)\,\mathbf{1}$.
Here Q_n denotes n-ubic form [1].
$C_k^{(n)}$ generalized Clifford algebra [2] are generated by matrices satisfying:

$$\gamma_i\gamma_j = \omega\gamma_j\gamma_i\,, \quad i < j\,, \quad \gamma_i^n = 1\,, \quad i, j = 1, 2, ..., k\,. \tag{1.1}$$

($\{\gamma_i\}_{i=1}^k$ are α_0 - images of the vector space $\mathcal{V}$ basis) is the epimorphic image of the universal $k - C_n$ algebra [1].
Naturally $n - C_k = T(\mathcal{V})/I(Q_n)$, where $T(\mathcal{V})$ is the tensor algebra and $I(Q_n)$ denotes the ideal of T generated by elements
$\{x \otimes x \otimes x \otimes ... \otimes x - I(Q_n)\mathbf{1}\}_{x \in \mathcal{V}}$; here n-th tensor power of x is understood.
Universal $k - \mathcal{G}_n$ Grassmann algebra is defined analogously [1] and similarly $G_k^{(n)}$ denotes the associative generalized Grassmann algebra $G_k^{(n)}$ which by definition [1] is generated by the generators $\{\theta_i\}_{i=1}^k$ subjected to the following relations

$$\{\theta_{i_1}, \theta_{i_2}, ...\theta_{i_n}\} = 0; \quad i_1, ...i_n = 1, 2, ..., k \quad ; \tag{1.2}$$

here the "n-anticommutator" [1] is just the symmetrizer operator

$$\{a_1, a_2, ..., a_n\} = \sum_{\sigma \in S_n} a_{\sigma(1)}...a_{\sigma(n)}$$

with S_n denoting the group of permutations.
Realizations of $G_k^{(n)}$ algebras are obtained from $C_{2k}^{(n)}$ generalized Clifford algebras in a similar way as in the case of "usual" $G_k^{(2)}$ and $C_{2k}^{(2)}$ algebras. $G_k^{(n)}$ algebras include various para-Grassmann generalizations of Grassmann variables - now widely spread out trough the vast literature which is too big to be quoted here representatively enough.

Anyhow a paper close in the spirit to what we are going to discuss briefly here is for example the work of Filippov, Isaev and Kurolikov [18]. The authors of [18] use their own notation for general Grassmann algebras [1] $G_N^{(p+1)} \equiv \Gamma_p(N)$ and en passant are calling the general Clifford algebras $C_k^{(n)}$ as "non-commutative spaces satisfying the commutation relations $\theta_i \theta_j = q \theta_j \theta_i \quad i < j$" ($q \equiv \omega$). Filippov, Isaev and Kurolikov have introduced in [18] a generalized derivative on $G_k^{(n)}$ after giving the commutation relations for the Lie algebra of the automorphism group of $G_k^{(n)}$ generalized Grassmann algebra. The authors of [18] established deep relations of their (FIK) Para-Grassmann analysis to ω-deformed structures ("quantum groups"). The FIK Para-Grassmann analysis was afterwards applied also in [19].

2. Generalized Grassmann Algebras and q-Deformed Quantum Oscillator

The authors of [19] work in the general Grassmann algebras and also generalized Clifford algebras setting although they seem to have not realized that they were dealing with generalized Grassmann algebras defined in [1].

The authors of [19] consider also Para-Grassmann ω-deformed harmonic oscillator as an example and its coherent states are given by the following vectors

$$|k(\mathbf{z})\rangle = \sum_{k \in Z_n} \frac{1}{(k)_\omega!} \mathbf{z}^n \otimes |n\rangle \qquad (2.1)$$

where $(k)_\omega = 1 + \omega + ... + \omega^{k-1}$; $k \in Z_n$ is the Gauss number and $(k)_\omega! = (1)_\omega (2)_\omega \cdot ... \cdot (k)_\omega$. The formula (2.1) mimics perfectly the corresponding expression for standard quantum harmonic oscillator with $\mathbf{z}$ being a generator of $G_1^{(n)}$ generalized Grassmann algebra. Another type of "ω-deformed" quantum harmonic oscillator was proposed by Biedenharn [20]. In this connection it is to be noticed that the q-deformed Heisenberg algebra for $q \equiv \omega$ was constructed with help of generalized Pauli algebra generators by M. Rausch de Traubenberg in [21]. As for the further references concerning generalization of Grassmann algebras one may consult [22] apart from [1].

3. Generalized Clifford Algebras and Potts Models

Generalized Clifford algebras generators had appeared in [23,24] where Potts models were considered; however it seems that the authors had not realized that they were dealing with Generalized Clifford Algebras. Afterwards Kwaśniewski [3] reduced the problem of finding of the partition function

$$Z = \sum_{(s_{i,k}) \in S} \exp\left\{ -\frac{E\left[\{s_{i,k}\}\right]}{kT} \right\} \tag{3.1}$$

where

$$-\frac{E\left[(s_{i,k})\right]}{-kT} = a \sum_{i,k=1}^{p,q} \left(s_{i,k}^{-1} s_{i,k+1} + s_{i,k+1}^{-1} s_{i,k} \right)$$

$$+ b \sum_{i,k=1}^{p,q} \left(s_{i,k}^{-1} s_{i+1,k} + s_{i+1,k}^{-1} s_{i,k} \right)$$

with $S = \{(s_{i,k}) = (p \times q); s_{i,k} \in Z_n\}$ for the Potts model to the calculation of $\overline{Tr}(P_1, ..., P_s)$, where $\overline{Tr}$ is the normalized trace while P's are linear combinations of γ matrices - generators of a generalized Clifford algebra naturally assigned to the lattice - the $p \times q$ torus lattice (p rows, q columns; $pq = N$). There in [3] Z_n-vector Potts model is introduced in a form resembling the Ising model without external field and then Kwaśniewski represents the transfer matrix (an element of $C_N^{(n)}$) as a sum of expressions proportional to $Tr(\gamma_{i_1}...\gamma_{i_s})$, where γ's are generalized γ matrices. Then he derives the formula for $Tr(\gamma_{i_1}...\gamma_{i_s})$. The very formula is crucial for getting the complete partition function for Potts models. In general the problem of explicit trace formula for $M^q \in C_N^{(n)}$, is decisive in the calculation of Z function for those models on the lattice in which the transfer matrix is an element of $C_N^{(n)}$.

4. Generalized Clifford Algebras and Z_n-Quantum Mechanics

Generalized Clifford algebras - (although these were not named so and neither were investigated) have appeared already in [9] Weyl's book *Theory of Groups and Quantum Mechanics* in 1931 as the English translation. Neither the authors of "*Observations sur la mécanique quantique finie*" [25] seem to have realized that they were dealing with generalized Clifford Algebras. Afterwards Z_n-Quantum Mechanics model was studied by J. M. Levi-Leblond , T. S. Santhanam, A. K. Kwaśniewski and others [8]. In Kwaśniewski et al. papers, one degree of freedom is represented by a toroidal grid $Z_k \times Z_k$ i.e. classical phase space. At the same time the group $Z_k \times Z_k$ is the grading group for the resulting generalized Clifford algebra [8]. Generators of these generalized Clifford algebras serve as realization of Weyl relations interpreted quantum-mechanically via corresponding transitive system of imprimitivities. Toroidal grid $Z_k \times Z_k$-phase space was treated in as a stage for linear map's quantization. Further details and review of references on quantum mechanics and generalized Clifford Algebras see [26] and [13].

5.   Quasinumbers and special functions of Tchebyshev-type

Clifford numbers from $C_1^{(n)}$ are perfectly suited for the development of generalized "hyperbolon and ellipton" trigonometries as developed in [11] and [13]. These quasinumber algebras [13] enable explicit construction of specific generalizations of Tchebyshev polynomials [14]. These are Z_n labeled sets of solutions of n-th order recurrences with parameter dependent coefficients and are Tchebyshev-like special functions which are polynomials in coordinates of a certain curve on hypersurface in R^n determined by quasi-numbers of determinant one [11]. If $n = 2$ then one gets classical Tchebyshev polynomials of both kinds. And here is an example taken from the Kwaśniewski Mexican paper [14]; namely for Tchebyshev 3-polynomials of zero-th kind we have

$$T_n(x,y,z) \equiv \sum_{k=0}^{n} \binom{n}{k} \sum_{k=0}^{n} \binom{n-k}{i\delta} (i-k)\, x^{n-k-i} y^i z^k \ , \qquad (5.1)$$

where x, y, z are coordinates of a point from the surface given by the group of hyperbolons of volume one [13]

$$\det \begin{pmatrix} h_0(\alpha) & h_1(\alpha) & h_2(\alpha) \\ h_2(\alpha) & h_0(\alpha) & h_1(\alpha) \\ h_1(\alpha) & h_2(\alpha) & h_0(\alpha) \end{pmatrix} \equiv \det \begin{pmatrix} x & y & z \\ z & x & y \\ y & z & x \end{pmatrix} = 1 \ . \qquad (5.2)$$

This surface defined by the equation $x^3 + y^3 + z^3 - 3xyz = 1$. Tchebyshev 3-polynomials of the first and second kind are also functions on the one-parameter subgroup of hyperbolons of volume one. Here we use generalizations of cosh and sinh hyperbolic functions defined as:

$$h_i(x) = \frac{1}{m} \sum_{k \in Z_n} \omega^{-ki} \exp\left\{\omega^k x\right\} \; ; \quad i \in Z_n \; ; \quad \omega = \exp\left\{i\frac{2\pi}{n}\right\} . \quad (5.3)$$

6. Explicit construction of ε Lie Γ graded Algebras

Generalized Lie-like algebras considered in [27] and constructed in [28] were further developed in [1,4,7] due to the application of algebra extensions of abelian groups. (The authors of [27,28] seem not to realize that they had been dealing with generalized Clifford algebras.) In [1,4,7] cohomological classification of the introduced structures is given, the Poincare-Birkhoff-Witt theorem for ε Lie Γ graded algebras is proved and other related topics are discussed. Generalized Clifford algebras appear in explicit constructions as special cases of algebra extensions for $\Gamma = Z_k \otimes Z_k \otimes \ldots \otimes Z_k$ grading group. Because of lack of space let us give at least the definition of Γ graded ε skew symmetric algebra L.

Definition 6.1 The map $\varepsilon : \Gamma \times \Gamma \longrightarrow C^*$ is said to be a commutation factor if (a) ε is a bimorphism; (b) $\forall \alpha, \beta \in \Gamma \quad \varepsilon(\alpha, \beta) \varepsilon(\beta, \alpha) = 1$.

Definition 6.2 Let L be any Γ graded algebra (associative or not). A linear mapping $\Delta : L \longrightarrow L$ homogeneous of degree α satisfying $\Delta \langle x_\alpha, y_\gamma \rangle = \langle \Delta x_\beta, y_\gamma \rangle + \varepsilon(\alpha, \beta) \langle x_\beta, \Delta y_\gamma \rangle$ is said to be an ε derivation.

Definition 6.3 Let L be Γ graded, ε skew symmetric algebra. L is said to be ε Lie Γ graded algebra if
a) ε is a commutation factor and
b) $\forall \alpha, \beta, \gamma \in \Gamma \quad \langle x_\alpha, \langle y_\beta, z_\gamma \rangle \rangle = \langle \langle x_\alpha, y_\beta \rangle, z_\gamma \rangle + \varepsilon(\alpha, \beta) \langle y_\beta, \langle x_\alpha, z_\gamma \rangle \rangle$
i.e. the Jacobi identity holds or, equivalently stated, ε Lie multiplication is an ε derivation of the ε skew symmetric algebra L.

7. On the Last Fermat Theorem and Generalized Clifford Algebras

The authors of [15] showed that the Last Fermat Theorem is equivalent to the statement that all rational solutions of $x^k + y^k = 1$ equation ($k \geq 2$) are provided by an orbit of rationally parameterized subgroup of a group preserving .k-ubic form. This very group according to K. Morinaga and T. Nono [2], [1] is given by

$$\mathfrak{G}(n; C) = \left\{ \omega^l \delta_{i,\sigma(j)} ; \ l \in Z_k , \ \sigma \in S_n \right\} \quad k \geq 3 \qquad (7.1)$$

where $\omega = \exp\left\{\frac{2\pi i}{k}\right\}$. Naturally $|\mathfrak{G}(n; C)| = k^n n!$, hence every "$k$-Fermat group" orbit of solutions of $x^k + y^k = 1$ counts $2k^2$ elements. One readily notices that the orbit $\mathfrak{G}(2; C) \begin{pmatrix} 1 \\ 0 \end{pmatrix}$ does not exhibit any nontrivial rational solution, as the k-Fermat group, $k \geq 3$ i.e. $\mathfrak{G}(n; C)$ contains as the only rationally parameterized subgroup, the matrix permutation subgroup $\simeq S_n$. Thus we arrive at the

Conclusion: The Last Fermat Theorem is equivalent to the statement, that all available rational solutions of $x^k + y^k = 1$ $k \geq 2$ are provided by the orbit $\mathfrak{G}(2; \overline{Q}) \begin{pmatrix} 1 \\ 0 \end{pmatrix}$; $\mathfrak{G}(n; Q) \subset \mathfrak{G}(n; C)$.

REFERENCES

1. A. K. Kwaśniewski: J. Math. Phys., **26** (9) (1985), 2234.

2. K. Morinaga and T. Nono: J. Sci. Hiroshima Univ., Ser. A, Math. Phys. Chem., **16** (1952), 13. K. Yamazaki: J. Fac. Sci. Univ. Tokyo, Sect. 1, **10** (1964), 147. I. Popovici and C. Gheorge: C. R. Acad. Sci. Paris, **262** (1966), 682. A. O. Morris: Q. J. Math. Oxford, Ser. (2) **18** (1967), 7; **19** (1968), 289. E. Thomas: Glasgow Math. J., **15** 74 (1974).

3. A. K. Kwaśniewski: J. Phys. **A**: Math. Gen., **19** (1986), 1469. T. T. Truong in "Clifford Algebras and their Applications in Mathematical Physics" ed. J. S. R. Chisholm and A. K.Commmon; Nato ASI Series, Series C: Mathematical and Physical Sciences **183** (1986), 541-548 T. T. Truong , H. J. de Vega; Phys. Letters **151 B** (2) (1985), 135-141, A. K. Kwaśniewski in "Clifford Algebras and their Applications in Mathematical Physics", ed. J. S. R. Chisholm and A. K. Commmon; Nato ASI Series, Series C: Mathematical and Physical Sciences **183** (1986), 549-554 A. K. Kwaśniewski Supplemento ai Rendiconti del Circolo Matematico di Palermo; Serie II, (1985), 137-139

4. A. K. Kwaśniewski: Supplemento ai Rendiconti del Circolo Matematico di Palermo Serie II No 9 (1985), 141-155.

5. J. Lukierski, V. Rittenberg: Phys. Rev. **D 18** (1975), 385. M. Scheunert, *Generalized Lie Algebras*, J. Math. Phys., **20** (1979), 712-720.

6. M. Scheunert, *Generalized Lie Algebras*, J. Math. Phys., **20** (1979), 712-720.

7. A. K. Kwaśniewski, *On Generalized Lie-Like Algebras*, Preprint N.574 10 Novembre 1987; Dipartimento di Fisica; Universita di Roma "La Sapienza" see also: Differential Geometry and Its Applications - Szczecin September 1988; Szczecin Technical University Publications No11 (1988), 163-172.

8. A. K. Kwaśniewski, W. Bajguz Journal of Group Theory in Physics **3** (2) (1995), 107-113; W. Bajguz, A. K. Kwaśniewski Advances in Applied Clifford Algebras, **4** (1) (1994), 73-88; J. M. Levi-Leblond, Rev. Mexi. Fis., **22** 15 (1973); T. S. Santhanam, Foundations of Physics, Vol. 7, Nos. 1/2, Feb. 1977; T. S. Santhanam, Lettere al Nuovo Cimento, **20** (1) (1977), 13-16; T. S. Santhanam, Physica, **114 A** (1982), 445-447; R. Jagannathan, T. S. Santhanam, Int. J. of Theor. Phys., **21** (5) (1982), p. 151;

9. H. Weyl, *Theory of Groups and Quantum Mechanics, Dover Publications*, inc. 1931. J. Schwinger: Proceedings of the National Academy of Sciences (U.S.A.), 46 (1960), 570.

10. K. Weierstrass: *Zur Theorie der aus n Haupteinheiten gebildeten Grössen*, Leipzig, 1884.

11. N. Fleury, M. Rauch de Trautenberg and R. M. Yamaleev: Université Louis Pasteur, Strasbourg, CRN-PHTH/91-08 (1991).
N. Fleury, M. Rauch de Trautenberg and R. M. Yamaleev: Journal of Mathematical Analysis and Applications **180**, pp.431-457, (1993)
N. Fleury, M. Rauch de Trautenberg; Journal of Mathematical Analysis and Applications **191**, pp.118-136, (1995)

12. L. Bruwier: Bull. Soc. Roy. Sci. Liege, **18** (1949), 72-82, 169-183 J. G. Mikusiäski: Annales de la Société Polonaise de Mathématique, **19** (1946), 165-205; Ann. Soc. Polon. Math., **21** (1948), 46-51. T. Oniga: C. R. Acad. Sci. Paris, **227** (1948), 1138-1140. Y. Lehrer: Riveon Lematematica, **7** (1953), 71-76. L. L. Silverman: Riveon Lematematica, **6** (1953), 53-60. L. Poli: Cahier Rhodaniens, **1** (1949), 1-15.

13. A. K. Kwaśniewski, R. Czech: Reports on Math. Phys., **31** (No 3) (1992), 341-351.

14. A. K. Kwaśniewski Advances in Applied Clifford Algebras, Vol. **9** No. 1 (1999) pp. 41-54 W. Bajguz, A. K. Kwaśniewski Reports on Math. Phys.**43** No. 3 (1999) pp. 367-376; W. Bajguz, A. K. Kwaśniewski Integral Transforms and Special Functions Vol. **8** No 3-4 pp.165-174 (1999) W. Bajguz Integral Transforms and Special Functions Vol. **9** No 2 pp 91-98 (2000)

15. A. K. Kwaśniewski, W. Bajguz, Advances in Applied Clifford Algebras, Vol. **6** (1) (1996), 49-54.

16. A. K. Kwaśniewski; Czech J. Phys. 50 No.1 (2000) 123-127; A. K. Kwaśniewski; Advances in Applied Clifford Algebras, Vol. **9**, (2) pp. 249-260 (1999) in press

17. E. H. L. Khinani, A.Quarab; Advances in Applied Clifford Algebras Vol. **9** (1), (1999) pp. 103-108

18. A. T. Filippov, A. P. Isaev, A. B. Kurolikov, Mod. Phys. Lett. **A 7** (1992), 2129.

19. G. Chadzitaskos, A. Odzijewicz, Letters in Mathematical Physics **43**: 199-209 (1988).

20. L. C. Biedenharn , J. Phys. **A**: Math. Gen. **22** (1989), L873-L878.

21. M. Rausch de Traubenberg Advances in Applied Clifford Algebras, **4** No 2 (1994), pp.131-144.

22. P. Revoy Advances in Applied Clifford Algebras, **3** No 1 (1993), pp.39-54.

23. L. Mittag, M. J. Stephen, J. Math. Phys. **12**, 441 (1971)

24. Yu. A. Bashilov, S. V. Pokrovsky Commun. Math. Phys. **76**, (1980) pp. 129-141

25. R. Balian, C. Itzykson C. R. Acad. Sc. Paris, t. 303, Serie I, No 16, (1986) pp. 773-778

26. A. Kwaśniewski et all. Advances in Applied Clifford Algebras **8** (2) pp.417-432 (1998)

27. J. Lukierski, V. Rittenberg: Phys. Rev. **D 18** (1975), 385.

28. M. Scheunert, "Generalized Lie Algebras" J. Math. Phys., **20** (1979), pp. 712- 720.

Is the Visual Cortex a "Clifford Algebra Quantum Computer"?

Valeri Labunets (lab@cs.tut.fi), Ekaterina Labunets-Rundblad (karen@cs.tut.fi) and Jaakko Astola (jta@cs.tut.fi)
Signal Processing Laboratory, Tampere University of Technology

Abstract. We propose a novel method to calculate invariants of color and multi-color nD images. It employs an idea of multidimensional hypercomplex numbers and combines it with the idea of Fourier–Clifford–Galois Number Theoretical Transforms over hypercomplex algebras, which reduces the computational complexity of a global recognition algorithm from $\mathcal{O}(knN^{n+1})$ to $\mathcal{O}(kN^n \log N)$ for nD k–multispectral images. From this point of view the visual cortex of a animal's brain can by considered as a "Fast Clifford Algebra Quantum Computer".

Keywords: Fast Fourier–Clifford–Galois Transforms, Pattern Recognition, Multi-color Images

Mathematics Subject Classification: 15A66, 68T45, 68H35

Dedicated to Richard Delanghe on the occasion of his 60th birthday.

1. Introduction

One of the main and interesting problems of informatics is the clarification of how human eyes and brain recognize objects in the real world. Practice shows that they successfully cope with problems of recognizing objects at different locations, of different views and elimination, and in different orders of blurring. But how is this done by the brain? How do we see? How do we recognize constantly moving and changing objects of the surrounding world? The phenomenon of moving objects recognition is as obvious as it is incomprehensible, because a moving object is fixed in the retina in the form of the sequence of images each of which in its own right does not permit one to conclude the true shape of the object. But it is beyond question that this sequential set of images appearing in the retina must contain something constant, thanks to which we see and realize the object as something constant. This "something" constant is called *invariant*.

The present work describes new methods of image recognition based on algebraic–geometrical theory of invariants. Changes in surrounding world which cause object shape and color transformations can be

F. Brackx et al. (eds.), Clifford Analysis and Its Applications, 173–182.
© 2001 *Kluwer Academic Publishers. Printed in the Netherlands.*

treated as the action of some Clifford numbers in physical and perceptual spaces. Our main hypothesis is: the brain of primates calculates hypercomplex–valued invariants of an image during recognizing [1]–[8]. Visual systems of animals with different evolutionary history use different hypercomplex algebras. For example, the human brain uses 3D hypercomplex (triplet) numbers to recognize color (RGB)–images and mantis shrimps use 10D multiplet numbers to recognize multicolor images. From this point of view the visual cortex of a animal's brain can be considered as a "Clifford algebra quantum computer". We propose a novel method to calculate invariants of color and multicolor images. It employs an idea of multidimensional hypercomplex numbers and combines it with the idea of number theoretical transforms over hypercomplex algebras, which reduces the computational complexity of the global recognition algorithm from $\mathcal{O}(knN^{n+1})$ to $\mathcal{O}(kN^n \log N)$ for nD k–multispectral images.

2.   Clifford algebras as models of geometrical spaces

We suppose that a brain calculates some hypercomplex–valued invariant of an image when recognizing it. Of course the algebraic nature of hypercomplex numbers must correspond to the spaces with respect to geometrically perceivable properties. For recognition of 2–D, 3–D and n–D images we turn the spaces $\mathbf{R}^2, \mathbf{R}^3, \mathbf{R}^n$ into corresponding algebras of hypercomplex numbers. Here, we present a brief introduction to the conventions of geometrical algebra that are used in this paper. More comprehensive introduction can be found in [9].

Let us consider an "small" n–D space $\mathbf{R}^n$ spanned on the orthonormal basis of n hyperimaginary units I_i, $i = 1, 2, \ldots n$. We suppose $I_i^2 = +1$ for $i = 1, 2, \ldots, p$, $I_i^2 = -1$ for $i = p+1, 2, \ldots, p+q$, $I_i^2 = 0$ for $i = p + q + 1, 2, \ldots, p + q + r = n$ and $I_i I_j = -I_j I_i$. Now, we construct the "big" 2^n–D space $\mathbf{R}^{2^n}$ as a direct sum of subspaces of dimensions $C_n^0, C_n^1, \ldots C_n^n$: $\mathbf{R}^{2^n} = \mathbf{R}^{C_n^0} \oplus \mathbf{R}^{C_n^1} \oplus \ldots \oplus \mathbf{R}^{C_n^n}$, where subspaces $\mathbf{R}^{C_n^p}$ ($p = 0, 1, \ldots, n$) are spanned on the p–products of units $I_{i_1} I_{i_2} \ldots I_{i_p}$ ($i_1 < i_2 < \ldots < i_p$). By definition, we suppose that $I_{i_0} := 1$ is the classical real unit 1. Hence, $\mathbf{R}^{C_n^0} := \{x I_{i_0} \mid x \in \mathbf{R}\}$, $\mathbf{R}^{C_n^1} := \{x_1 I_1 + x_2 I_2 + \ldots + x_n I_n \mid x_i \in \mathbf{R}\}$, $\mathbf{R}^{C_n^2} := \{x_{12} I_1 I_2 + x_{13} I_1 I_3 + \ldots + x_{n-1,n} I_{n-1} I_n \mid x_{i_1 i_2} \in \mathbf{R}\}, \ldots$

Every element of $\mathbf{R}^{2^n}$ has the following representation $\mathcal{C} = \text{Vec}^0(\mathcal{C}) + \text{Vec}^1(\mathcal{C}) + \ldots + \text{Vec}^n(\mathcal{C})$, where $\text{Vec}^0(\mathcal{C}) \equiv \text{Sc}(\mathcal{C}) \in \mathbf{R}^{C_n^0}$ is the scalar part of the Clifford numbers, $\text{Vec}^1(\mathcal{C}) \in \mathbf{R}^{C_n^1}$ is its the vector part, $\text{Vec}^2(\mathcal{C}) \in \mathbf{R}^{C_n^2}$ is its the bivector part, $\ldots$ $\text{Vec}^n(\mathcal{C}) \in \mathbf{R}^{C_n^n}$ is its the n–vector part. If $C_1, C_2 \in \mathbf{R}^{2^n}$, then we can define their product $C_1 C_2$. There 3^n possibilities for $I_i^2 = +1, 0, -1, \forall i = 1, 2, \ldots, n$. Every possibility generates one algebra. Consequently, the space $\mathbf{R}^{2^n}$ with 3^n rules of the multiplication forms 3^n different 2^nD algebras $\mathcal{A}_{2^n}^{p,q,r}(\mathbf{R})$, which are called *Clifford algebras*.

EXAMPLE 1. For $n = 1$ the space $\mathbf{R}^{2^n} = \mathbf{R}^2$ has the algebraic frame of the algebra of generalized complex numbers: $\mathbf{R}^2 \longrightarrow \mathcal{A}_2^{p,q,r}(\mathbf{R}) :=$ $\mathbf{R} + \mathbf{R}I = \{z = x_1 + Ix_2 \mid x_1, x_2 \in \mathbf{R}\}$, where I is a generalized imaginary unit. If $I^2 = -1$, i.e. $I = i$, then $\mathcal{A}_2^{1,1,0} := \{x_1 + ix_2 \mid x_1, x_2 \in \mathbf{R}; \ i^2 = -1\}$ is *the field of complex numbers*. If $I^2 = +1$, i.e $I = e$, then $\mathcal{A}_2^{2,0,0} := \{x_1 + ex_2 \mid x_1, x_2 \in \mathbf{R}; \ e^2 = 1\}$ is *the ring of double numbers*. If $I^2 = 0$, i.e $I = \varepsilon$ then $\mathcal{A}_2(\mathbf{R}) := \{x_1 + \varepsilon x_2 \mid x_1, x_2 \in \mathbf{R}; \ \varepsilon^2 = 0\}$ is *the ring of dual numbers*. □

In $\mathcal{A}_{2^n}^{p,q,r}(\mathbf{R})$ we introduce the conjugation operation which maps every Clifford number C to the number $\overline{C}$. If the pseudodistance between two Clifford numbers $\mathcal{A}$ and $\mathcal{B}$ is defined as the modulus of their difference $\rho(\mathcal{A}, \mathcal{B}) = |\mathcal{A} - \mathcal{B}|$, then the algebras $\mathcal{A}_{2^n}^{p,q,r}(\mathbf{R})$ are transformed into 2^n-D pseudometric spaces designed as $C\mathcal{L}_{2^n}^{p,q,r}$. Subspaces of pure vector Clifford numbers $x_1 I_1 + \ldots + x_n I_n \in \mathrm{Vec}^1\left(\mathcal{A}_{2^n}^{p,q,r}(\mathbf{R})\right)$ are n-D spaces $\mathbf{R}^n := \mathcal{G}\mathcal{R}_n^{p,q,r}$. The pseudometrics constructed in $C\mathcal{L}_{2^n}^{p,q,r}$ induces in $\mathcal{G}\mathcal{R}_n^{p,q,r}$ the corresponding pseudometrics.

EXAMPLE 2. In $\mathcal{A}_2^{p,q,r}(\mathbf{R})$ we introduce a conjugation operation which maps every element $z = x_1 + Ix_2$ to the element $\overline{z} = x_1 - Ix_2$. Now, generalized complex plane is turned into pseudometric space: $\mathcal{A}_2(\mathbf{R}) \longrightarrow \mathcal{G}\mathcal{C}_2^{p,q,r}$ if one defines pseudodistance $\rho(z_1, z_2) = |z_2 - z_1|$. In this case the 2D plane of the classical complex numbers is the Euclidean space $\mathcal{G}\mathcal{C}_2^{2,0,0}$, the double numbers plane is the Minkowskian space $\mathcal{G}\mathcal{C}_2^{1,1,0}$ and the dual numbers plane is the Galilean space $\mathcal{G}\mathcal{C}_2^{1,0,1}$. When one speaks about all three algebras (or geometries) simultaneously then the corresponding algebra (or geometry) is that of *generalized complex numbers*, which is denoted as $\mathcal{A}_2^{p,q,r}$ (or $\mathcal{G}\mathcal{C}_2^{p,q,r}$). □

Every algebra $\mathcal{A}_{2^n}^{p,q,r}(\mathbf{R})$ has even subalgebra $^{ev}\mathcal{A}_{2^n}^{p,q,r}(\mathbf{R})$, spanned on products of even number of hyperimaginary units. All Clifford numbers $\mathcal{E} \in {}^{ev}\mathcal{A}_{2^n}^{p,q,r}$ of unit modulus represent the rotation group of the corresponding space $\mathcal{G}\mathcal{R}_n^{p,q,r}$ which is called *spinor group* and is denoted as $\mathrm{Spin}(\mathcal{A}_{2^n}^{p,q,r})$. Generalized complex numbers and quaternions of unit modulus have the form $z = e^{I\varphi} = \cos\varphi + I\sin\varphi$, $q = e^{u_0\varphi} = \cos\varphi + u_0 \sin\varphi$, where $\cos\varphi$ and $\sin\varphi$ are trigonometric functions in corresponding geometries, φ is a rotation angle around vector-valued quaternion $\mathbf{u}_0$ of unit modulus ($|\mathbf{u}_0| = 1$, $\mathbf{u}_0 = -\overline{\mathbf{u}_0}$). Clifford spinors $\mathcal{E} \in \mathrm{Spin}(\mathcal{A}_{2n}^{p,q,r})$ with unit modulus have the analogous form $\mathcal{E} = e^{\mathbf{u}_0\varphi} = \cos\varphi + \mathbf{u}_0\sin\varphi \in \mathrm{Spin}(\mathcal{A}_{2n}^{p,q,r})$.

THEOREM 1. [9]. All motions of n–D spaces $\mathcal{GR}_n^{p,q,r}$ are represented in the form $\mathbf{x}' = \mathcal{E}^{1/2}\mathbf{x}\mathcal{E}^{-1/2} + \mathbf{y}$, $\quad \mathbf{x},\mathbf{y} \in \mathcal{GR}_n^{p,q,r}$, where $\mathbf{e} := e^{I\varphi}$, $|\mathbf{q}| = 1$, $|\mathcal{E}| = 1$.

If $|\mathbf{e}|, |\mathbf{q}|, |\mathcal{E}| \neq 1$ then the latter transformations form the affine groups $\mathbf{Aff}(\mathcal{GR}_2^{p,q,r})$, $\mathbf{Aff}(\mathcal{GR}_3^{p,q,r})$, $\mathbf{Aff}(\mathcal{GR}_n^{p,q,r})$.

3. Clifford algebras as models of perceptual spaces

The color vision model of primates is based on the existence of three different types of photoreceptors in the eye. However, for example, mantis shrimps have at least ten spectral types of photoreceptors in its eyes giving them capability to recognize fine spectral details. The multicomponent color image is measured as k–component vector

$$\mathbf{f}_{Mcol}(\mathbf{x}) := \begin{bmatrix} f_1(\mathbf{x}) \\ f_2(\mathbf{x}) \\ \ldots \\ f_k(\mathbf{x}) \end{bmatrix} = \begin{bmatrix} \int_\lambda S^{obj}(\lambda, \mathbf{x})H_1(\lambda)d\lambda \\ \int_\lambda S^{obj}(\lambda, \mathbf{x})H_2(\lambda)d\lambda \\ \ldots \\ \int_\lambda S^{obj}(\lambda, \mathbf{x})H_k(\lambda)d\lambda \end{bmatrix},$$

where $S^{obj}(\lambda, \mathbf{x})$ is the color spectrum received from the point $\mathbf{x}$ of the object, $H_1(\lambda)$, $H_2(\lambda)$, $\ldots$, $H_k(\lambda)$ are sensor sensitivity functions. Usually, the three primary colors are chosen as Red, Green, Blue for the primate visual system.

We will interpret such images as hypercomplex–valued signals

$$\mathbf{f}_{Mcol}(\mathbf{x}) = f_0 1(\mathbf{x}) + f_1(\mathbf{x})\varepsilon_{col}^1 + \ldots + f_{k-1}(\mathbf{x})\varepsilon_{col}^{k-1}, \quad \mathbf{x} \in \mathbf{R}^n,\ n = 2, 3, \ldots$$

which takes values in k–cycle algebra $\mathcal{A}_k^{Mcol} := \mathbf{R}1 + \mathbf{R}\varepsilon_{Mcol}^1 + \ldots + \mathbf{R}\varepsilon_{Mcol}^{k-1}$ (where $\varepsilon_{Mcol}^k = 1$) and which we will call *multicolor algebras*. Such algebras generalize the classical HSV–model perceptual color space on multispectral spaces. In particular, RGB–color images are represented as triplet–valued functions: $\mathbf{f}_{col}(\mathbf{x}) = f_R(\mathbf{x})1_{col} + f_G(\mathbf{x})\varepsilon_{col} + f_B(\mathbf{x})\varepsilon_{col}^2$.

One can show that the multiplet algebra $\mathcal{A}_k^{Mcol}$ is the direct sum of real and complex fields: $\mathcal{A}_k^{Mcol} = \sum_{i=1}^{k_{lu}}[\mathbf{R} \cdot \mathbf{e}_{lu}^i] + \sum_{j=1}^{k_{Ch}}[\mathbf{C} \cdot \mathbf{E}_{Ch}^j]$, where $k_{lu} = 1$ or $= 2$ and $k_{Ch} = \frac{k}{2}$ or $\frac{k-1}{2}$ if k is odd or even, respectively, and $\mathbf{e}_{lu}^i$ and $\mathbf{E}_{Ch}^j$ are orthogonal idempotent units such that $(\mathbf{e}_{lu}^i)^2 = \mathbf{e}_{lu}^i$,

$(\mathbf{E}_{Ch}^j)^2 = \mathbf{E}_{Ch}^j$ and $\mathbf{e}_{lu}^i \mathbf{E}_{Ch}^j = \mathbf{E}_{Ch}^j \mathbf{e}_{lu}^i = 0$, for all i, j. Every multiplet C can be represented as a linear combination of k_{lu} "scalar" parts and k_{Ch} "complex" parts: $C = \sum_{i=1}^{k_{lu}} (a_i \cdot \mathbf{e}_{lu}^i) + \sum_{j=1}^{k_{Ch}} (\mathbf{z}_j \cdot \mathbf{E}_{Ch}^j)$. The real numbers $a_i \in \mathbf{R}$ are called *intensity numbers* and complex numbers $\mathbf{z}_j = b + ic \in \mathbf{C}$ are called *multichromaticity numbers*. For example, triplet color algebra is the direct sum of real and complex fields:

$$(1) \qquad \mathcal{A}_3^{col} = \mathcal{A}_3^{col}(\mathbf{R} \mid 1, \varepsilon_{col}, \varepsilon_{col}^2) := \mathbf{R}1_{col} + \mathbf{R}\varepsilon_{col} + \mathbf{R}\varepsilon_{col}^2 =$$
$$\mathbf{R} \cdot e_{lu} + \mathbf{C} \cdot \mathbf{E}_{Ch}.$$

Now we will interpret the multicolor nD image as multiplet–valued nD signal of the following form

$$(2) \qquad \mathbf{f}_{Mcol}(\mathbf{x}) = \sum_{i=1}^{k_{lu}} [f_{lu}^i(\mathbf{x}) \cdot \mathbf{e}_{lu}^i] + \sum_{j=1}^{k_{Ch}} [f_{Ch}^j(\mathbf{x}) \cdot \mathbf{E}_{Ch}^j], \qquad \mathbf{x} \in \mathcal{GR}_n^{p,q,r},$$

which takes values in multiplet algebras $\mathcal{A}_k^{Mcol}$ and the argument $\mathbf{x}$ belongs to the vector part $\mathcal{GR}_n^{p,q,r}$ of the space algebra $\mathcal{A}_{2^n}^{p,q,r}$, i.e. $\mathbf{x} \in \mathcal{GR}_n^{p,q,r} = \mathbf{Vec}\left(\mathcal{A}_{2^n}^{p,q,r}\right)$.

A more complete model of the visual perception is based on the lateral geniculate nucleus (LGN), on which the optical nerve terminates. There are many types of color sensitive cells in LGN. Spectrally opponent cells respond to wide uniform fields by increasing their firing rate within the same wavelength region, and by decreasing the firing rate for other wavelengths. Depending on the firing threshold, these cells are called: "–Red+Blue", "+Red–Blue", "–Green+Red", "+Green–Red", "–Green+Blue", "+Green– Blue", where "+" means excitation and "–" means inhibition.

Let us consider 3–D color space $\mathbf{R}_{col}^3 := \mathbf{R}I_R + \mathbf{R}I_G + \mathbf{R}I_B$ spanned on three units I_R (red unit), I_G (green unit), I_B (blue unit). We suppose $I_R^2 = +1, 0, -1$, $I_G^2 = +1, 0, -1$, $I_B^2 = +1, 0, -1$. Now, we construct a new color Clifford algebra

$$\mathcal{A}_8^{col} := \mathbf{R}I_{Bl} + (\mathbf{R}I_R + \mathbf{R}I_G + \mathbf{R}I_B) + (\mathbf{R}I_{RG} + \mathbf{R}I_{RB} + \mathbf{R}I_{GB}) + \mathbf{R}I_{Wh},$$

where $I_{Bl} = 1$, $I_{Wh} = I_R I_G I_B$ are "black" and "white" units, $I_{RG} := I_R I_G$, $I_{RB} := I_R I_B$, $I_{GB} := I_G I_B$.

DEFINITION 1. Functions of the form $\mathbf{f}_{col} : \mathbf{R}^n \longrightarrow \mathcal{A}_8^{col}$ will be called $\mathcal{A}_8^{col}$–*valued color n–D images.*

To form the algebraic model of multicolor images we consider the kD color space $\mathbf{R}_{col}^k := \mathbf{R}I_1 + \mathbf{R}I_2 + \ldots + \mathbf{R}I_k$ spanned by the basis I_i, $i = 1, 2, \ldots, k$. This multicolor space generates the multicolor Clifford algebra $\mathcal{A}_{2^k}^{Mcol} := {}^{p,q,r}\mathcal{A}_{2^k}^{Mcol}(I_1, I_2, \ldots, I_k | \mathbf{R})$.

DEFINITION 2. $\mathcal{A}_{2^k}^{Mcol}$-valued $f_{Mcol} : \mathbf{R}_{Sp}^n \longrightarrow \mathcal{A}_{2^k}^{Mcol}$ will be called *clifford–valued multicolor nD images.*

Further, we interpret an image as an embedding of a manifold in a higher dimensional spatial–color Clifford algebra. The embedding manifold is a "hybrid" (spatial–color) space that includes spatial coordinates as well as color coordinates. For example, a 2D color image is accordingly considered as a 3–D manifold in the 5–D spatial–color space $\mathbf{R}_5^{SpCol}(I_1^{Sp}, I_2^{Sp}; I_R^{Col}, I_G^{Col}, I_B^{Col}) = (\mathbf{R}I_1^{Sp} + \mathbf{R}I_2^{Sp}) + (\mathbf{R}I_R^{Col} + \mathbf{R}I_G^{Col} + \mathbf{R}I_B^{Col}) = \mathbf{R}_{Sp}^2 \oplus \mathbf{R}_{col}^3$, whose coordinates are (x, y, f_R, f_G, f_B), where $x \in \mathbf{R}I_1^{Sp}$, $y \in \mathbf{R}I_2^{Sp}$ are spatial coordinates and $f_R \in \mathbf{R}I_R^{Col}$, $f_G \in \mathbf{R}I_R^{Col}$, $f_B \in \mathbf{R}I_R^{Col}$ are color coordinates. For nD k–multicolor images we have an $n + k$ spatial–color space
$$\mathbf{R}_{n+k}^{SpMcol}(I_1^{Sp}, \ldots, I_n^{Sp}; I_1^{Mcol}, \ldots, I_k^{Mcol}) = (\mathbf{R}I_1^{Sp} + \ldots + \mathbf{R}I_n^{Sp})$$
$$\oplus(\mathbf{R}I_1^{Mcol} + \mathbf{R}I_2^{Mcol} + \ldots + \mathbf{R}I_k^{Mcol}) = \mathbf{R}_{Sp}^n \oplus \mathbf{R}_{Col}^k,$$ whose coordinates are $(x_1, \ldots, x_n; f_1, f_2, \ldots, f_k)$, where $x_i \in \mathbf{R}I_i^{Sp}$, $i = 1, \ldots, n$ are spatial coordinates and $f_j \in \mathbf{R}I_j^{Mcol}$, $j = 1, 2, \ldots, k$. It is clear that the geometrical, color and spatial–multicolor spaces $\mathbf{R}_{Sp}^n$, $\mathbf{R}_{Mcol}^k$, $\mathbf{R}_{SpMcol}^{n+k}$ generate *spatial, color and spatial–multicolor algebras Clifford algebras* $\mathcal{A}_{2^n}^{Sp}$, $\mathcal{A}_{2^k}^{Mcol}$, $\mathcal{A}_{2^{n+k}}^{SpMcol} = \mathcal{A}_{2^n}^{Sp} \otimes \mathcal{A}_{2^k}^{Col}$, respectively, where $\otimes$ is the symbol of the tensor product. We propose that all spatial hyperimaginary units commute with all color units.

4. Hypercomplex–valued invariants of n–D images

Let us assume that $f_{Mcol}(\mathbf{x}) : \mathbf{R}_{Sp}^n \longrightarrow \mathbf{R}_{Mcol}^k$ is an image of some a k-multicolor nD object. This image $f_{Mcol}(\mathbf{x})$ can be considered as a function of Clifford variables: $f_{Mcol}(\mathbf{x})$, $\mathbf{x} \in \mathrm{Vec}^1(\mathcal{A}_{2^n}^{Sp}) := \mathcal{G}\mathcal{R}_n$ with values in multicolor algebra $\mathcal{A}_{2^k}^{Mcol}$. Changes in the surrounding world can be treated in the language of the multicolor algebra as action of two groups: the space affine group $\mathrm{Aff}(\mathcal{G}\mathcal{R}_n^{p,q,r})$, acting on the physical space $\mathrm{Vec}^1(\mathcal{A}_{2^n}^{Sp}) := \mathcal{G}\mathcal{R}_n$ and the multicolor affine group $\mathrm{Aff}(\mathcal{A}_{2^k}^{Mcol})$, acting on the perceptual space $\mathcal{A}_{2^k}^{Mcol}$.

Let $\mathbf{G}_{n,k}^{SpMcol} : \mathrm{Aff}(\mathcal{G}\mathcal{R}_n^{p,q,r}) \times \mathrm{Aff}(\mathcal{A}_{2^k}^{Mcol})$ be the spatial–multicolor group, and $(\mathbf{g}^{Sp}, \mathbf{g}^{Mcol}) \in \mathbf{G}_{n,k}^{SpMcol}$, where $\mathbf{g}^{Sp} \in \mathrm{Aff}(\mathcal{G}\mathcal{R}_n^{p,q,r})$, $\mathbf{g}^{Mcol} \in \mathrm{Aff}(\mathcal{A}_{2^k}^{Mcol})$.

DEFINITION 3. An $\mathcal{A}^{SpMcol}$-valued functional $\mathcal{J} = \mathcal{F}[f_{Mcol}(\mathbf{x})]$ of image $f_{Mcol}(\mathbf{x})$ is *called a relative* $\mathbf{G}_{n,k}^{SpMcol}$*-invariant* if
$$\mathcal{J} = \mathcal{F}[(\mathbf{g}^{Sp}, \mathbf{g}^{Mcol}) \circ f(\mathbf{x})] = \mathcal{F}[\mathbf{g}^{Mcol} \circ f(\mathbf{g}^{Sp} \circ \mathbf{x})] =$$

$$= \mathcal{C}(\mathbf{g}^{Sp}, \mathbf{g}^{Mcol}) \cdot \mathcal{F}[f(\mathbf{x})] \cdot \mathcal{C}^{-1}(\mathbf{g}^{Sp}, \mathbf{g}^{Mcol}), \quad \forall \mathbf{g} \in \mathbf{G}_{n,k}^{SpMcol},$$

where $\mathcal{C}, \mathcal{C}^{-1}$ are left and right $\mathcal{A}^{SpMcol}$-valued multiplicators. If $\mathcal{C} = 1$ then $\mathcal{J}$ is called *absolute invariant*.

Here, $\mathcal{A}^{SpMcol} := \mathcal{A}_{2^n k}^{SpMcol} = \mathcal{A}_{2^n}^{Sp} \otimes \mathcal{A}_k^{Mcol}$ if we use a k–cycle model of the multicolor image and $\mathcal{A}^{SpMcol} := \mathcal{A}_{2^{n+k}}^{SpMcol} = \mathcal{A}_{2^n}^{Sp} \otimes \mathcal{A}_{2^k}^{Mcol}$ if we use a clifford–valued model of the multicolor image.

DEFINITION 4. If $\mathbf{c}$ is the centroid of the image $\mathbf{f}_{Mcol}$ then the functionals

$$\mathfrak{M}_p := \int_{\mathbf{x} \in \mathcal{GR}_n^{Sp}} (\mathbf{x} - \mathbf{c})^p \mathbf{f}_{Mcol}(\mathbf{x}) d\mathbf{x}$$

are called *central $\mathcal{GA}^{SpMcol}$–valued moments* of the multicolor nD image $\mathbf{f}_{Mcol}(\mathbf{x})$.

Here all products are of the type $\mathbf{x}^p \mathbf{f}_{Mcol}$, where $\mathbf{x} \in \mathcal{GR}_n^{Sp}$, $\mathbf{f}_{Mcol} \in \mathcal{GA}_k^{Mcol}$ are spatial–multicolor numbers belonging to the generalized spatial–color algebra. Let us clarify the rules of moment transformation under geometrical and multicolor distortions of the initial images: $\mathbf{f}_{\lambda, Q, \mathbf{a}}^A(\mathbf{x}^*) = A \left\{ \mathbf{f}_{Mcol} \left(\lambda Q(\mathbf{x} + \mathbf{a}) Q^{-1} \right) \right\} A^{-1}$, where λ is a scale factor, $\mathbf{x}^* = r Q(\mathbf{x} + \mathbf{a}) Q^{-1}$, $|Q| = 1$, $|A| \neq 0$.

THEOREM 2. (**Main theorem**). The central moments $\mathfrak{M}_p$ of the multicolor images $\mathbf{f}_{Mcol}(\mathbf{x})$ are relative $\mathcal{GA}^{SpMcol}$–valued invariants

$$(3) \quad \mathfrak{I}_p\{\lambda_Q \mathbf{a} \mathbf{f}_{Mcol}^A\} := \mathfrak{M}_p\{\lambda_Q \mathbf{a} \mathbf{f}_{Mcol}^A\} = \lambda^{p+3} A Q^p [\mathfrak{M}_p\{\mathbf{f}_{Mcol}\}] A^{-1} Q^{-p}$$

with respect to the spatial–multicolor group $\mathbf{G}_{n,k}^{SpMcol}$.

Obviously, the following ratios $\mathfrak{N}_p := \mathfrak{M}_p / \mathfrak{M}_0^{\frac{p+3}{3}}$ are normalized moments. They are *respective* $\mathcal{GA}^{SpMcol}$–valued invariants $\mathfrak{N}_p\{\lambda_Q \mathbf{a} \mathbf{f}_{Mcol}^A\}$:= $Q^p \mathfrak{N}_p\{\mathbf{f}_{Mcol}\} Q^{-p}$. with respect to the spatial–multicolor group $[\text{aff}(\mathcal{GC}_2^{Sp})] \times [(\mathbf{M} \times \mathbf{SO})(\mathcal{GA}^{Mcol})]$ with left Q^p and right Q^{-p} multiplicators, respectively.

THEOREM 3. Modules of unary moments $|\mathfrak{N}_p\{\lambda_Q \mathbf{a} \mathbf{f}_{Mcol}^A\}| = |\mathfrak{N}_p\{\mathbf{f}_{Mcol}\}|$ are absolute scalar–valued invariants with respect to the spatial–multicolor group $[\text{aff}(\mathcal{GC}_3^{Sp})] \times [(\mathbf{M} \times \mathbf{SO})(\mathcal{GA}^{Mcol})]$.

5. Fast Calculation Algorithms of Multiplet Invariants

As every term of a 2D discrete multicolor image $f_{Mcol}(m, n) = \sum_{i=0}^{k-1} f_i(m,n)\varepsilon_{Mcol}^i$ has 2^n gray–levels that there are no principal limits for considering the mathematical model of every term as function which has the values in the finite field $\mathbf{GF}(Q)$: $f_i(m, n)$: $[0, N-1]^2 \longrightarrow \mathbf{GF}(Q)$, $i = 0, 1, ..., k-1$, if $Q > 2^n$. In this case numbers of the form $a_0 + a_1\varepsilon_{Mcol}^1 + A_{k-1}\varepsilon_{Mcol}^{k-1}$, where $a_i \in \mathbf{GF}(Q)$, are called *modular multiplets*. They form the *modular multiplet algebra*:

$$\mathcal{A}^{Mcol}(\mathbf{GF}(Q) \mid 1, \varepsilon_{Mcol}^1, \ldots, \varepsilon_{Mcol}^{k-1}) :=$$

$$\mathbf{GF}(Q) + \mathbf{GF}(Q)\varepsilon_{Mcol}^1 + \ldots + \mathbf{GF}(Q)\varepsilon_{Mcol}^{k-1}.$$

One can show that for special cases, the Q modular multiplet algebra is the direct sum of Galois fields $\mathbf{GF}(Q)$ and complex Galois fields $\mathbf{GF}(Q^2)$:

$$\mathcal{A}^{Mcol}(\mathbf{GF}(Q)) := \sum_{i=1}^{k_{lu}} \mathbf{GF}(Q) \cdot e_{lu} + \sum_{j=1}^{k_{Ch}} \mathbf{GF}(Q^2) \cdot \mathbf{E}_{Ch}^j.$$

A 2D discrete image of the form $f_{Mcol}(m, n) : \mathcal{GC}_2^{Sp} \to \mathcal{A}^{Mcol}(\mathbf{GF}(Q))$ is called *a modular multiplet–valued image*.

DEFINITION 5. Functionals $\mathcal{M}_p\{f_{Mcol}\} := \mathfrak{M}_p\{f_{Mcol}\} \,(\text{mod}\,Q) :=$

$$= \sum_{m=0}^{Q-1} \sum_{n=0}^{Q-1} (m + In)^p f_{Mcol}(m + In) =$$

$$\sum_{i=1}^{k_{lu}} \left(\sum_{m=0}^{Q-1}\sum_{n=0}^{Q-1} (m + In)^p f_l^i(m + In) \right) \cdot e_{lu}^i +$$

$$+ \sum_{j=1}^{k_{Sh}} \left(\sum_{m=0}^{Q-1}\sum_{n=0}^{Q-1} (m + In)^q f_{Ch}^j(m + In) \right) \cdot \mathbf{E}_{Ch}^j \,(\text{mod}\,Q)$$

are called *modular $\mathcal{A}^{SpCol}(\mathbf{GF}(Q))$–valued moments* of the 2D image $f_{Mcol}^k(m, n)$.

Let $\mathcal{E}$ be a primitive root in the Galois field $\mathbf{GF}(Q^2)$ then $m + In = \mathcal{E}^k$, and

$$\mathcal{M}_p\{f_{Mcol}\} = \sum_{i=1}^{k_{lu}} \left(\sum_{i=0}^{Q^2-1} \mathcal{E}^{pk} f_{lu}^i(\mathcal{E}^k) \right) e_{lu}^i +$$

$$\sum_{j=1}^{k_{Sh}} \left(\sum_{k=0}^{Q^2-1} \mathcal{E}^{pk} f_{Ch}^j (\mathcal{E}^k) \right) \mathbf{E}_{Ch}^j \ (mod\, Q).$$

We obtain an algorithm for calculating modular moments $\mathcal{M}_p$ as the $(k_{lu} + k_{Ch})$ *Fourier–Clifford–Galois NTTs*. Its computational complexity of this algorithm is the defined complexity of $(k_{lu} + k_{Ch})$ fast Fourier–Clifford–Galois Transforms: $[k_{lu}+k_{Ch}][N^2 \log_2 N]$ additions and multiplications, if we use the classical computer. Computational complexity of this algorithm can be reduced by special choice of primitive root $\mathcal{E}$. Indeed, if $\mathcal{E} = \pm 2, \pm 2I, \pm(1 \pm I)$ or $2(1 \pm I)$ then Fourier–Clifford–Galois transform is reduced to the computation of 2–D fast Radon–transform, which can be done without multiplication. Computational complexity of such a computational scheme is only $[k_{lu} + k_{Ch}]N^2 \log_2 N$ additions.

But we can use a quantum computer for the calculation of the hypercomplex moments. Note, that every nD multicolor discrete image can be described as a set of $N = N^{Sp} N^{Mcol} = Q^n Q^k = Q^n Q^{k_{lu}+k_{Ch}}$ real and complex numbers, corresponding to the intensity of light in each multicolor pixel (picture element), i.e as some vector $f_{Mcol} \in \mathbb{C}^N$. In order to store this image we use N-qbit register (simple quantum Read–Only–Memory). Using fast quantum Fourier number theoretical transform we obtain moments $\mathcal{M}_p$. For certain values N there are very efficient algorithms with computer complexity $\log_2 N$ [10].

6. Conclusion

We have presented a novel algebraic tool for the integration of data from multiple sensors into a uniform representation. We have provided an explicit expressions for relative and absolute hypercomplex–valued invariants of color and k–multispectral 2D and 3D images with respect to geometrical and color distortions. The behavior of relative invariants with respect to the more important subgroups of the spatial–color groups is studied in detail. Our technique uses high–dimensional hypercomplex algebras and reduces the computational complexity of the global recognition algorithm from $\mathcal{O}(kN^3)$ to $\mathcal{O}(kN^2 \log N)$ for 2–D k–multicolor images.

Note that digital computers use Boolean algebra. This algebra is the Clifford algebra $\mathcal{A}_{2^n}(\mathbf{GF}(2))$ over field Galois $\mathbf{GF}(2)$. But are there analog computers working with the Clifford algebra $\mathcal{A}_{2^n}^{p,q,r}(\mathbf{R})$? Now we will try to answer to this question by remembering the question in the title: "*Is the Brain a Clifford Algebra Quantum Computer*"? Yes, it can be! If this brain belongs to a human being living even in n–D non–Euclidean space $\mathcal{GR}_n^{p,q,r}$. But, as a fast calculator of invariants? The full answer to this question the interested reader can find in [1],[2], [4]–[8], where we use for this purpose the Fourier–Clifford and the Fourier–Hamilton Number Theoretical Transforms.

References

1. Labunets–Rundblad E.V. (2000) Fast Fourier–Clifford Transforms Design and Application in Invariant Recognition. Thesis for degree of Doctor of Technology. Publication 292, Tampere University Technology, p. 262

2. Labunets–Rundblad E.V., Labunets V.G. (2000): Spatial–Colour Clifford Algebra for Invariant Image Recognition. (Geometric Computing with Clifford Algebra), Springer, Berlin Heideberg, be published

3. Labunets V.G, Labunets E.V., Egiazarian, K., Astola, J. (1988): Hypercomplex moments application in invariant image recognition. Proc. of IEEE Int. Conf. on Image Processing, Chicago, Illinois, 2, 257–261

4. Labunets E.V. (1996): Group–Theoretical Methods in Image Recognition. Report No LiTH–ISY–R–1855. Linköping University, 1–281

5. Assonov M.B., Labunets E.V., Labunets V.G., Lenz, R. (1996): Fast spectral algorithms for invariant pattern recognition and image matching based on modular invariants. ICIP'96, Switzerland, 284–288

6. Labunets E.V., Labunets V.G., Egiazarian, K., Astola, J. (1999): Fast spectral algorithms of invariants calculation. Proc. 10th Inter. Conf. on Image Analysis and Processing ICIAP'99, Venice, Italy, 203–208

7. Rundblad–Labunets E.V., Labunets V.G. (1999): Fast invariant recognition of multicolour images based on Triplet–Fourier–Gauss transform. Second Int. Workshop on Transforms and Filter Banks, Tampere, Finland, TICSP Series, 4, 405–438

8. Rundblad–Labunets E.V., Labunets V.G., Astola, J., Egiazarian, K., Polovnev S.V. (1999): Fast invariant recognition of colour images based on Triplet–Fourier–Gauss transform. Proc. of Int. Conf. Computer Science and Information Technologies, Yerevan, Armenia, 265–268

9. Hestenes D., Sobczyk G. (1984): Clifford Algebra to Geometric Calculus. D. Reidel Publishing

10. Deutsch D. (1989) Quantum computational networks. Proc. Poy. Soc. Lond. A 425, 73

The $C\ell_n$-Valued Robin Boundary Value Problem on Lipschitz Domains in $\mathbb{R}^n$

Loredana Lanzani * (lanzani@comp.uark.edu)
Department of Mathematics, University of Arkansas

Abstract. We show that the L^2-Robin boundary value problem for Laplace's equation on a Lipschitz domain in $\mathbb{R}^n$, with $C\ell_n$-valued boundary data (see (R) below), is explicitly solvable via layer potential operators. Corresponding results for the L^p-Robin problem are stated without proof.

Keywords: Robin condition, single layer potential, Lipschitz domains, Clifford algebras

1991 Mathematics Subject Classification: 31B10, 31B20, 35C15

Dedicated to Richard Delanghe on the occasion of his 60th birthday.

1. Introduction

In this note we present the solution in $W^{1,2}$ of the Robin boundary value problem for the Laplacian on a Lipschitz domain $\Omega \subset \mathbb{R}^n$ with $C\ell_n$-valued datum $f \in L^2(b\Omega)$ (see (R) below). This work originates from [6], where we considered the case of scalar-valued datum $f \in L^p(b\Omega)$, $1 < p \leq 2$. In the present context of the Clifford algebra $C\ell_n$, the direct relationship between the Clifford derivatives of the single layer potential and left Clifford-Cauchy integral operators allows for a more unified and direct approach to the solution of the problem. Because we are choosing the Robin coefficient b in the space $L^s(b\Omega)$ with s greater than the critical exponent $n - 1$, the solution operator for the Robin problem turns out to be a compact perturbation of the solution operator of the Neumann problem. In this respect, the situation we present here bears a close affinity with the classical study of the Neumann problem for C^1-domains (see [3]). The treatment of the critical exponent case (namely, $b \in L^{n-1}(b\Omega)$) requires a different approach, which has been developed in [6]. The structure of this paper is as follows. In sections 2 and 3 we describe and summarize the features of the Clifford algebras, the function spaces and the singular integral operators that are involved in this work. In Section 4 we present a

---

* Research supported by CNR grant no. 203.01.71.

F. Brackx et al. (eds.), Clifford Analysis and Its Applications, 183–191.
© 2001 *Kluwer Academic Publishers. Printed in the Netherlands.*

simple proof of the L^2-solution of the Robin problem with non-critical Robin coefficient, and we state without proof the corresponding result in L^p, with critical Robin coefficient.

2. Clifford Algebras: Notation and Basic Properties

The *real Clifford Algebra associated with the Euclidean space* $\mathbb{R}^n$, denoted $C\ell_n$, is defined as the minimal enlargement of $\mathbb{R}^n$ to a unitary algebra not generated by any proper subspace of $\mathbb{R}^n$, with the property that

$$x^2 = -|x|^2 = -\sum_{j=0}^{n-1} x_j^2 \tag{1}$$

for any $x \in \mathbb{R}^n$. This implies that

$$e_j e_k + e_k e_j = -2\delta_{jk}, \quad j,k \geq 1 \tag{2}$$

where $\{e_j\}_{j=0}^{n-1}$ denote the generating elements of $C\ell_n$, which are usually identified with the standard orthonormal basis of $\mathbb{R}^n$. In particular, e_0 is identified with the unit of the algebra. $C\ell_n$ is thus a 2^n-dimensional vector space over $\mathbb{R}$ and any element $a \in C\ell_n$ can be uniquely represented as

$$a = \sum_{l=0}^{n-1} \sum_{|I|=l} a_I\, e_I, \quad a_I \in \mathbb{R}, \quad \text{where} \tag{3}$$

$$e_I = e_{i_1} e_{i_2} \ldots e_{i_l}, \; 0 \leq i_1 < i_2 < \ldots i_l \leq n-1, \; I = (i_1, i_2, \ldots i_l) \tag{4}$$

In particular, we single out the *Scalar part of* a, denoted $\mathrm{Sc}(a)$, defined as

$$\mathrm{Sc}(a) = a_0\, e_0 \tag{5}$$

which we consider to be equivalent to a_0. *Clifford conjugation in* $C\ell_n$ is defined as the unique (real-)linear involution on $C\ell_n$ with

$$\bar{e}_I\, e_I = e_I\, \bar{e}_I = 1 \quad \text{for all } I \tag{6}$$

Thus

$$\bar{a} = \sum_{l=0}^{n-1} \sum_{|I|=l} a_I\, \bar{e}_I, \quad \text{and} \quad \bar{e}_I = (-1)^{\frac{l(l+1)}{2}} e_I, \quad |I| = l. \tag{7}$$

In particular, we have

$$\mathrm{Sc}(a\,\bar{a}) = \mathrm{Sc}(\bar{a}\,a) = |a|^2 = \sum_{l=0}^{n-1} \sum_{|I|=l} (a_I)^2 \tag{8}$$

It is customary to view $\mathbb{R}^n$ as embedded into $C\ell_n$ via the obvious identification

$$x = \sum_{j=0}^{n-1} x_j\, e_j \tag{9}$$

A $C\ell_n$-valued function f on an open set $\Omega \subset \mathbb{R}^n$ is defined via:

$$f(x) = \sum_{l=0}^{n-1} \sum_{|I|=l} f_I(x)\, e_I, \quad \text{with } f_I : \Omega \to \mathbb{R} \tag{10}$$

Any continuity, differentiability or integrability property which is ascribed to f has to be possessed by all components f_I. In particular, the Lebesgue and Sobolev spaces of $C\ell_n$-valued functions $L^2(b\Omega, C\ell_n)$, $W^{1,2}(b\Omega, C\ell_n)$ are defined by requiring that each component f_I belong to $L^2(b\Omega)$ (resp. $W^{1,2}(b\Omega)$), with norm in L^2 defined via: $\|f\|_2^2 :=$ Sc$(\int_{b\Omega} f\, \overline{f}\, d\sigma)$ (see (8)). The $W^{1,2}$-norm is defined similarly.

The *Left* and *Right Dirac derivatives* of a (differentiable) $C\ell_n$-valued function f are defined respectively as

$$Df := \sum_{j=0}^{n-1} \left(\sum_{l=0}^{n-1} \sum_{|I|=l} \left(\frac{\partial}{\partial x_j} f_I \right) e_j\, e_I \right); \tag{11}$$

$$fD := \sum_{j=0}^{n-1} \left(\sum_{l=0}^{n-1} \sum_{|I|=l} \left(\frac{\partial}{\partial x_j} f_I \right) e_I\, e_j \right) \tag{12}$$

Similarly, we define

$$\overline{D}f := \sum_{j=0}^{n-1} \left(\sum_{l=0}^{n-1} \sum_{|I|=l} \left(\frac{\partial}{\partial x_j} f_I \right) \overline{e}_j\, e_I \right); \tag{13}$$

$$f\overline{D} := \sum_{j=0}^{n-1} \left(\sum_{l=0}^{n-1} \sum_{|I|=l} \left(\frac{\partial}{\partial x_j} f_I \right) e_I\, \overline{e}_j \right) \tag{14}$$

It is immediate to check that

$$\overline{(Df)} = \overline{f}\, \overline{D} \tag{15}$$

Moreover, we have

$$D\overline{D}f = \overline{D}Df = \Delta f := \sum_{l=0}^{n-1} \sum_{|I|=l} (\Delta f_I)\, e_I. \tag{16}$$

We conclude by recalling *Stokes' Formula:*

$$\int_{b\Omega} u(x)\, n(x)\, v(x)\, d\sigma(x) =$$

$$\iint_{\Omega} (uD)(x)\, v(x)\, dx + \iint_{\Omega} u(x)\, (Dv)(x)\, dx, \tag{17}$$

provided that the functions above are integrable.

3. Singular Integrals: Notations and Main Properties

The operators we will be mainly concerned with are the (non-tangential) boundary values of the *Left Cauchy Integral C*, the *Single Layer Potential S* and its left Dirac derivative DS; here we recall their definitions and their basic properties. The fundamental result in the context of Lipschitz domains is the following theorem, due to A. Calderon (for domains with small Lipschitz constant) and R. Coifman, A. McIntosh and Y. Meyer (for arbitrary Lipschitz domains) (see [1], [2], [4], [5], [7]):

THEOREM 3.1. *Let $\Omega \subset \mathbb{R}^n$ denote a bounded, connected domain with Lipschitz boundary.*
Then, for any $f \in L^2(b\Omega, C\ell_n)$ the Left Clifford-Cauchy Integral of f:

$$Cf(y) := \frac{1}{\omega_n} \int_{b\Omega} \frac{q-y}{|q-y|^n}\, \bar{n}(q)\, f(q)\, d\sigma(q), \qquad y \in \Omega \tag{18}$$

has the following properties:

(i)   *The Non-Tangential Maximal Function of Cf, $(Cf)^*$ (see [4], [5], [7]) is square-integrable on $b\Omega$, and*

$$\| (Cf)^* \|_2 \leq C \|f\|_2 \tag{19}$$

(ii)   *Cf has Non-Tangential Limit $(Cf)^+(p)$ (see[4], [5], [7]) at almost every $p \in b\Omega$ and*

$$(Cf)^+(p) = \frac{1}{2}\left(-f(p) + \frac{2}{\omega_n}\, \text{p.v.} \int_{b\Omega} \frac{q-p}{|q-p|^n}\, \bar{n}(q)\, f(q)\, d\sigma(q) \right) \tag{20}$$

(Here, $\bar{n}$ denotes the Clifford-conjugate of the outer normal unit vector).

For $f \in L^2(b\Omega, C\ell_n)$ the *Single·Layer Potential of f*, denoted Sf, is defined as

$$Sf(y) = \frac{-1}{\omega_n(n-1)} \int_{b\Omega} \frac{1}{|y-q|^{n-2}} f(q) \, d\sigma(q), \qquad y \in \Omega \qquad (21)$$

Since

$$D_y\left(\frac{-1}{(n-1)|y-q|^{n-2}}\right) = \frac{q-y}{|y-q|^n} \qquad (22)$$

it follows that DS and C are related via

$$D(Sf)(y) = C(nf)(y), \quad y \in \Omega, \quad f \in L^2(b\Omega, C\ell_n) \qquad (23)$$

where n denotes the outer unit normal vector to $b\Omega$. The following corollary is thus a direct consequence of Theorem 3.1 (see [4], [7]):

COROLLARY 3.2. *With the same notations and hypotheses as Theorem (3.1), for any $f \in L^2(b\Omega, C\ell_n)$ we have:*

$$\|(DSf)^*\|_2 \le C\|f\|_2 \quad and \qquad (24)$$

$$\overline{n}(p)\,(D(Sf))^+(p) = \frac{1}{2}(-I+\mathcal{K}^*)f(p), \quad a.e.\ p \in b\Omega \qquad (25)$$

where $\mathcal{K}^*$ denotes the $L^2(b\Omega, C\ell_n)$-adjoint of the Left Clifford-Hilbert transform $\mathcal{K}$ (see, e.g., [4] or [7]):

$$\mathcal{K}f(p) = \frac{2}{\omega_n} p.v. \int_{b\Omega} \frac{p-q}{|p-q|^n} n(q)\, f(q)\, d\sigma(q) \qquad (26)$$

The following result, essentially due to G. Verchota, will be of great importance to us (see [4], [7], [8]):

THEOREM 3.3. *With the same notations and hypotheses as above, we have:*

(i) $S: L^2(b\Omega, C\ell_n) \to W^{1,2}(b\Omega, C\ell_n)$ *is invertible;*

(ii) $-I + \mathcal{K}^*$ *is invertible on $L_0^2(b\Omega, C\ell_n)$ and, moreover*

$$\mathrm{Ind}_{L^2}(-I+\mathcal{K}^*) = 0 \qquad (27)$$

(Here, $\mathrm{Ind}_{L^2}(T) := \dim \mathrm{Ker}(T) - \dim(L^2 \setminus \mathrm{Range}(T))$, see [9]).

In addition, we have

$$\|(I \pm \mathcal{K}^*)g\|_2 \le$$

$$\le C(\|(I \mp \mathcal{K}^*)g\|_2 + \|Sg\|_2) \qquad (28)$$

4.　The Robin Problem in L^p

We are finally ready to state and prove the main result of this note:

THEOREM 4.1.　*Let $\Omega \subset \mathbb{R}^n$ denote a bounded, connected domain with Lipschitz boundary. Let b denote a given, scalar-valued function, $b \in L^s(b\Omega, \mathbb{R})$, $s > n - 1$, $b \geq 0$ (b positive on some subset of $b\Omega$ with positive measure).*
With the same notations as in Section 3, for $f \in L^2(b\Omega, C\ell_n)$ define

$$Tf(p) := \left(\frac{1}{2}(-I + \mathcal{K}^*)f \right)(p) + b(p)\,\mathcal{S}f(p), \quad a.e. \ p \in b\Omega \qquad (29)$$

Then, we have that T is bounded and invertible in $L^2(b\Omega, C\ell_n)$.
Moreover, the Robin Problem for $C\ell_n$-valued harmonic functions:

$$(R) \quad \begin{cases} \Delta u = 0 & in \ \Omega \\ (Du)^* \in L^2(b\Omega) \\ \bar{n}Du + bu = f & on \ b\Omega, \end{cases}$$

is uniquely solvable in $L^2(b\Omega, C\ell_n)$, and the solution u is represented via

$$u(x) = \mathcal{S}\left(T^{-1}f \right)(x), \qquad x \in \Omega \qquad (30)$$

Proof. By (i) in Theorem 3.3 and the Rellich-Kondrachev compact embedding theorem (see [9]) it follows that the point-wise multiplication operator $b\,\mathcal{S}(f)(p) := b(p)\,\mathcal{S}f(p)$ is compact in $L^2(b\Omega, C\ell_n)$. This, and (ii) in Theorem 3.3 imply at once that T is bounded in $L^2(b\Omega, C\ell_n)$. The fact that (30) gives the solution of (R) (provided T is invertible in $L^2(b\Omega, C\ell_n)$) is an immediate consequence of (24) and (25).

We are thus left to show that T is invertible in $L^2(b\Omega, C\ell_n)$. We begin by showing that T is one-to-one. Indeed, if we let

$$Tf = 0 \quad \text{for some} \quad f \in L^2(b\Omega, C\ell_n) \qquad (31)$$

and apply Stokes' formula (17) to $v := \mathcal{S}f$ and $u := \overline{Dv} = \bar{v}\overline{D}$ (see (15)) we obtain (as $uD = \bar{v}\,\overline{D}\,D = \Delta\bar{v} = \Delta v = 0$)

$$\iint_\Omega (\overline{Dv})Dv = \int_{b\Omega} \overline{Dv}\,nv = \int_{b\Omega} \overline{(\bar{n}\,Dv)}\,v = -\int_{b\Omega} b\,\bar{v}\,v \qquad (32)$$

In particular, by considering the scalar components of (32) we obtain (see (8))

$$\iint_\Omega |Dv|^2 = -\int_{b\Omega} b\,|v|^2 \qquad (33)$$

It follows $Dv = 0$ in Ω, i.e. $v = const$ and, since $b \geq 0$ (and b is positive on a subset of $b\Omega$ with positive measure) it must be $v = 0$ in Ω. Since we have set $v = Sf$, by (i) in Theorem 2.3 we conclude $f = 0$.

Next, we show that $\mathcal{T}$ has dense range in $L^2(b\Omega, C\ell_n)$. Indeed, the compactness of bS and (27) imply (see [9])

$$\text{Ind}_{L^2}(\mathcal{T}) = \text{Ind}_{L^2}(-I + \mathcal{K}^*) = 0 \tag{34}$$

But we just proved that $\mathcal{T}$ is one-to-one, thus

$$\text{Ind}_{L^2}(\mathcal{T}) = \dim\left(L^2(b\Omega, C\ell_n)/\text{Range}\,(\mathcal{T})\right) \tag{35}$$

and the dense range property is proved.

Finally, we show that $\mathcal{T}$ has closed range. To this end, have

$$(y_n)_{n\in\mathbb{N}} \subset L^2(b\Omega, C\ell_n), \quad y_n \to y \text{ in } L^2(b\Omega, C\ell_n), \quad y_n = \mathcal{T} x_n \tag{36}$$

We distinguish two cases: if $\|x_n\|_2 \leq C$ for each n, then by the Banach-Alaoglou theorem (see [9]) we have (modulo a subsequence)

$$x_n \rightharpoonup x, \quad \text{for some } x \in L^2(b\Omega, C\ell_n) \quad \text{(weak convergence)} \tag{37}$$

By the uniqueness of weak limits (and the boundedness of $\mathcal{T}$) we conclude $y = \mathcal{T}x$.

If, instead, $(x_n)_{n\in\mathbb{N}}$ contains an unbounded subsequence, we consider

$$z_n := \frac{x_n}{\|x_n\|_2} \tag{38}$$

In this case it is not difficult to show that (modulo a subsequence)

$$z_n \rightharpoonup 0 \text{ (weak)}, \quad \|\mathcal{T}z_n\|_2 \to 0, \quad \|bSz_n\|_2 \to 0 \text{ and } \|Sz_n\|_2 \to 0 \tag{39}$$

By combining (28) with the triangle inequality we obtain:

$$\|(I + \mathcal{K}^* + bS)z_n\|_2 \leq$$
$$\leq C\left(\|\mathcal{T}z_n\|_2 + \|Sz_n\|_2 + 2\|bSz_n\|_2\right) \to 0 \tag{40}$$

This leads to the following contradiction:

$$1 = \|z_n\|_2 \leq \|\mathcal{T}z_n\|_2 + \|(I + \mathcal{K}^* + bS)z_n\|_2 \to 0 \tag{41}$$

The proof of Theorem 4.1 is concluded. QED

Even though Corollary 3.2 and Theorem 3.3 extend to the case $f \in L^p(b\Omega)$, $1 < p \leq 2$, the solution of the Robin problem in the general case: $b \in L^{n-1}(b\Omega)$, $f \in L^p(b\Omega)$ requires a more sophisticated approach than Theorem 4.1 because the operator bS now fails to be compact in $L^p(b\Omega)$ (even though it still bounded, by the Sobolev embedding theorem). Moreover, Stokes' formula can no longer be applied to show uniqueness since, in this case, the functions involved may not be integrable. Nevertheless, the result is maintained. We have:

THEOREM 4.2. *Let* $\Omega \subset \mathbb{R}^n$ *denote a bounded, connected domain with Lipschitz boundary. Let* b *denote a given, scalar-valued function,* $b \in L^{n-1}(b\Omega)$, $b \geq 0$ (b *positive on some subset of* $b\Omega$ *with positive measure). With the same notations as in Section 3, for* $f \in L^p(b\Omega)$, $1 < p \leq 2$, *define*

$$\mathcal{T}f(p) := \left(\frac{1}{2}(-I + \mathcal{K}^*)f\right)(p) + b(p)\,\mathcal{S}f(p), \quad a.e.\ p \in b\Omega \quad (42)$$

Then, we have that $\mathcal{T}$ *is bounded and invertible in* $L^p(b\Omega)$.
Moreover, the Robin Problem for $C\ell_n$-*valued harmonic functions:*

$$(R) \quad \begin{cases} \Delta u = 0 & in\ \Omega \\ (Du)^* \in L^p(b\Omega) & \\ \bar{n}Du + bu = f & on\ b\Omega, \end{cases}$$

is uniquely solvable in $L^p(b\Omega)$, *and the solution* u *is represented via*

$$u(x) = \mathcal{S}\left(\mathcal{T}^{-1}f\right)(x), \quad x \in \Omega \quad (43)$$

The proof of Theorem 4.2 in the case of scalar-valued coefficients will appear in [6].

References

1. A.P. Calderòn, *Cauchy Integrals on Lipschitz Curves and Related Operators*, Proc. Nat. Acad. Sci. USA, **74** (1977), 1324-7.
2. R.R. Coifman, A. McIntosh, Y. Meyer, *L'Intègrale de Cauchy Definit un Opèrateur Borné sur L^2 Pour les Courbes Lipschitziennes*, Annals of Math., **116** (1982), 361-87.
3. E. Fabes, M. Jodeit, N. Riviére *Potential Techniques for Boundary Value Problems on C^1 Domains*, Acta Math.,**141** (1978), 165-186.
4. J.E. Gilbert, M.A.M. Murray, *Clifford Algebras and Dirac Operators in Harmonic Analysis*, Cambridge Studies in Advanced Mathematics **26** (1991).
5. C.E. Kenig, *Harmonic Analysis Techniques for Second Order Elliptic Boundary Value Problems*, AMS-CBMS **83** (1994).
6. L. Lanzani, Z. Shen, *On the Robin Problem for the Laplace Equation in Lipschitz Domains*, preprint.
7. M. Mitrea, *Clifford Wavelets, Singular Integrals, and Hardy Spaces*, Lecture Notes in Mathematics **1575** (1991).

8. G. C. Verchota, *Layer Potentials and Boundary value Problems for Laplace's Equation in Lipschitz Domains*, J. Funct. Anal., **59** (1984), 572-611.
9. A.E. Taylor, D.C. Lay *Introduction to Functional Analysis*, John Wiley & Sons (1980).

Quaternionic Analysis in $\mathbb{R}^3$ Versus Its Hyperbolic Modification

Heinz Leutwiler
Mathematisches Institut, Universität Erlangen-Nürnberg

Abstract. Two generalizations of the Cauchy-Riemann system in $\mathbb{R}^3$ are compared.

Keywords: generalized function theory, quaternions, hyperbolic metric, Dirac operator

1991 Mathematics Subject Classification: 30G35

Dedicated to Richard Delanghe on the occasion of his 60th birthday.

1. Introduction

At this stage of the theory one may ask: Among the two generalizations of classical complex analysis to $\mathbb{R}^3$, which one is worthwhile to be further investigated? Is it the usual quaternionic analysis or rather its hyperbolic modification, introduced in [8]? The goal of this paper is to convince the reader that there are good reasons for both theories.

The ambiguity in the analysis over quaternions already starts when one is looking for an appropriate embedding of $\mathbb{R}^3$ into $\mathbb{H}$, the algebra of quaternions (with generalized imaginary units i, j, k, satisfying $i^2 = j^2 = k^2 = -1$, $ij = -ji = k$, $jk = -kj = i$, $ki = -ik = j$). Identifying the points $(x, y, t) \in \mathbb{R}^3$ with the so-called *"reduced quaternions"*

$$z = x + iy + jt \tag{1.1}$$

has the advantage that the powers z^n ($n \in \mathbb{Z}$) remain reduced, i.e. are of the form

$$f = u + iv + jw, \tag{1.2}$$

with u, v, w real valued. We therefore prefer this identification to the more common one, where the point $(x, y, t) \in \mathbb{R}^3$ is identified with the "pure quaternion" $ix + jy + kt$ (see, e.g., [6]). In order to check that for any $n \in \mathbb{Z}$ the power z^n remains reduced, it is best to introduce polar coordinates r, φ, ϑ ($0 \leq r < \infty$, $0 \leq \varphi \leq \pi$, $0 \leq \vartheta \leq 2\pi$), i.e. set

$$z = r(\cos \varphi + i \sin \varphi \cos \vartheta + j \sin \varphi \sin \vartheta), \tag{1.3}$$

F. Brackx et al. (eds.), *Clifford Analysis and Its Applications*, 193–211.
© 2001 *Kluwer Academic Publishers. Printed in the Netherlands.*

and verify that

$$z^n = r^n(\cos n\varphi + i \sin n\varphi \cos \vartheta + j \sin n\varphi \sin \vartheta), \qquad (1.4)$$

an extension of De Moivre's identity.

From a "*mathematical*" point of view one expects that the generalized function theory in $\mathbb{R}^3$ contains the power functions z^n ($n \in \mathbb{Z}$). In particular the kernel

$$z^{-1} = \frac{\bar{z}}{|z|^2} \qquad (z \neq 0), \qquad (1.5)$$

where $\bar{z} = x - iy - jt$ and $|z|^2 = z\bar{z} = x^2 + y^2 + t^2$, should be a solution. From a "*physicists*" point of view, however, the kernel at 0 should probably represent the electrical field of a unit charge at that point, i.e. should be of the form

$$\frac{\bar{z}}{|z|^3} = \frac{z^{-1}}{|z|} \qquad (z \neq 0), \qquad (1.6)$$

provided one represents - as usual - the electrical field by $\bar{f}$ rather than by f. It now happens that these two different points of view lead to the generalizations of complex analysis mentioned above, namely the modified quaternionic analysis on one hand and the usual quaternionic one on the other. Surprisingly this splitting can also be interpreted geometrically, as we shall see now. Whereas the "mathematical" extension is related to "hyperbolic" (non-Euclidean) geometry, the "physical" one is connected with the Euclidean one. To see this, let us first recall the classical Cauchy-Riemann equations, satisfied by the holomorphic functions $f = u + iv$, defined on some open set $\omega \subset \mathbb{C}$:

$$\frac{\partial u}{\partial x} = \frac{\partial v}{\partial y}, \qquad \frac{\partial u}{\partial y} = -\frac{\partial v}{\partial x}. \qquad (1.7)$$

Passing from f to its conjugate $\bar{f} = u - iv$, and interpreting $\bar{f}$ as a vector field, it satisfies the system $\operatorname{div} \bar{f} = 0$ and $\operatorname{rot} \bar{f} = 0$ (some authors write "curl" instead of "rot"). This system has a natural extension to $\mathbb{R}^3$: On an arbitrary open set $\Omega \subset \mathbb{R}^3$ one considers the continuously differentiable vector fields $\vec{g}$, which satisfy the system

$$\operatorname{div} \vec{g} = 0 \qquad \text{and} \qquad \operatorname{rot} \vec{g} = 0, \qquad (1.8)$$

sometimes called the *M.Riesz system*. It occurs in physics when one looks at the velocity field of a steady, incompressible fluid flow, or when one is interested in the behavior of the electrical field $\vec{E}$ outside of its charged body. Furthermore it is satisfied by the magnetic field $\overrightarrow{H}$ generated by a current in a closed wire. Written in terms of the conjugate $\overline{f}$ of the function $f = u+iv+jw$, i.e. setting $\overrightarrow{g} = (u, -v, -w)$, the system (1.8) reads as follows:

$$(R) \quad \begin{cases} \dfrac{\partial u}{\partial x} - \dfrac{\partial v}{\partial y} - \dfrac{\partial w}{\partial t} = 0 \\[2mm] \dfrac{\partial u}{\partial y} = -\dfrac{\partial v}{\partial x}, \quad \dfrac{\partial u}{\partial t} = -\dfrac{\partial w}{\partial x}, \quad \dfrac{\partial v}{\partial t} = \dfrac{\partial w}{\partial y}. \end{cases} \qquad (1.9)$$

It can also be written, more compact, either in the form

$$D_\ell f = \frac{\partial f}{\partial x} + i\frac{\partial f}{\partial y} + j\frac{\partial f}{\partial t} = 0, \quad \text{or} \quad D_r f = \frac{\partial f}{\partial x} + \frac{\partial f}{\partial y}i + \frac{\partial f}{\partial t}j = 0, \quad (1.10)$$

demonstrating its connection with the Dirac-Fueter operator (see, e.g., [5], [13], [6], [2]).

Another way to write down the system (R) is by looking at the differential $1-$form $\sigma = udx - vdy - wdt$. Obviously the last three equations of (R) are fulfilled if and only if the outer derivative $d\sigma$ vanishes. In order to write down the first equation one has to introduce - besides $d\sigma$ - its adjoint $1-$form $\quad \delta\sigma := \star d(\star\sigma)$, where $\star$ denotes the famous Hodge star (see, e.g., [14]), which maps $k-$forms to $(3-k)-$forms. The system (R) can then be written in the invariant form

$$\delta\sigma = 0 \quad \text{and} \quad d\sigma = 0, \qquad (1.11)$$

showing that the $1-$form σ is harmonic (in the sense of Hodge).

The first equation in (1.11), in contrast to the second one, depends on the metric of the underlying manifold, which in the above case is the Euclidean one. Replacing this metric by the *hyperbolic* one, more precisely considering the upper half space $\mathbb{R}^3_+$, endowed with the Riemannian metric

$$ds^2 = \frac{dx^2 + dy^2 + dt^2}{t^2}, \qquad (1.12)$$

the equations (1.11) yield the following modification of the system (R):

$$(H) \quad \begin{cases} t\left(\dfrac{\partial u}{\partial x} - \dfrac{\partial v}{\partial y} - \dfrac{\partial w}{\partial t}\right) + w = 0 \\[2mm] \dfrac{\partial u}{\partial y} = -\dfrac{\partial v}{\partial x}, \quad \dfrac{\partial u}{\partial t} = -\dfrac{\partial w}{\partial x}, \quad \dfrac{\partial v}{\partial t} = \dfrac{\partial w}{\partial y}. \end{cases} \qquad (1.13)$$

Solutions $f = u + iv + jw$ of this system, defined on some open set $\Omega \subset \mathbb{R}^3$, are called (H)-*solutions* in [8] (in honor of W.V.D. Hodge).

Written in terms of $\vec{g} = \overline{f}$, the modified system (1.13) reads as follows:

$$\operatorname{div}(\rho\,\vec{g}) = 0 \qquad \text{and} \qquad \operatorname{rot}\vec{g} = 0, \qquad (1.14)$$

where $\rho(z) = \frac{1}{t}$ plays the role of a density.

Drawing on the polar representation (1.4), a tedious calculation shows that the power-function

$$z \to z^n \quad (n \in \mathbb{Z}) \qquad (1.15)$$

defines an (H)-solution (see [8] for a shorter proof). In particular the kernel (1.5) satisfies the system (H), whereas (1.6) fulfills (R). The systems (H) and (R) can thus be viewed as the "*mathematical*", resp. the "*physical*", extension of the Cauchy-Riemann system.

2.　A comparison of the systems (R) and (H)

We start with the observation that holomorphic functions naturally lead to (H)-solutions. Indeed, given any holomorphic function $g = u+iv$ on an open set $\omega \subset \mathbb{C}^+$, the upper half plane, the function

$$f(z) = u\left(x, \sqrt{y^2 + t^2}\right) + \frac{iy + jt}{\sqrt{y^2 + t^2}}\, v\left(x, \sqrt{y^2 + t^2}\right), \qquad (2.1)$$

defined on the open set $\Omega = \left\{ (x, y, t) : \left(x, \sqrt{y^2 + t^2}\right) \in \omega \right\}$, yields an (H)-solution (see, e.g., [8]). Geometrically this construction, already considered by R. Fueter [5], means that the holomorphic function g is "rotated" around the real axis. Starting with the power-function $g(x + iy) = (x + iy)^n$, the construction (2.1) leads to the function z^n, with $z = x + iy + jt$, showing again that z^n is an (H)-solution. Consequently elementary functions, defined by power series with real coefficients, like $e^z = \sum_{n=0}^{\infty} \frac{z^n}{n!}, \log z, \sin z, \cos z, \sqrt{z}, \dots$ yield (H)-solutions.

An analogous result, based on the above construction, also exists for the system (R), provided we modify the Cauchy-Riemann system (1.7). Choosing the function $g = u + iv$ to be a solution of the so-called Stokes-Beltrami system·

$$y\left(\frac{\partial u}{\partial x} - \frac{\partial v}{\partial y}\right) - v = 0, \qquad \frac{\partial u}{\partial y} = -\frac{\partial v}{\partial x}, \qquad (2.2)$$

the function f, associated with $g = u + iv$ by (2.1), defines a solution of the system (R).

Next we compare the local representations for the solutions of the systems (R) and (H). Since in the two systems the last three equations are identical and in fact represent the "integrability conditions", there is, at least locally, a real function h such that

$$f = \overline{\text{grad}}\, h = \frac{\partial h}{\partial x} - i\frac{\partial h}{\partial y} - j\frac{\partial h}{\partial t}. \tag{2.3}$$

Inserting this representation into the first equation of (R), we see that h is *harmonic*, i.e. satisfies

$$\Delta h = 0, \quad \Delta = \frac{\partial^2}{\partial x^2} + \frac{\partial^2}{\partial y^2} + \frac{\partial^2}{\partial t^2}. \tag{2.4}$$

Inserting it into the first equation of (H), we get the equation

$$t\Delta h - \frac{\partial h}{\partial t} = 0, \tag{2.5}$$

showing that h is *hyperbolic harmonic*. Note that both equations, (2.4) and (2.5), are Laplace-Beltrami equations, (2.4) with respect to the Euclidean metric $ds^2 = dx^2 + dy^2 + dt^2$, (2.5) with respect to the hyperbolic metric (1.12). From the representation (2.3), with h satisfying (2.4), resp. (2.5), one sees at once that in case of (R) the components u, v, w of f are harmonic, whereas in case of (H), only the first two components u and v are hyperbolic harmonic. The third component w satisfies the equation

$$t^2 \Delta w - t\frac{\partial w}{\partial t} + w = 0, \tag{2.6}$$

i.e. is an eigenfunction with eigenvalue -1 of the Laplace-Beltrami operator

$$\Delta_{hyp} = t^2 \Delta - t\frac{\partial}{\partial t}, \tag{2.7}$$

associated with the metric (1.12).

Recall that the harmonic functions (2.4) are characterized by their *mean-value property*: If h is harmonic on a neighborhood of the closure $\overline{B}$ of a ball B, the value of h at the center of B is equal to the average of h over the boundary ∂B. A similar result also holds for the hyperbolic harmonic functions (2.5), as we are going to see now. Here , of course, the Euclidean ball B has to be replaced by the non-euclidean ball $B(\zeta, R_{hyp}) \subset \mathbb{R}^3_+$ of n.e. center $\zeta = \xi + i\eta + j\tau$ and n.e. radius R_{hyp}, computed with respect to the hyperbolic metric (1.12). As a set, it agrees with the Euclidean ball with Euclidean center $\tilde{\zeta} = \xi + i\eta +$

$j\tau \cosh(R_{hyp})$ and Euclidean radius $R_e = \tau \sinh(R_{hyp})$. Its boundary, the sphere $\partial B(\zeta, R_{hyp})$ may be described by the equation

$$\left\{ z \in \mathbb{R}^3_+ : \frac{|z - \zeta|}{|z - \hat{\zeta}|} = c \right\}, \tag{2.8}$$

where $\hat{\zeta} = \xi + i\eta - j\tau$ and $c = \tanh\left(\frac{R_{hyp}}{2}\right)$. Since the Green function $G(z, \zeta)$ for the equation (2.5), with respect to $\mathbb{R}^3_+$ and $\zeta \in \mathbb{R}^3_+$, is given by (see, e.g., [1])

$$G(z, \zeta) = \frac{|z - \zeta|}{|z - \hat{\zeta}|} + \frac{|z - \hat{\zeta}|}{|z - \zeta|} - 2, \tag{2.9}$$

it is clear that $G(z, \zeta)$ is constant on each n.e. sphere ∂B of n.e. center ζ. This result will now be combined with the following "hyperbolic version" of Green's 2nd identity, to be found, e.g., in [1]: For arbitrary, twice continuously differentiable functions u and v, defined on some open set $\Omega \subset \mathbb{R}^3_+$ and an arbitrary, smoothly bounded compact set $K \subset \Omega$ with outer unit normal field $\nu = \nu_0 + i\nu_1 + j\nu_2$ we have

$$\int_K (u\Delta_{hyp}v - v\Delta_{hyp}u)dV_h = \int_{\partial K} \left(u\frac{\partial v}{\partial \nu_h} - v\frac{\partial u}{\partial \nu_h}\right)d\sigma_h, \tag{2.10}$$

where $\frac{\partial}{\partial \nu_h} = t\frac{\partial}{\partial \nu}$, $d\sigma_h = \frac{d\sigma}{t^2}$ and $dV_h = \frac{dV}{t^3}$. Choosing $u = h$ to be hyperharmonic and $v \equiv 1$, we infer that

$$\int_{\partial K} \frac{\partial h}{\partial \nu_h} d\sigma_h = 0. \tag{2.11}$$

A second application of (2.10), this time to the hyperbolic function h, Green's function $G(\cdot, \zeta)$ and the set difference $K = \overline{B}(\zeta, R_{hyp}) \setminus B(\zeta, r_{hyp})$, yields

$$\int_{S_R} h\frac{\partial G}{\partial \nu_h} d\sigma_h = \int_{S_r} h\frac{\partial G}{\partial \nu_h} d\sigma_h, \tag{2.12}$$

where $S_R = \partial B(\zeta, R_{hyp})$ and $S_r = \partial B(\zeta, r_{hyp})$. Here we have used the fact that $G(\cdot, \zeta)$ is constant on each n.e. sphere with n.e. center ζ and combined it with (2.11). Now, by (2.8) and (2.9), $G = 2(\coth R_{hyp} - 1)$, $\frac{\partial G}{\partial \nu_h} = \frac{\partial G}{\partial R} = 2\frac{\partial}{\partial R}(\coth R - 1) = -\frac{2}{\sinh^2 R}$ and so, by (2.12),

$$\frac{1}{\sinh^2 R}\int_{S_R} h d\sigma_h = \frac{1}{\sinh^2 r}\int_{S_r} h d\sigma_h. \tag{2.13}$$

Since for $r \to 0$ the integral on the right hand side tends to $4\pi h(\zeta)$, we have proved

Theorem 2.1 *Let h be hyperbolic harmonic on a neighborhood of the closure $\overline{B}$ of a non-euclidean ball $B = B(\zeta, R_{hyp}) \subset \mathbb{R}^3_+$ with n.e. center ζ and n.e. radius R_{hyp}. Then*

$$h(\zeta) = \frac{1}{4\pi \sinh^2 R_{hyp}} \int_{S_R} h \, d\sigma_h, \tag{2.14}$$

where S_R denotes the non-euclidean sphere, with n.e. center ζ and n.e. radius $R = R_{hyp}$, and $d\sigma_h = \frac{d\sigma}{t^2}$ the n.e. area element.

A similar result holds for the solutions w of the equation (2.6). Here, however the constant function is no longer a solution and has to be replaced by the following one

$$w_0(z, \zeta) = \left(\frac{|z - \hat{\zeta}|}{|z - \zeta|} - \frac{|z - \zeta|}{|z - \hat{\zeta}|} \right) \tanh^{-1} \left(\frac{|z - \zeta|}{|z - \hat{\zeta}|} \right). \tag{2.15}$$

On the other hand, the Green function for (2.6), with respect to $\mathbb{R}^3_+$ and $\zeta = \xi + i\eta + j\tau \in \mathbb{R}^3_+$, is defined by

$$\tilde{G}(z, \zeta) = \frac{|z - \hat{\zeta}|}{|z - \zeta|} - \frac{|z - \zeta|}{|z - \hat{\zeta}|} \qquad (\hat{\zeta} = \xi + i\eta - j\tau). \tag{2.16}$$

Applying the solutions w and w_0 of (2.6) to (2.10) and observing that, by (2.8), on the n.e. sphere with n.e. center ζ,

$$w_0 = \frac{R_{hyp}}{\sinh R_{hyp}}, \quad \frac{\partial w_0}{\partial \nu_h} = \frac{\partial w_0}{\partial R} = \frac{\partial}{\partial R} \left(\frac{R}{\sinh R} \right) = \frac{\sinh R - R \cosh R}{\sinh^2 R}, \tag{2.17}$$

one obtains the equality

$$\frac{\partial w_0}{\partial \nu_h} \int_{S_R} w \, d\sigma_h = w_0 \int_{S_R} \frac{\partial w}{\partial \nu_h} d\sigma_h. \tag{2.18}$$

Then, applying (2.10) to w, a solution of (2.6), $\tilde{G}(\cdot, \zeta)$ and the above set difference $K = \overline{B}(\zeta, R_{hyp}) \setminus B(\zeta, r_{hyp})$, we infer that

$$\int_{S_R} (w \frac{\partial \tilde{G}}{\partial \nu_h} - \tilde{G} \frac{\partial w}{\partial \nu_h}) d\sigma_h = \int_{S_r} (w \frac{\partial \tilde{G}}{\partial \nu_h} - \tilde{G} \frac{\partial w}{\partial \nu_h}) d\sigma_h. \tag{2.19}$$

Since, by (2.8),

$$\tilde{G} = \frac{2}{\sinh R_{hyp}}, \quad \frac{\partial \tilde{G}}{\partial \nu_h} = \frac{\partial \tilde{G}}{\partial R} = \frac{\partial}{\partial R} \left(\frac{2}{\sinh R} \right) = -\frac{2 \cosh R}{\sinh^2 R},$$

(2.17) - (2.19) yield

$$\frac{1}{R \sinh R} \int_{S_R} w d\sigma_h = \frac{1}{r \sinh r} \int_{S_r} w d\sigma_h. \qquad (2.20)$$

Letting $r \to 0$, the right hand side tends to $4\pi w(\zeta)$, and hence we have proved

Theorem 2.2 *Let w be a solution of the equation (2.6), considered on a neighborhood of the closure $\overline{B}$ of some non-euclidean ball $B = B(\zeta, R_{hyp}) \subset \mathbb{R}^3_+$ with n.e. center ζ and n.e. radius R_{hyp}. Then*

$$w(\zeta) = \frac{1}{4\pi R_{hyp} \sinh R_{hyp}} \int_{S_R} w d\sigma_h, \qquad (2.21)$$

where S_R denotes the non-euclidean sphere, with n.e. center ζ and n.e. radius $R = R_{hyp}$, and $d\sigma_h = \frac{d\sigma}{t^2}$ the n.e. area element.

3. Polynomial solutions of the system (H)

As pointed out in sections 1 and 2, the power-function $z \to z^n$ is an (H)-solution. Combined with the observation that for any (H)-solution f and any $a \in \mathbb{C}$ the function $z \to af(aza)a$ $(z \in \mathbb{R}^3)$ is again an (H)-solution, we infer that $z \to a(aza)^n a = (a^2 z)^n a^2$ is also an (H)-solution. Choosing now $a \in \mathbb{C}$ in such a way that for a given $\lambda \in \mathbb{R}$, $a^2 = 1 + i\lambda$, we get the result that for any $\lambda \in \mathbb{R}$ the function $F_n(z, \lambda) = [(1+i\lambda)z]^n (1+i\lambda)$ is an (H)-solution. Expanding in terms of λ (keeping fixed $z \in \mathbb{R}^3$), we find that

$$F_n(z, \lambda) = [(1 + i\lambda)z]^n (1 + i\lambda) = \sum_{k=0}^{n+1} E_n^k(z) \lambda^k, \qquad (3.1)$$

with some polynomials $E_n^k(z)$ of degree n. The *generating function* $F_n(z, \lambda)$ uniquely determines the polynomials E_n^k; in fact

$$E_n^k(z) = \frac{1}{k!} \frac{d^k}{d\lambda^k} F_n(z, \lambda) |_{\lambda=0}. \qquad (3.2)$$

On account of this formula it is clear that the E_n^k's $(k = 0, 1, \ldots, n+1)$ are (H)-solutions. In order to deduce the recursion formula for these polynomials, observe first that $F_n(z, \lambda) = (1+i\lambda)z F_{n-1}(z, \lambda)$ and then plug in (3.1) on both sides. Comparing coefficients one then finds the recursion formula:

$$E_n^k(z) = z E_{n-1}^k(z) + iz E_{n-1}^{k-1}(z), \qquad (3.3)$$

where by definition $E_n^k = 0$, if $k \notin \{0, 1, \ldots, n+1\}$. With this formula at our disposal it is easy to verify (see [8]) that E_n^k has the following explicit expression:

$$E_n^k(z) = \sum_{\substack{\mu_0 + \mu_1 + \cdots + \mu_n = k \\ (\mu_\nu \in \{0,1\})}} i^{\mu_0} z i^{\mu_1} z i^{\mu_2} z \ldots i^{\mu_{n-1}} z i^{\mu_n} \qquad (z \in \mathbb{R}^3). \quad (3.4)$$

In [8] it has also been shown that the E_n^k's, for $k = 0, 1, \ldots, n+1$, form a *basis* for the vector space V_n (over $\mathbb{R}$) of homogeneous polynomial (H)-solutions of degree n.

Another way to construct (H)-polynomials is by differentiation of the power-function (1.15), making use of the fact that for any (H)-solution f the partial derivatives $\frac{\partial f}{\partial x}$ and $\frac{\partial f}{\partial y}$ are also (H)-solutions (but not $\frac{\partial f}{\partial t}$). It is therefore quite natural to also introduce, beside the E_n^k's, the polynomial (H)-solutions (see [9] and [4])

$$L_n^k : z \to \frac{1}{k!} \frac{\partial^k z^{n+k}}{\partial y^k} \qquad (k, n \in \mathbb{N}_0). \quad (3.5)$$

It will be convenient to put $L_n^k = 0$ for $k < 0$. Clearly the L_n^k's are homogeneous (H)-polynomials of degree n. In addition they obey the recursion formula

$$L_n^k(z) = z L_{n-1}^k(z) + i L_n^{k-1}(z), \quad (3.6)$$

as can be seen by induction on k (assuming the formula to hold true for all n). From (3.6) it immediately follows that the L_n^k's have the following explicit form (see [4]):

$$L_n^k(z) = \sum_{\substack{\mu_0 + \mu_1 + \cdots + \mu_n = k \\ \mu_\nu \in \{0,1,\ldots,k\}}} i^{\mu_0} z i^{\mu_1} z i^{\mu_2} z \ldots i^{\mu_{n-1}} z i^{\mu_n} \qquad (z \in \mathbb{R}^3). \quad (3.7)$$

As a simple application of (3.7) we note that for all $n, k \in \mathbb{N}_0$ and $z \in \mathbb{R}^3$:

$$\left| L_n^k(z) \right| \le \binom{n+k}{k} |z|^n. \quad (3.8)$$

Surprisingly the polynomials L_n^k can be written down explicitly in terms of the components x, y, t of z. Indeed we have

$$L_n^k(z) = \sum_{p+q+r=n} \binom{n+k}{p} \binom{k+q}{q} x^p y^q t^r L_r^{k+q}(j), \quad (3.9)$$

for all $n, k \in \mathbb{N}_0$ and all $z = x + iy + jt \in \mathbb{R}^3$. Thus the coefficients $L_r^{k+q}(j)$ are explicitly known (see [4]).

Our next goal is to determine the generating function for the polynomials L_n^k, i.e. the function

$$\widetilde{F}_n(z, \lambda) = \sum_{k=0}^{\infty} L_n^k(z) \lambda^k. \qquad (3.10)$$

The series converges for $\lambda \in \mathbb{R}$ with $|\lambda| < 1$, since on account of (3.8)

$$\left| \sum_{k=0}^{\infty} L_n^k(z) \lambda^k \right| \leq \sum_{k=0}^{\infty} \left| L_n^k(z) \right| \left| \lambda^k \right| \leq |z|^n \sum_{k=0}^{\infty} \binom{n+k}{k} \left| \lambda^k \right|;$$

the convergence thus follows from the well-known expansion

$$\frac{1}{(1-x)^{n+1}} = \sum_{k=0}^{\infty} \binom{n+k}{k} x^k \quad (|x| < 1), \qquad (3.11)$$

obtained by differentiating the geometric series. Inserting the recursion relation (3.6) we now conclude that $\widetilde{F}_n = z\widetilde{F}_{n-1} + i\lambda\widetilde{F}_n$, and hence that

$$\widetilde{F}_n(z, \lambda) = (1 - i\lambda)^{-1} z\widetilde{F}_{n-1}(z, \lambda) = \frac{1 + i\lambda}{1 + \lambda^2} z\widetilde{F}_{n-1}(z, \lambda). \qquad (3.12)$$

Based on this recursion formula for $\widetilde{F}_n$ it is now easy to check (using induction on n) that

$$\widetilde{F}_n(z, \lambda) = \frac{1}{(1+\lambda^2)^{n+1}} [(1 + i\lambda)z]^n (1 + i\lambda) = \frac{1}{(1+\lambda^2)^{n+1}} F_n(z, \lambda), \qquad (3.13)$$

with F_n as in (3.1). Applying now the binomial theorem to the left hand side of the equality $(1 + \lambda^2)^{n+1} \widetilde{F}_n(z, \lambda) = F_n(z, \lambda)$, we find that it equals

$$\sum_{p=0}^{n+1} \binom{n+1}{p} \lambda^{2p} \sum_{\ell=0}^{\infty} L_n^\ell \lambda^\ell = \sum_{k=0}^{\infty} \left[\sum_{p=0}^{\left[\frac{n+1}{2}\right]} \binom{n+1}{p} L_n^{k-2p} \right] \lambda^k.$$

Comparing it with the expansion (3.1) of the right hand side, we conclude that

$$E_n^k = \sum_{p=0}^{\left[\frac{k}{2}\right]} \binom{n+1}{p} L_n^{k-2p}, \text{ for } 0 \leq k \leq n+1. \qquad (3.14)$$

this formula expresses the E_n^k's for $0 \leq k \leq n+1$ in terms of the L_n^k's. For $k \geq n+2$ we get

$$\sum_{p=0}^{n+1} \binom{n+1}{p} L_n^{k-2p} = 0, \qquad \text{resp.} \qquad L_n^k = -\sum_{p=1}^{m} \binom{n+1}{p} L_n^{k-2p},$$

(3.15)

where $m = \min\left\{\left[\frac{k}{2}\right], n+1\right\}$. Inserting on the other hand the expansion (3.11) for $x = -\lambda^2$ into the right hand side of equation (3.13) we find that

$$\sum_{k=0}^{\infty} L_n^k \lambda^k = \sum_{p=0}^{\infty} (-1)^p \binom{n+p}{p} \sum_{k=0}^{\infty} E_n^{k-2p} \lambda^k$$

$$= \sum_{k=0}^{\infty} \left[\sum_{p=0}^{\left[\frac{k}{2}\right]} (-1)^p \binom{n+p}{p} E_n^{k-2p} \right] \lambda^k,$$

and hence comparing coefficients,

$$L_n^k = \sum_{p=0}^{\left[\frac{k}{2}\right]} (-1)^p \binom{n+p}{p} E_n^{k-2p}, \quad \text{for all } k, n \in \mathbb{N}_0.$$

(3.16)

The formulas (3.14) and (3.16) have already been deduced in [8], but (3.15) is new. They show, in particular, that the sets $\left\{E_n^k : 0 \leq k \leq n+1\right\}$ and $\left\{L_n^k : 0 \leq k \leq n+1\right\}$ can be interchanged; thus the linear independence of the E_n^k's yields the linear independence of the L_n^k's, as long as we restrict k to $0 \leq k \leq n+1$. Summarizing, we have therefore shown

Theorem 3.1 *The sets $\left\{E_n^k : 0 \leq k \leq n+1\right\}$ and $\left\{L_n^k : 0 \leq k \leq n+1\right\}$ both form a basis for the $\mathbb{R}$-vector space V_n of all homogeneous polynomial (H)-solutions of degree n.*

On account of Theorem 3.1, each (H)-solution f, defined on some open set containing 0, can be represented, at least near 0, in terms of the E_n^k's, resp. L_n^k's (see [8],[9]). We thus conclude that the (H)-solutions yield the smallest $\mathbb{R}$-vector space of functions, closed under uniform limits, which contains the powers z^n $(n \in \mathbb{N})$ as well as its partial derivatives with respect to y.

We note that there is a further way to obtain a basis for the homogeneous (H)-polynomials of degree n, by looking at the (H)-polynomials $(a_k z)^n a_k$, $k = 0, 1, \ldots, n+1$, defined by some complex numbers $a_0, a_1, \ldots, a_{n+1}$. Based on Theorem 3.1 of [9] one can easily show

Theorem 3.2 *The complex number $\zeta = e^{i\varphi}$ ($0 \leq \varphi \leq 2\pi$) has the property that for each $n \in \mathbb{N}_0$ the functions $(\zeta^k z)^n \zeta^k$ ($0 \leq k \leq n+1$) represent a basis for the $\mathbb{R}$-vector space V_n if and only if the real number $\frac{\varphi}{\pi}$ is irrational.*

4. Polynomial solutions of the system (R)

Fueter has already recognized that in case of the system (R) the so-called *hypercomplex variables*

$$z_1 = y - ix = -\frac{1}{2}(iz + zi), \qquad z_2 = t - jx = -\frac{1}{2}(jz + zj), \quad (4.1)$$

rather than $z = x + iy + jt$, play the essential role (see also [13]). Starting with the function

$$S_n(z, \lambda) = (z_1 + \lambda z_2)^n \qquad (\lambda \in \mathbb{R}, \, n \in \mathbb{N}), \tag{4.2}$$

one quickly verifies that the function $z \to S_n(z, \lambda)$ is an (R)-solution. Indeed, since (by induction on n) $\frac{\partial S_n}{\partial y} = n S_{n-1}$ and $\frac{\partial S_n}{\partial t} = n\lambda S_{n-1}$, the equation (1.10) is fulfilled:

$$x D_\ell S_n = x \left[\frac{\partial S_n}{\partial x} + i\frac{\partial S_n}{\partial y} + j\frac{\partial S_n}{\partial t} \right]$$

$$= x\frac{\partial S_n}{\partial x} + (y - z_1)\frac{\partial S_n}{\partial y} + (t - z_2)\frac{\partial S_n}{\partial t}$$

$$= \left(x\frac{\partial S_n}{\partial x} + y\frac{\partial S_n}{\partial y} + t\frac{\partial S_n}{\partial t} \right) - z_1\frac{\partial S_n}{\partial y} - z_2\frac{\partial S_n}{\partial t}$$

$$= n\left[S_n - (z_1 + \lambda z_2)S_{n-1} \right] = 0.$$

Here we have made use of Euler's relation.

By analogy with (3.1), we now expand S_n in terms of the powers of λ:

$$S_n(z, \lambda) = (z_1 + \lambda z_2)^n = \sum_{k=0}^{n} P_n^k(z) \lambda^k. \tag{4.3}$$

From

$$P_n^k(z) = \frac{1}{k!}\frac{d^k}{d\lambda^k} S_n(z, \lambda)|_{\lambda=0}, \qquad k = 0, 1, \ldots, n, \tag{4.4}$$

we conclude that the P_n^k's are homogeneous (R)-polynomials of degree n. Furthermore, from $S_n(z, \lambda) = (z_1 + \lambda z_2)S_{n-1}(z, \lambda)$ we infer that

$$P_n^k(z) = z_1 P_{n-1}^k(z) + z_2 P_{n-1}^{k-1}(z), \tag{4.5}$$

provided we set $P_n^k = 0$ for $k \notin \{0, 1, \ldots, n\}$. With this formula at hand, it is now easy to check (by induction on n) that

$$P_n^k(z) = \sum_{(\sigma_1, \ldots, \sigma_n) \in \sigma(k)} z_{\sigma_1} z_{\sigma_2} \cdots z_{\sigma_n}, \qquad (4.6)$$

where $\sigma(k)$ denotes the set of all permutations $(\sigma_1, \ldots, \sigma_n)$ of n elements with $n - k$ elements equal to 1 and k elements equal to 2. We thus end up with the polynomials introduced by Fueter [5] (see also [13]). An inductional argument, based on (4.5), also shows that

$$\frac{\partial P_n^k}{\partial y} = n P_{n-1}^k, \qquad \frac{\partial P_n^k}{\partial t} = n P_{n-1}^{k-1}, \qquad (4.7)$$

and hence, by (1.10), that

$$\frac{\partial P_n^k}{\partial x} = -n \left(i P_{n-1}^k + j P_{n-1}^{k-1} \right) = -n \left(P_{n-1}^k i + P_{n-1}^{k-1} j \right). \qquad (4.8)$$

Observing that for an arbitrary (R)-solution f, and its real part Ref, $\overline{\mathrm{grad}}\,(Ref) = \frac{\partial f}{\partial x}$, we get

$$P_n^k i + P_n^{k-1} j = -\frac{1}{n+1} \overline{\mathrm{grad}}\,(Re P_{n+1}^k), \quad \text{for } k = 0, 1, \ldots, n+1. \quad (4.9)$$

In order to find an explicit expression for the polynomials P_n^k let us recall that

Lemma 4.1 *Given the real homogeneous polynomials $g_0(y, t)$, $g_1(y, t)$ of degree $n + 1$, resp. n, there is a harmonic polynomial h on $\mathbb{R}^3$, uniquely determined by the conditions*

$$h(0, y, t) = g_0(y, t) \quad \text{and} \quad \frac{\partial h}{\partial x}(0, y, t) = g_1(y, t), \qquad (4.10)$$

namely

$$h(x, y, t) = \sum_{k=0}^{\left[\frac{n+1}{2}\right]} (-1)^k \frac{x^{2k}}{(2k)!} \Delta_2^k g_0(y, t) + \sum_{k=0}^{\left[\frac{n}{2}\right]} (-1)^k \frac{x^{2k+1}}{(2k+1)!} \Delta_2^k g_1(y, t),$$
$$(4.11)$$

where Δ_2^k denotes the kth power of the two-dimensional Laplace operator $\Delta_2 = \frac{\partial^2}{\partial y^2} + \frac{\partial^2}{\partial t^2}$.

According to (2.3) there is a harmonic function $h_n^k : \mathbb{R}^3 \to \mathbb{R}$ such that

$$P_n^k = \overline{\mathrm{grad}}\, h_n^k. \qquad (4.12)$$

By (4.6),

$$P_n^k(0, y, t) = \binom{n}{k} y^{n-k} t^k, \tag{4.13}$$

so that we have

$$\frac{\partial h_n^k}{\partial x}(0, y, t) = \binom{n}{k} y^{n-k} t^k, \quad \frac{\partial h_n^k}{\partial y}(0, y, t) = 0, \quad \frac{\partial h_n^k}{\partial t}(0, y, t) = 0. \tag{4.14}$$

From the last two equations it follows that $h_n^k(0, y, t) = $ constant. But h_n^k in (2.3) being determined only up to a constant, it may be chosen in such a way that $h_n^k(0, y, t) = 0$ for all y, t. Together with (4.14), Lemma 4.1 then yields the representation

$$h_n^k(x, y, t) = \binom{n}{k} \sum_{\ell=0}^{[\frac{n}{2}]} (-1)^\ell \frac{x^{2\ell+1}}{(2\ell+1)!} \Delta_2^\ell \{y^{n-k} t^k\}. \tag{4.15}$$

A similar formula holds for $P_n^k i + P_n^{k-1} j$. Indeed, setting $\tilde{h}_n^k = \frac{-1}{n+1} Re P_{n+1}^k$, then by (4.9)

$$P_n^k i + P_n^{k-1} j = \overline{\text{grad } \tilde{h}_n^k}. \tag{4.16}$$

By (4.15), $Re P_n^k \, (= \frac{\partial h_n^k}{\partial x})$ - and hence also $\tilde{h}_n^k$ - is even in x, forcing $\frac{\partial \tilde{h}_n^k}{\partial x}(0, y, t) = 0$. Since in addition, due to (4.13), $\tilde{h}_n^k(0, y, t) = \frac{-1}{n+1} \binom{n+1}{k} y^{n+1-k} t^k$, Lemma 4.1 applied to $\tilde{h}_n^k$ yields the representation

$$\tilde{h}_n^k(x, y, t) = -\frac{\binom{n+1}{k}}{n+1} \sum_{\ell=0}^{[\frac{n+1}{2}]} (-1)^\ell \frac{x^{2\ell}}{(2\ell)!} \Delta_2^\ell \{y^{n+1-k} t^k\}. \tag{4.17}$$

Next observe that (by induction on ℓ)

$$\Delta_2^\ell \{y^{n-k} t^k\} = (2\ell)! \sum_{p=0}^{\ell} \frac{\binom{\ell}{p} \binom{n-k}{2\ell-2p} \binom{k}{2p}}{\binom{2\ell}{2p}} y^{n-k-2\ell+2p} t^{k-2p}, \tag{4.18}$$

provided we set $\binom{n}{k} = 0$, for $n < k$. Inserted in (4.15) one finds, keeping in mind (4.12):

Theorem 4.2 *For* $P_n^k = u_n^k + i v_n^k + j w_n^k$, $0 \le k \le n$, *we have*

$$u_n^k(z) = \binom{n}{k} \sum_{q=0}^{\left[\frac{n}{2}\right]} \sum_{p=0}^{q} (-1)^q \frac{\binom{q}{p}\binom{n-k}{2p}\binom{k}{2q-2p}}{\binom{2q}{2p}} x^{2q} y^{n-k-2p} t^{k-2q+2p}$$

$$v_n^k(z) = -\binom{n}{k} \sum_{q=0}^{\left[\frac{n}{2}\right]} \sum_{p=0}^{q} (-1)^q \frac{\binom{q}{p}\binom{n-k}{2p+1}\binom{k}{2q-2p}}{\binom{2q+1}{2p+1}} x^{2q+1} y^{n-k-1-2p} t^{k-2q+2p}$$

$$w_n^k(z) = -\binom{n}{k} \sum_{q=0}^{\left[\frac{n}{2}\right]} \sum_{p=0}^{q} (-1)^q \frac{\binom{q}{p}\binom{n-k}{2p}\binom{k}{2q-2p+1}}{\binom{2q+1}{2p}} x^{2q+1} y^{n-k-2p} t^{k-2q+2p-1}$$

From (4.8) and Theorem 4.2 one obtains
(with the convention $\binom{n}{-1}/\binom{m}{-1} = (m+1)/(n+1)$)

Theorem 4.3 *For* $P_n^k i + P_n^{k-1} j = \tilde{u}_n^k + i \tilde{v}_n^k + j \tilde{w}_n^k$, $0 \le k \le n$, *we have*

$$\tilde{u}_n^k(z) = -\binom{n}{k} \sum_{q=1}^{\left[\frac{n+1}{2}\right]} \sum_{p=0}^{q} (-1)^q \frac{\binom{q}{p}\binom{n-k}{2p-1}\binom{k}{2q-2p}}{\binom{2q-1}{2p-1}} x^{2q-1} y^{n+1-k-2p} t^{k-2q+2p}$$

$$\tilde{v}_n^k(z) = \binom{n}{k} \sum_{q=0}^{\left[\frac{n+1}{2}\right]} \sum_{p=0}^{q} (-1)^q \frac{\binom{q}{p}\binom{n-k}{2p}\binom{k}{2q-2p}}{\binom{2q}{2p}} x^{2q} y^{n-k-2p} t^{k-2q+2p}$$

$$\tilde{w}_n^k(z) = \binom{n}{k} \sum_{q=0}^{\left[\frac{n+1}{2}\right]} \sum_{p=0}^{q} (-1)^q \frac{\binom{q}{p}\binom{n-k}{2p-1}\binom{k}{2q-2p+1}}{\binom{2q}{2p-1}} x^{2q} y^{n+1-k-2p} t^{k-2q+2p-1}$$

As an application, we now show that the polynomials $\{P_n^k : k = 0, \ldots, n\}$ and $\{P_n^k i + P_n^{k-1} j : k = 0, \ldots, n+1\}$ together form a basis (over $\mathbb{R}$) for the vector space of homogeneous (R)-polynomials of degree n. The linear independence immediately follows from the fact that the P_n^k's are linearly independent over $\mathbb{H}$ (see, e.g. [13]). Hence it remains to be shown that the polynomials under consideration generate the vector space. To check this statement, let f be an arbitrary polynomial (R)-solution, homogeneous of degree n. Then, by (2.3), there is a harmonic polynomial, homogeneous of degree $n+1$, such that $f = \overline{\mathrm{grad}}\, h$. Applying Lemma 4.1 to h, (4.11) yields a representation for h in terms of $g_0(y,t) = h(0,y,t)$ and $g_1(y,t) = \frac{\partial h}{\partial x}(0,y,t)$. Since g_0 and g_1 are homogeneous polynomials of degree $n+1$, respectively n, we have $g_0(y,t) = \sum_{k=0}^{n+1} \alpha_k y^{n+1-k} t^k$ and $g_1(y,t) = \sum_{k=0}^{n} \beta_k y^{n-k} t^k$, with $\alpha_k, \beta_k \in \mathbb{R}$. Inserted into (4.11), taking into consideration (4.12),

(4.15), (4.16) and (4.17), we get

$$f = \overline{\mathrm{grad}}\, h = -\frac{n+1}{\binom{n+1}{k}} \sum_{k=0}^{n+1} \alpha_k \left(P_n^k i + P_n^{k-1} j \right) + \frac{1}{\binom{n}{k}} \sum_{k=0}^{n} \beta_k\, P_n^k,$$

as desired. Summarizing, we thus have proved

Theorem 4.4 *The polynomials $\{P_n^k : k = 0, 1, \ldots, n\}$, together with their linear combinations $\{P_n^k i + P_n^{k-1} j : k = 0, 1, \ldots, n+1\}$, form a basis for the $\mathbb{R}-$ vector space of homogeneous, polynomial (R)-solutions of degree n.*

Finally let us note that with the same technique of proof one shows that

Theorem 4.5 *The polynomials $\{P_n^k : k = 0, 1, \ldots, n\}$ form a basis for the subspace of those homogeneous, polynomial (R)-solutions of degree n, whose values on the plane $x = 0$ are real.*

5. An integral representation for the M. Riesz system

Combining the identity

$$g(D_\ell f) + (D_r g)f = \frac{\partial}{\partial x}(gf) + \frac{\partial}{\partial y}(gif) + \frac{\partial}{\partial t}(gjf), \qquad (5.1)$$

valid for arbitrary continuously differentiable functions $f, g : \Omega \to \mathbb{R}^3$, Ω open in $\mathbb{R}^3$, with Gauss's divergence theorem, whereupon for any compact set $K \subset \Omega$ with smooth boundary ∂K

$$\int_K [\frac{\partial}{\partial x}(gf) + \frac{\partial}{\partial y}(gif) + \frac{\partial}{\partial t}(gjf)]dV = \int_{\partial K} [gf\nu_0 + gif\nu_1 + gjf\nu_2]dV =$$

$$\int_{\partial K} g\nu f dS,$$

$\nu = \nu_0 + i\nu_1 + j\nu_2$, the outer unit normal field, one obtains the formula

$$\int_K [g(D_\ell f) + (D_r g)f]dV = \int_{\partial K} g\nu f dS. \qquad (5.2)$$

In particular, if f and g are solutions of (R), then by (1.10),

$$\int_{\partial K} g\nu f dS = 0. \qquad (5.3)$$

Choosing now for g the kernel (1.6) with singularity at the point $z \in \overset{\circ}{K}$ (the interior of K), i.e. the function

$$\zeta \;\rightarrow\; \frac{(\zeta - z)^{-1}}{|\zeta - z|} = \frac{\overline{\zeta - z}}{|\zeta - z|^3}, \tag{5.4}$$

we conclude, if (5.3) is applied to $\Omega \setminus \{z\}$ and $K_\epsilon = K \setminus B(z, \epsilon)$ ($B(z, \epsilon)$ denoting the Euclidean ball of center z and radius ϵ), that

$$\int\limits_{\partial K} \frac{\overline{\zeta - z}}{|\zeta - z|^3} \nu(\zeta) f(\zeta) \, dS = - \int\limits_{\partial B(z,\epsilon)} \frac{\overline{\zeta - z}}{|\zeta - z|^3} \nu(\zeta) f(\zeta) \, dS$$

$$= \frac{1}{\epsilon^2} \int\limits_{\partial B(z,\epsilon)} f(\zeta) \, dS.$$

Letting $\epsilon \to 0$, the last integral tends to $4\pi f(z)$ so that we get the following, well-known Cauchy-type formula:

Theorem 5.1 *Let $\Omega \subset \mathbb{R}^3$ be an open set, $K \subset \Omega$ a compact set with smooth boundary ∂K, and let $\nu = \nu_0 + i\nu_1 + j\nu_2$ be its outer unit normal field. Then for any solution $f : \Omega \to \mathbb{R}^3$ of the system (R) we have*

$$f(z) = \frac{1}{4\pi} \int\limits_{\partial K} \frac{(\zeta - z)^{-1}}{|\zeta - z|} \nu(\zeta) f(\zeta) \, dS \quad (z \in \overset{\circ}{K}). \tag{5.5}$$

This elegant formula however has the disadvantage that its integrand is no longer in $\mathbb{R}^3$. Drawing on the following algebraic identity, valid for arbitrary $a, b, c \in \mathbb{R}^3$,

$$a\overline{b}c = (b, c)a - a \times (b \times c) - (a, b \times c)k, \tag{5.6}$$

involving the Euclidean scalar product as well as the cross-product (vector product) in $\mathbb{R}^3$, one can easily remove this obstacle. Indeed, setting $a = (\zeta - z)^{-1}/|\zeta - z|$, $b = \overline{\nu}(\zeta)$, $c = f(\zeta)$ in (5.6), the representation (5.5) yields (since there is no k-term on the left hand side !) $f(z) =$

$$\frac{1}{4\pi} \int\limits_{\partial K} \left\{ \frac{(\zeta - z)^{-1}}{|\zeta - z|} (\overline{\nu}(\zeta), f(\zeta)) - \frac{(\zeta - z)^{-1}}{|\zeta - z|} \times (\overline{\nu}(\zeta) \times f(\zeta)) \right\} dS,$$

$$\tag{5.7}$$

as well as

$$\int\limits_{\partial K} \left(\frac{(\zeta - z)^{-1}}{|\zeta - z|}, \overline{\nu}(\zeta) \times f(\zeta) \right) dS(\zeta) = 0, \quad \text{for all } z \in \overset{\circ}{K}. \tag{5.8}$$

Passing to the conjugate value on both sides of (5.7) we get, recalling that f satisfies (R) if and only if $\overrightarrow{g} = \overline{f}$ satisfies (1.8):

Theorem 5.2 *Let $\Omega \subset \mathbb{R}^3$ be an open set, $K \subset \Omega$ a compact set with smooth boundary ∂K, and let $\nu = \nu_0 + i\nu_1 + j\nu_2$ be its outer unit normal field. Then for any solution $\overrightarrow{g} : \Omega \to \mathbb{R}^3$ of the M.Riesz system (1.8) we have*

$$\overrightarrow{g}(z) = \frac{1}{4\pi} \int\limits_{\partial K} \frac{\zeta - z}{|\zeta - z|^3} \left(\nu(\zeta), \overrightarrow{g}(\zeta) \right) dS$$

$$- \frac{1}{4\pi} \int\limits_{\partial K} \frac{\zeta - z}{|\zeta - z|^3} \times \left(\nu(\zeta) \times \overrightarrow{g}(\zeta) \right) dS,$$

for all $z \in \overset{\circ}{K}$.

This beautiful representation theorem for the solutions of the M. Riesz system (1.8) deserves to be better known. To our knowledge it is mentioned so far only in the Russian literature. It occurs in the book [3] of Dzhuraev, but the proof given there is different from ours, as it is based on the solution of Poisson's equation. Our proof is similar to the one contained in the book [15] of Zhdanov, although this author identifies $\mathbb{R}^3$ with the pure quaternions, rather than with the reduced ones, as we do. Theorem 5.2 is by the way also mentioned, without proof, in Presa-Sage and Havin's interesting article [12].

A corresponding representation theorem for the solutions of the system (H) is, unfortunately, not yet known. We notice however that since the (H)-solutions are contained in the larger class of "Holomorphic Cliffordian functions", studied by Laville and Ramadanoff in [7] (see also Pernas [11]), there is hope for such a formula.

References

[1] Ahlfors, L.V.: Möbius transformations in several dimensions. Ordway Lectures in Mathematics, University of Minnesota, 1981.
[2] Brackx, F., Delanghe, R., Sommen, F.: Clifford analysis. Pitman, Boston-London-Melbourne, 1982.
[3] Dzhuraev, A.D.: Methods of singular integral equations. Pitman

monographs 60, Lengman Sci. & Techn., Harlow, New York 1992 (Russian original: Nauka, Moscow 1987)

[4] Eriksson-Bique, S.L. and Leutwiler, H.: On modified quaternionic analysis in $\mathbb{R}^3$. Arch. Math. 70 (1998), 228-234.

[5] Fueter, R.: Die Funktionentheorie der Differentialgleichungen $\Delta u = 0$ und $\Delta\Delta u = 0$ mit vier reellen Variablen. Comment. Math. Helv. 7 (1934/35), 307-330.

[6] Gürlebeck, K. and Sprößig, W.: Quaternionic analysis and elliptic boundary values problems. Birkhäuser, Basel, 1990

[7] Laville, G. and Ramadanoff, I.: Holomorphic Cliffordian functions. Advances in Clifford algebra (1999)

[8] Leutwiler, H.: Modified quaternionic analysis in $\mathbb{R}^3$. Complex Variables Theory Appl. 20 (1992), 19-51.

[9] Leutwiler, H.: Rudiments of a function theory in $\mathbb{R}^3$. Expositiones Math. 14 (1996), 97-123.

[10] Leutwiler, H.: More on modified quaternionic analysis in $\mathbb{R}^3$. Forum Math. 7 (1995), 279-305

[11] Pernas, L.: Holomorphie quaternionienne (preprint Nov. 1997).

[12] Presa Sague, A. and Khavin, V.P.: Uniform approximation by harmonic differential forms on Euclidean space. St. Petersbg. Math. J. 7, No.6 (1996), 943-977 (translation from Algebra Anal. 7, No.6 (1995), 104-152).

[13] Sudbery, A.: Quaternionic analysis. Math. Proc. Camb. Philos. Soc. 85 (1979), 199-225.

[14] Warner, F.W.: Foundations of differentiable manifolds and Lie groups. Scott, Foresman and Company, Glenview 1971.

[15] Zhdanov, M.S.: Integral transforms in geophysics. Springer, Heidelberg 1988

Contributions to a Geometric Function Theory in Higher Dimensions by Clifford Analysis Methods: Monogenic Functions and M-conformal mappings

Helmuth R. Malonek (`hrmalon@mat.ua.pt`)
Departamento de Matemática, Universidade de Aveiro

Abstract. We introduce the concept of M(onogenic)-conformal mappings realized by functions which are defined in an open subset of $\mathbb{R}^{n+1}$ and with values in a Clifford algebra. The relation of this concept to the geometric interpretation of the hypercomplex derivative (Gürlebeck and Malonek, 1999) allows us to complete the theory of monogenic functions by providing a still missing geometric characterization of those functions. We also show that M-conformal mappings are generalizations of conformal mappings in the plane, but different from conformal mappings (in the sense of Gauss) in higher dimensions, the latter being restricted for $n \geq 2$ to the set of Möbius transformations (Liouville's Theorem).

Keywords: monogenic functions, hypercomplex derivative, M-conformal mappings

Mathematics Subject Classification: 30G35, 28A75, 26B15

Dedicated to Richard Delanghe on the occasion of his 60th birthday.

1. Introduction

The books *Clifford Analysis* (1982) and *Clifford algebra and spinor valued functions* (1992) written by Richard Delanghe and his disciples and collaborators are well known to those who treat analytic problems in higher dimensions with Clifford algebras and are thereby concerned with some relations to complex function theory. The first work appeared after Delanghe had spent a decade setting up enthusiastically the fundamentals and cornerstones for this new discipline, which soon has been named after the title of the book: *Clifford Analysis*. The second one, with the subtitle *A function theory for the Dirac operator*, marked ten years later a successful further development of the theory with important applications in physics and other areas. Meanwhile the recognition of the work of Richard Delanghe and his research group in Gent is worldwide.

But it was not only his mathematics that made Richard Delanghe one of the founders of Clifford Analysis. His passion for new ideas, his

213

humanity and his enthusiastic support to young researchers have also contributed to make him the main pillar of this field.

Looking back to last thirty years we can say that time has shown that Clifford Analysis is an independent mathematical discipline with its own goals and characteristic tools. Some of the most important indicators for such a development (behind all concrete results) are the capacity of generalizing, unifying and simplifying several other analytic theories developed for solving problems in higher dimensional spaces, its overwhelming variety of fields of applications and, last but not least, the existence of still a lot of interesting open and unsolved problems. This also concerns the different approaches to the fundamentals of Clifford Analysis.

Although the general fundamentals of Clifford Analysis are rooted in almost all classical areas, it seems to be appropriate to consider complex function theory of one complex variable as one of the most important stimulus for that theory. Moreover, in very interesting philosophical (and sometimes controversial) discussions, results in Clifford Analysis are often recognized as the truly appropriate generalizations or extensions of classical theorems of complex analysis to higher dimensions. All this is the reason for the constant interest in the relations between both theories and contributes for sometimes surprising new insights.

In this line of reasoning, we will characterize monogenic functions by an intrinsic geometric property which reduces in the complex case to the property of conformal mapping.

The second chapter in the book *Clifford Analysis* is dealing with the general theory of monogenic functions in $\mathbb{R}^{n+1}$ as solutions of a generalized Cauchy-Riemann system. At that time (1982) the characterization of monogenic functions by a corresponding differentiability concept or by the existence of a well defined derivative was still missing. The same happened with the characterization of these functions via a suitable generalized conformality concept. Meanwhile the former characterizations have been found, c.f. (Malonek, 1990), (Gürlebeck and Malonek, 1999), but nothing was known up to now about the latter.

It is clear from the situation in the plane that both characterizations should be intimately related to each other, because conformal mappings in the plane are realized in the neighborhood of a point by all complex functions for which the complex derivative exists and is not equal to zero. But the usual definition of conformal mappings (in the sense of Gauss) in $\mathbb{R}^{n+1}$ with $n \geq 2$ applies only to the restricted set of Möbius transformations (Liouville's Theorem). These transformations can algebraically be expressed and studied by using Clifford algebraic and analytic methods in very beautiful ways, but they are *not* monogenic functions.

Therefore it seems to be natural to ask whether or not there exists another mapping property than conformality for all monogenic functions, independently of the dimension n (but coinciding with conformality in $\mathbb{R}^2$) if the hypercomplex derivative exists and is not equal to zero.

The answer can be found by analyzing the geometric meaning of the hypercomplex derivative defined in (Gürlebeck and Malonek, 1999). Our purpose is to perform this analysis in the most elementary way to keep as much as possible clear the connections with the complex case and to show the essentials. A general and rigorous exposition with more technical details and applications will be published elswhere. To be short we will use the notations of (Gürlebeck and Malonek, 1999) without repeating them here; these notations coincide mainly with the notations already used in (Brackx, Delanghe and Sommen, 1982).

2. The complex case revisited

A starting point for the classical geometric function theory in the plane is the theory of conformal mappings based on the Riemann mapping theorem (which unfortunately is not valid in higher dimensions). More generally, all consequences related to the geometric interpretation of properties of holomorphic functions and their composition with other special function classes (for instances with homoemorphisms of Beltrami equations which leads to quasiconformal mappings) are important subjects in this field.

A simple method of studying conformal mappings (and also quasiconformal mappings, c.f. (Ahlfors, 1966)) is in the infinitesimal form, i.e., by looking to the differential of a non-constant real differentiable function $f : \Omega \longrightarrow \mathbb{C}$ defined in the domain $\Omega \subset \mathbb{C}$. This differential is given in the neighborhood of some point $z_0 \in \Omega$ in the form

$$df = f_z dz + f_{\bar{z}} d\bar{z}. \qquad (1)$$

If f is holomorphic, i.e., $f_{\bar{z}} = 0$, then (1) reduces to

$$df = f_z dz = \frac{df}{dz} dz. \qquad (2)$$

This implies that f realizes (locally) a conformal mapping in the sense of Gauss, namely there exists a positive function $\lambda(z) = |f'(z)|^2$ such that

$$|df|^2 = \lambda(z)|dz|^2. \qquad (3)$$

It is not difficult to show that the last property guarantees (locally) the conservation of angles between two paths passing through z_0 and between their images passing through $f(z_0)$.

This usual treatment is suggested by the special role of z and $\bar{z}$ and the application of the corresponding Wirtinger derivatives in formulas (1) or (2) of the differential. However, for generalizations to higher dimensions the corresponding integral relation is more interesting. In view of (2), the (trivial) relation

$$\int_{z_0}^z df = \int_{z_0}^z \frac{df}{d\zeta}(\zeta)d\zeta. \tag{4}$$

holds for some path between $z_0, z \in \Omega$. Dividing (4) by the (complex valued) integral $\int_{z_0}^z d\zeta$ we obtain with the help of the mean value theorem

$$\lim_{z \to z_0} \frac{\int_{z_0}^z df}{\int_{z_0}^z d\zeta} = \lim_{z \to z_0} \frac{f(z) - f(z_0)}{z - z_0} = \frac{df}{dz}(z_0). \tag{5}$$

Here the simple line integral $\int_{z_0}^z d\zeta$ is written in the form of a definite complex Riemann integral which is typical for dealing with line integrals in complex form. But all the relations (1) - (2) are also true if we multiply them by $-i$. The corresponding rotation implies in (4) and (5) the change to line integrals with elements defined by the (positive) normal to the integration path (e.g., instead of $dz = dx + idy$ we have now $-idz = dy - idx$). Concerning the conformal mapping property we could think that instead of observing the conservation of the angles between paths we now get the expression for the conservation of the angles between the normals to the paths. This makes no difference in the plane. But thinking in terms of surface integrals (which we have to expect in higher dimensions) this is the key for understanding all what follows.

Before ending this section let us still remark that $-idz = dy - idx$ coincides with the form of the usually applied hypercomplex surface element, (c.f. (Gürlebeck and Malonek, 1999)),

$$d\sigma = d\hat{x}_0 - e_1 d\hat{x}_1 + e_2 d\hat{x}_2 + \cdots + (-1)^n e_n d\hat{x}_n, \tag{6}$$

restricted to the complex case $n = 1$ whereby $x := x_0$, $y := x_1$, $i := e_1$. Moreover, using the nomenclature introduced in (Delanghe, 1970) $z_1 = -iz = x_1 - e_1 x_0$ is one of the *totally regular variables*. Such variables are of fundamental importance for the approach to differentiability of monogenic functions (Malonek, 1990) and appear in several other circumstances, c.f. (Brackx, Delanghe and Sommen, 1982).

3. The basic relations in terms of differential forms

We first introduce some prerequisites. Let us normalize the generalized Cauchy-Riemann operators D and $\overline{D}$ of Clifford analysis by

$$\partial_{\overline{z}} := \frac{1}{2}D \quad \text{and} \quad \partial_z := \frac{1}{2}\overline{D}.$$

This changes the usual notation but allows us to avoid any changes if we are considering the restriction to the complex case. With the help of the totally regular variables $z_k := x_k - e_k x_0$, $k = 1, \ldots n$, we introduce also the new main variable z by defining

$$z := \frac{1}{n}(z_1 e_1 + \ldots + z_n e_n) = \tag{7}$$

$$= x_0 + \frac{1}{n}(x_1 e_1 + \ldots + x_n e_n)$$

and the $(n-1)$-differential form ($n \geq 2$; if $n = 1$ we have simply to choose the constant form $d\mu = -e_1$):

$$d\mu := (-1)^n e_1 dz_2 \wedge dz_3 \wedge \cdots \wedge dz_n + \tag{8}$$

$$(-1)^{n+1} e_2 d\overline{z}_1 \wedge dz_3 \wedge \cdots \wedge dz_n + \ldots +$$

$$(-1)^{2n-1} e_n d\overline{z}_1 \wedge d\overline{z}_2 \wedge \cdots \wedge d\overline{z}_{n-1}$$

$$= \sum_{k=1}^{n} (-1)^{n-1+k} e_k d\hat{x}_{0,k}.$$

Here the notation $d\hat{x}_{0,m}$ ($m = 1, \ldots, n$) means that in the ordered outer product of the 1-forms dx_k ($k = 0, \ldots, n$) the factors dx_0 and dx_m are absent (c.f. (Gürlebeck and Malonek, 1999) in what concerns the (elementary) calculus of the introduced differential forms).

Further, let dF be the left [1] differential of an arbitrary real differentiable function

$$F : \Omega \subset \mathbb{R}^{n+1} \longrightarrow Cl_{0,n}.$$

This differential can be written, c. f. (Malonek, 2000), in the form:

$$dF = dz\, \partial_z F + d\overline{z}\, \partial_{\overline{z}} F + \tag{9}$$

$$+ \sum_{k=2}^{n} \frac{1}{n}(dx_k - e_k e_1 dx_1)\left[e_k(\partial_{\overline{z}} - \partial_z)F + n\frac{\partial F}{\partial x_k}\right].$$

Since the differential form $d\mu$ is constructed in such a way that it annihilates the 1-forms $(dx_k - e_k e_1 dx_1)$, $k = 2, \ldots, n$, we obtain as

---

[1] To be short we consider in general only left monogenic functions, left derivative etc. The right version is dual.

the outer product of $d\mu$ and dF the n-form:

$$d\mu \wedge dF = d\mu \wedge dz\, \partial_z F + d\mu \wedge d\bar{z}\, \partial_{\bar{z}} F. \qquad (10)$$

We also have the very useful factorization of $d\sigma$ and $\overline{d\sigma}$ in the form

$$d\sigma = dz_1 \wedge \cdots \wedge dz_n = d\mu \wedge dz = -dz \wedge d\mu \quad \text{and} \qquad (11)$$
$$\overline{d\sigma} = d\overline{z_1} \wedge \cdots \wedge d\overline{z_n} = -d\mu \wedge d\bar{z} = d\bar{z} \wedge d\mu$$

It is worthwile noticing that the outer product of $d\mu$ and dz is alternating. Moreover, c.f. (Brackx, Delanghe and Sommen, 1982), that if ds stands for the classical (and real) surface measure and

$$N = \sum_{k=0}^{n} e_k N_k$$

where N_k is the k-th component of the (outward pointing) normal, then the $Cl_{0,n}$-valued surface element $d\sigma$ can be written as $d\sigma = N ds$.

Stokes' theorem for manifolds of dimension n implies that the hypercomplex derivative $\partial_z F$ of a monogenic function, i. e. with $\partial_{\bar{z}} F = 0$, has to be understood as the limit of the quotient of two integrals, c.f. (Sudbery, 1979). Suppose we consider in Ω a positively oriented differentiable n-dimensional hypersurface S with coherent oriented boundary ∂S. Taking into account that (c.f. (Gürlebeck and Malonek, 1999))

$$(-1)^{n-1} d(d\mu F) = d\mu \wedge dF = d\mu \wedge dz\, \partial_z F \qquad (12)$$

then it follows by Stokes' theorem that

$$(-1)^{n-1} \int_{\partial S} (d\mu F) = \int_{S} d\mu \wedge dz\, \partial_z F. \qquad (13)$$

Let $z^* \in S$ be a fixed point in S. Consider now a so called *regular sequence* of subdomains $\{S_m\}$ which is shrinking to z^* as m tends to infinity and whereby z^* belongs to all S_m. Dividing both sides of (13) by

$$\int_{S} d\mu \wedge dz$$

and applying the mean value argument, in case the limit for all possible regular sequences exists, we get the following integral representation of $\partial_z F$

$$\partial_z F(z^*) = (-1)^n \lim_{m \to \infty} \left[\int_{S_m} d\mu \wedge dz \right]^{-1} \int_{\partial S_m} (d\mu F). \qquad (14)$$

Notice that by means of a suitable parametrization of S_m the surface integral in (14) can be transformed into a surface integral of the first kind

$$\int_{S_m} d\mu \wedge dz = \int_{S_m} d\sigma = \int_{S_m} N ds \qquad (15)$$

(c.f.(11) and formula (5) in the complex case). This means that this integral is related to a direction (the integral-direction of the varying over the surface normal) and the limit of the quotient (14) depends a priori from that direction. Contrarily, with respect to (10), we see that a quotient of the form (14) for an arbitrary real differentiable function and all possible regular sequences S_m will only be independent from the (integral-)direction of convergence if the second expression on the right hand side of (10) vanishes, i.e. if F is a monogenic function.

4. The definition of M-conformal mappings and its relation to monogenic functions

To simplify the exposition we start with several definitions.

Definition 1. Let S be a positively oriented differentiable n-dimensional hypersurface with coherent oriented boundary in Ω. Then we will call the integral

$$\mathcal{M}_S = \int_S d\mu \wedge dz = \int_S N ds \qquad (16)$$

the Clifford measure of S. Given an arbitrary real differentiable function $F : \Omega \subset \mathbb{R}^{n+1} \longrightarrow Cl_{0,n}$ we will analogously call the integral

$$\mathcal{M}_S(F) = \int_S d\mu \wedge dF \qquad (17)$$

the (left) Clifford measure of $F(S)$

Notice that for the special complex case we have $d\mu = -e_1 \cong -i$ and the corresponding hypersurfaces are curves. Complex analysis normally deals with $\int_\gamma dz = \int_\gamma dx + i dy$, whereas the Clifford measure is equal to

$$\mathcal{M}_\gamma = -i \int_\gamma dz = \int_\gamma dy - i dx = \int_\gamma (\sin\alpha - i\cos\alpha) ds$$

$$= \int_\gamma (\cos\mu + i\cos\nu) ds, \quad \text{with } \alpha = \frac{\pi}{2} - \mu = \pi - \nu.$$

The last type of representation is the basis for the general treatment and shows the directional as well as the absolute part of the integral.

Definition 2. Let S_1 and S_2 be two positively oriented differentiable n-dimensional hypersurfaces with coherent oriented boundary in Ω. Consider also two regular sequences $\{S_{m,k}\}$, $k = 1, 2$ corresponding to the considered S_k, $k = 1, 2$ and some fixed point $z^* \in \Omega$ which belongs to all $S_{m,k}$ for $k = 1, 2$. Further we assume for simplicity that the euclidean measures of both $S_{m,k}$, $k = 1, 2$ are the same for all m. Then (in case of their existence) we will call the limits

$$\alpha^r_{1,2}(z^*) = \lim_{m \to \infty} \mathcal{M}^{-1}_{S_{m,1}} \mathcal{M}_{S_{m,2}} \text{ resp.} \tag{18}$$

$$\alpha^\ell_{1,2}(z*) = \lim_{m \to \infty} \mathcal{M}_{S_{m,2}} \mathcal{M}^{-1}_{S_{m,1}} \tag{19}$$

the right resp. left angle between S_1 and S_2 at the point z^*.
Substituting $\mathcal{M}_S$ by $\mathcal{M}_S(F)$ we use this notation also for describing angles between $F(S_1)$ and $F(S_2)$, but in that case we will assume that F is itself a para-vector and maps hypersurfaces in hypersurfaces.

Remark

As we previously mentioned the Clifford measure of a surface depends from its absolute part (its euclidean measure) as well as the directional part determined by the "field" of normals (which are now considered as unit para-vectors). The assumption about the regular sequences implies that the limits in Definition 2 are again unit para-vectors describing the relative position of two surfaces and therefore, in our opinion, really deserve their names. Of course, the assumption that F is a para-vector is restrictive but taking into account that it guarantees the same dimension in the domain and in the range of F it seems to be natural in mapping problems, which concern also the question of the existence of the inverse mapping. Moreover, this assumption is also important by algebraic reasons and can avoid zero-divisor problems. But in the following short introduction to M-conformal mappings we will not discuss such questions. Here we are only concerned with a brief introduction to this concept.

Corollaries:

1. With the help of Definition 1 we obtain from (14) that the (left) derivative is the limit of (left) quotient of the corresponding Clifford measures

$$\lim_{m \to \infty} \mathcal{M}^{-1}_{S_m} \mathcal{M}_{S_m}(F) \tag{20}$$

taken over all possible regular sequences $\{S_m\}$.

2. It is now obvious that a vanishing surface-measure, i.e. $\mathcal{M}_S(F) = 0$, for a monogenic function F is equivalent to the fact that $F_z = F_{\bar{z}} = 0$ but this means that F in fact depends only on x_k, $k = 1, \ldots, n$. But then F cannot be bijective. It is worthwhile noticing that the vanishing surface-measure relates to a generalized Cauchy theorem in domains of n dimensions complementary to the generalized Cauchy theorem valid for monogenic functions in the $(n + 1)$-dimensional domain Ω.

Definition 3. Let $\mathcal{S}_1$ and $\mathcal{S}_2$ be two positively oriented differentiable n-dimensional hypersurfaces with coherent oriented boundary in Ω. If there exists a mapping F defined on hypersurfaces in Ω which preserves the angle, i.e., such that

$$\alpha_{1,2}^{r}(z^*) = \lim_{m \to \infty} \mathcal{M}_{\mathcal{S}_{m,1}}^{-1} \, \mathcal{M}_{\mathcal{S}_{m,2}} = \lim_{m \to \infty} \mathcal{M}_{\mathcal{S}_{m,1}}(F)^{-1} \, \mathcal{M}_{\mathcal{S}_{m,2}}(F) \quad \text{or}$$

$$\alpha_{1,2}^{\ell}(z^*) = \lim_{m \to \infty} \mathcal{M}_{\mathcal{S}_{m,2}} \, \mathcal{M}_{\mathcal{S}_{m,1}}^{-1} = \lim_{m \to \infty} \mathcal{M}_{\mathcal{S}_{m,2}}(F) \, \mathcal{M}_{\mathcal{S}_{m,1}}(F)^{-1}, \quad (21)$$

then F is called a right resp. left M-conformal mapping.

Theorem 1. Let F be a para-vector valued real differentiable function in $\Omega \subset \mathbb{R}^{n+1}$. This function realizes locally in the neighborhood of a fixed point z^* a *left* M-conformal mapping if and only if F is *left* monogenic and its *left* derivative is different from zero.

Sketch of proof. The theorem can be proved very similarly to the complex case. First, let F be left monogenic in z^* and consider

$$\alpha_{1,2}^{\ell,F}(z^*) = \lim_{m \to \infty} \mathcal{M}_{\mathcal{S}_{m,2}}(F) \, \mathcal{M}_{\mathcal{S}_{m,1}}(F)^{-1}. \qquad (22)$$

Multiplying from the left and from the right by the number one in the form

$$1 = \lim_{m \to \infty} \mathcal{M}_{\mathcal{S}_{m,2}} \, \mathcal{M}_{\mathcal{S}_{m,2}}^{-1}, \quad \text{resp. } 1 = \lim_{m \to \infty} \mathcal{M}_{\mathcal{S}_{m,1}} \, \mathcal{M}_{\mathcal{S}_{m,1}}^{-1} \qquad (23)$$

an easy calculation leads to

$$\alpha_{1,2}^{\ell,F}(z^*) = \lim_{m \to \infty} \mathcal{M}_{\mathcal{S}_{m,2}} \, \partial_z^{(\ell)} F(z^*) \left[\partial_z^{(\ell)} F(z^*) \right]^{-1} \mathcal{M}_{\mathcal{S}_{m,1}}^{-1} \qquad (24)$$

$$= \alpha_{1,2}^{\ell}(z^*)$$

In order to prove the reciprocal implication we first multiply

$$\alpha_{1,2}^{\ell}(z^*) = \lim_{m \to \infty} \mathcal{M}_{\mathcal{S}_{m,2}} \, \mathcal{M}_{\mathcal{S}_{m,1}}^{-1}$$

from the left and from the right with

$$1 = \lim_{m \to \infty} M_{\mathcal{S}_{m,2}}(F) \, M_{\bar{\mathcal{S}}_{m,2}}^{-1}(F), \text{ resp. } 1 = \lim_{m \to \infty} M_{\mathcal{S}_{m,1}}(F) \, M_{\bar{\mathcal{S}}_{m,1}}^{-1}(F).$$

If follows that

$$\alpha_{1,2}^{\ell}(z^*) = \lim_{m \to \infty} M_{\mathcal{S}_{m,2}} \, M_{\bar{\mathcal{S}}_{m,1}}^{-1} =$$

$$= \lim_{m \to \infty} M_{\mathcal{S}_{m,2}}(F) \, M_{\bar{\mathcal{S}}_{m,2}}^{-1}(F) \, M_{\mathcal{S}_{m,2}} \times$$

$$\times M_{\bar{\mathcal{S}}_{m,1}}^{-1} \, M_{\mathcal{S}_{m,1}}(F) \, M_{\bar{\mathcal{S}}_{m,1}}^{-1}(F).$$

Since by the left M-conformality property the whole expression must be equal to

$$\lim_{m \to \infty} M_{\mathcal{S}_{m,2}}(F) \, M_{\mathcal{S}_{m,1}}(F)^{-1} = \alpha_{1,2}^{\ell,F}(z^*), \qquad (25)$$

the product of the four inner factors must be equal to 1. After changing the order of the second and third factor we see that therefore it has to be

$$\lim_{m \to \infty} M_{\bar{\mathcal{S}}_{m,2}}^{-1} \, M_{\mathcal{S}_{m,2}}(F)(z^*) = \lim_{m \to \infty} M_{\bar{\mathcal{S}}_{m,1}}^{-1} \, M_{\mathcal{S}_{m,1}}(F)(z^*). \qquad (26)$$

The facts that the (left) derivative of a real differentiable function F (in the sense of the given representation (14)) is uniquely defined only for a left monogenic function, together with the unicity of the inverse of a para-vector, finishes the proof.

References

Ahlfors, L.V. *Lectures on Quasiconformal Mappings*. Van Nostrand, Princeton 1966; Reprinted by Wadsworth Inc., Belmont, 1987.

Brackx, F., Delanghe, R. and Sommen, F. *Clifford Analysis*. Pitman **76**, Boston-London-Melbourne, 1982.

Delanghe, R. On regular-analytic functions with values in a Clifford algebra. *Math. Ann.*, **185**: 91–111, 1970.

Gürlebeck, K. and Malonek, H. A hypercomplex derivative of monogenic functions in $\mathbb{R}^{n+1}$ and its applications. *Complex Variables Theory Appl.*, Vol. **39**: 199–228, 1999.

Malonek, H. A new hypercomplex structure of the Euclidean space $\mathbb{R}^{m+1}$ and the concept of hypercomplex differentiability. *Complex Variables Theory Appl.*, Vol. **14**: 25–33, 1990.

Malonek, H. Function theoretic aspects in Clifford Analysis, submitted to *Proceedings of the International Conference on Clifford Analysis*, Beijing, August 2000.

Sudbery, A. Quaternionic analysis. *Math. Proc. Cambr. Phil. Soc.*, 85: 199–225, 1979

The Dirac Type Tensor Equation in Riemannian Space

Nikolai Marchuk (nmarchuk@mi.ras.ru) *
Steklov Mathematical Institute

Abstract. We suggest a tensor equation on Riemannian manifolds which can be considered as a generalization of the Dirac equation for the electron. The tetrad formalism is not used.

Keywords: tensor, spinor, Dirac equation, Riemannian manifolds

Mathematics Subject Classification: 35Q40

Dedicated to Richard Delanghe on the occasion of his 60th birthday.

1. Introduction

In this paper, following [1],[2], we consider the Dirac type tensor equation (7) on elementary Riemannian manifolds. The wave function of the electron in this equation has 16 real valued components, i.e., two times more than a bispinor in the Dirac equation but two times less than the wave function in the main equation in [1].

The research was carried out while the author was visiting at Bath University.

2. An elementary Riemannian manifold and differential forms

Let $\mathcal{M}$ be a four dimensional differentiable manifold covered by a system of coordinates x^μ. Greek indices run over (0,1,2,3). Summation convention over repeating indices is assumed. We consider atlases on $\mathcal{M}$ consisted of one chart. Suppose that there is a smooth twice covariant tensor field (a metric tensor) with components $g_{\mu\nu} = g_{\mu\nu}(x)$, $x \in \mathcal{M}$ such that

- $g_{\mu\nu} = g_{\nu\mu}$;

- $g = \det\|g_{\mu\nu}\| < 0$ for all $x \in \mathcal{M}$;

* Research supported by the Russian Foundation for Basic Research, grant 00-01-00224, and by the Royal Society.

F. Brackx et al. (eds.), *Clifford Analysis and Its Applications*, 223–230.

- The signature of the matrix $\|g_{\mu\nu}\|$ is equal to -2.

The matrix $\|g^{\mu\nu}\|$ composed from contravariant components of the metric tensor is the inverse matrix to $\|g_{\mu\nu}\|$. The full set of $\{\mathcal{M}, g_{\mu\nu}\}$ is called an *elementary Riemannian manifold* (with one chart atlases) and is denoted by $\mathcal{V}$.

Let $\Lambda_{\mathcal{V}}^k$ be the sets of exterior differential forms of rank $k = 0, 1, 2, 3, 4$ on $\mathcal{V}$ (covariant antisymmetric tensor fields) and

$$\Lambda_{\mathcal{V}} = \Lambda_{\mathcal{V}}^0 \oplus \ldots \oplus \Lambda_{\mathcal{V}}^4 = \Lambda_{\mathcal{V}}^{\text{even}} \oplus \Lambda_{\mathcal{V}}^{\text{odd}},$$

$$\Lambda_{\mathcal{V}}^{\text{even}} = \Lambda_{\mathcal{V}}^0 \oplus \Lambda_{\mathcal{V}}^2 \oplus \Lambda_{\mathcal{V}}^4, \quad \Lambda_{\mathcal{V}}^{\text{odd}} = \Lambda_{\mathcal{V}}^1 \oplus \Lambda_{\mathcal{V}}^3.$$

Elements of $\Lambda_{\mathcal{V}}$ are called (nonhomogeneous) *differential forms* and elements of $\Lambda_{\mathcal{V}}^k$ are called *k-forms* or differential forms of rank k. The set of smooth scalar functions on $\mathcal{V}$ (invariants) is identified with the set of 0-forms $\Lambda_{\mathcal{V}}^0$. A k-form $U \in \Lambda_{\mathcal{V}}^k$ can be written as

$$U = \frac{1}{k!} u_{\nu_1 \ldots \nu_k} dx^{\nu_1} \wedge \ldots \wedge dx^{\nu_k} = \sum_{\mu_1 < \cdots < \mu_k} u_{\mu_1 \ldots \mu_k} dx^{\mu_1} \wedge \ldots \wedge dx^{\mu_k}, \quad (1)$$

where $u_{\nu_1 \ldots \nu_k} = u_{\nu_1 \ldots \nu_k}(x)$ are real valued components of a covariant antisymmetric ($u_{\nu_1 \ldots \nu_k} = u_{[\nu_1 \ldots \nu_k]}$) tensor field. Differential forms from $\Lambda_{\mathcal{V}}$ can be written as linear combinations of the 16 basis differential forms

$$1, dx^{\mu}, dx^{\mu_1} \wedge dx^{\mu_2}, \ldots, dx^1 \wedge \ldots \wedge dx^n, \qquad \mu_1 < \mu_2 < \ldots . \quad (2)$$

The exterior multiplication of differential forms is defined in the usual way. If $U \in \Lambda_{\mathcal{V}}^r, V \in \Lambda_{\mathcal{V}}^s$, then

$$U \wedge V = (-1)^{rs} V \wedge U \in \Lambda_{\mathcal{V}}^{r+s}.$$

In this paper we consider changes of coordinates with positive Jacobian and do not distinguish tensors and pseudotensors.

Consider the Hodge star operator $\star : \Lambda_{\mathcal{V}}^k \to \Lambda_{\mathcal{V}}^{4-k}$. If $U \in \Lambda_{\mathcal{V}}^k$ has the form (1), then

$$\star U = \frac{1}{k!(4-k)!} \sqrt{-g} \, \varepsilon_{\mu_1 \ldots \mu_4} u^{\mu_1 \cdots \mu_k} dx^{\mu_{k+1}} \wedge \ldots \wedge dx^{\mu_4},$$

where $u^{\mu_1 \cdots \mu_k} = g^{\mu_1 \nu_1} \ldots g^{\mu_k \nu_k} u_{\nu_1 \ldots \nu_k}$, $\varepsilon_{\mu_1 \ldots \mu_4}$ is the sign of the permutation $(\mu_1 \ldots \mu_4)$, and $\varepsilon_{0123} = 1$. It is easy to prove that for $U \in \Lambda_{\mathcal{V}}^k$

$$\star(\star U) = (-1)^{k+1} U.$$

Further on we consider the bilinear operator Com : $\Lambda_\nu^2 \times \Lambda_\nu^2 \to \Lambda_\nu^2$ such that for basis 2-forms

$$\mathrm{Com}(dx^{\mu_1} \wedge dx^{\mu_2}, dx^{\nu_1} \wedge dx^{\nu_2}) =$$
$$-2g^{\mu_1\nu_1} dx^{\mu_2} \wedge dx^{\nu_2} - 2g^{\mu_2\nu_2} dx^{\mu_1} \wedge dx^{\nu_1}$$
$$+2g^{\mu_1\nu_2} dx^{\mu_2} \wedge dx^{\nu_1} + 2g^{\mu_2\nu_1} dx^{\mu_1} \wedge dx^{\nu_2}$$

Evidently, $\mathrm{Com}(U, V) = -\mathrm{Com}(V, U)$.

Now we define the Clifford multiplication of differential forms with the aid of the following formulas (see formulas for the space dimensions 2 and 3 in [1]):

$$\overset{0}{U}\overset{k}{V} = \overset{k}{V}\overset{0}{U} = \overset{0}{U} \wedge \overset{k}{V} = \overset{k}{V} \wedge \overset{0}{U},$$

$$\overset{1}{U}\overset{k}{V} = \overset{1}{U} \wedge \overset{k}{V} - \star(\overset{1}{U} \wedge \star \overset{k}{V}),$$

$$\overset{k}{U}\overset{1}{V} = \overset{k}{U} \wedge \overset{1}{V} + \star(\overset{k}{U} \wedge \star \overset{1}{V}),$$

$$\overset{2}{U}\overset{2}{V} = \overset{2}{U} \wedge \overset{2}{V} + \star(\overset{2}{U} \wedge \star \overset{2}{V}) + \frac{1}{2}\mathrm{Com}(\overset{2}{U}, \overset{2}{V}),$$

$$\overset{2}{U}\overset{3}{V} = \star \overset{2}{U} \wedge \star \overset{3}{V} - \star(\overset{2}{U} \wedge \star \overset{3}{V}),$$

$$\overset{2}{U}\overset{4}{V} = \star \overset{2}{U} \wedge \star \overset{4}{V},$$

$$\overset{3}{U}\overset{2}{V} = -\star \overset{3}{U} \wedge \star \overset{2}{V} - \star(\star \overset{3}{U} \wedge \overset{2}{V}),$$

$$\overset{3}{U}\overset{3}{V} = \star \overset{3}{U} \wedge \star \overset{3}{V} + \star(\overset{3}{U} \wedge \star \overset{3}{V}),$$

$$\overset{3}{U}\overset{4}{V} = \star \overset{3}{U} \wedge \star \overset{4}{V},$$

$$\overset{4}{U}\overset{2}{V} = \star \overset{4}{U} \wedge \star \overset{2}{V},$$

$$\overset{4}{U}\overset{3}{V} = -\star \overset{4}{U} \wedge \star \overset{3}{V},$$

$$\overset{4}{U}\overset{4}{V} = -\star \overset{4}{U} \wedge \star \overset{4}{V},$$

where ranks of differential forms are denoted as $\overset{k}{U} \in \Lambda_\nu^k$ and $k = 0, 1, 2, 3, 4$. From this definition we may obtain some properties of the Clifford multiplication of differential forms.

1. If $U, V \in \Lambda_\nu$, then $UV \in \Lambda_\nu$.

2. The axioms of associativity and distributivity are satisfied for the Clifford multiplication.

3. $dx^\mu dx^\nu = dx^\mu \wedge dx^\nu + g^{\mu\nu}, \quad dx^\mu dx^\nu + dx^\nu dx^\mu = 2g^{\mu\nu}.$

4. If $U, V \in \Lambda_\nu^2$, then $\mathrm{Com}(U, V) = UV - VU.$

Let us define the trace of a differential form as a linear operation $\mathrm{Tr} : \Lambda_\mathcal{V} \to \Lambda^0_\mathcal{V}$ such that

$$\mathrm{Tr}(1) = 1, \quad \mathrm{Tr}(dx^{\mu_1} \wedge \ldots \wedge dx^{\mu_k}) = 0 \quad \text{for} \quad k = 1, 2, 3, 4.$$

The reader can easily prove that

$$\mathrm{Tr}(UV - VU) = 0, \quad \mathrm{Tr}(V^{-1}UV) = \mathrm{Tr}\,U, \quad U, V \in \Lambda_\mathcal{V}.$$

The 4-form

$$I = \sqrt{-g}\,dx^0 \wedge dx^1 \wedge dx^2 \wedge dx^3$$

is called *the volume form*. It can be shown that $I^2 = -1$, $IU = UI$ for $U \in \Lambda^{\text{even}}_\mathcal{V}$, and $IV = -VI$ for $V \in \Lambda^{\text{odd}}_\mathcal{V}$.

Also, let us define an involution $* : \Lambda^k_\mathcal{V} \to \Lambda^k_\mathcal{V}$. By definition, put

$$U^* = (-1)^{\frac{k(k-1)}{2}}U, \quad U \in \Lambda^k_\mathcal{V}.$$

It is readily seen that

$$U^{**} = U, \quad (UV)^* = V^*U^*, \quad U, V \in \Lambda_\mathcal{V}.$$

Note that the Hodge $\star$ operator can be expressed with the aid of the volume form I, the involution $*$, and the Clifford multiplication

$$\star U = U^*I.$$

Now we are ready to define the spinor group

$$\mathrm{Spin}_\mathcal{V} = \{S \in \Lambda^{\text{even}}_\mathcal{V} : S^*S = 1\}$$

and the group

$$\mathrm{U}(1)_\mathcal{V} = \{\exp(\lambda I) : \lambda \in \Lambda^0_\mathcal{V}\},$$

where $\exp(\lambda I) = \cos\lambda + I\sin\lambda \in \Lambda^0_\mathcal{V} \oplus \Lambda^4_\mathcal{V}$. It is important to note, that elements of the group $\mathrm{Spin}_\mathcal{V}$ commute with elements of the group $\mathrm{U}(1)_\mathcal{V}$.

3. Tensors with values in $\Lambda^k_\mathcal{V}$

Let

$$u^{\lambda_1\ldots\lambda_r}_{\mu_1\ldots\mu_k\nu_1\ldots\nu_s}(x) = u^{\lambda_1\ldots\lambda_r}_{[\mu_1\ldots\mu_k]\nu_1\ldots\nu_s}(x), \quad x \in \mathcal{V}$$

be components of a tensor field of rank $(r, k + s)$ antisymmetric with respect to the first k covariant indices. One may consider the following objects:

$$U^{\lambda_1\ldots\lambda_r}_{\nu_1\ldots\nu_s} = \frac{1}{k!}u^{\lambda_1\ldots\lambda_r}_{\mu_1\ldots\mu_k\nu_1\ldots\nu_s}\,dx^{\mu_1} \wedge \ldots \wedge dx^{\mu_k} \tag{3}$$

which are formally written as k-forms. Under a change of coordinates $(x) \to (\tilde{x})$ the values (3) transform as components of a tensor field of rank (r, s)

$$\tilde{U}^{\alpha_1 \ldots \alpha_r}_{\beta_1 \ldots \beta_s} = q^{\nu_1}_{\beta_1} \cdots q^{\nu_s}_{\beta_s} p^{\alpha_1}_{\lambda_1} \cdots p^{\alpha_r}_{\lambda_r} U^{\lambda_1 \ldots \lambda_r}_{\nu_1 \ldots \nu_s}, \quad q^{\nu}_{\beta} = \frac{\partial x^{\nu}}{\partial \tilde{x}^{\beta}}, \quad p^{\alpha}_{\lambda} = \frac{\partial \tilde{x}^{\alpha}}{\partial x^{\lambda}}. \quad (4)$$

The objects (3) are called tensors of rank (r, s) with values in $\Lambda^k_{\mathcal{V}}$. We write this as

$$U^{\lambda_1 \ldots \lambda_r}_{\nu_1 \ldots \nu_s} \in \mathsf{T}^r_s \Lambda^k_{\mathcal{V}}.$$

Elements of $\mathsf{T}^r_s \Lambda^0_{\mathcal{V}}$ are ordinary tensors of rank (r, s) on $\mathcal{V}$. For $U_{\mu} \in \mathsf{T}_1 \Lambda^k_{\mathcal{V}}$ we have

$$dx^{\mu} U_{\mu} \in \Lambda^{k+1}_{\mathcal{V}} \oplus \Lambda^{k-1}_{\mathcal{V}}.$$

4. The covariant derivatives ∇_{μ}

On Riemannian manifold $\mathcal{V}$ the Christoffel symbols $\Gamma^{\lambda}_{\mu\nu} = \Gamma^{\lambda}_{\nu\mu}$ (Levi-Chivita connectedness components) are defined with the aid of the metric tensor by the well known formulas. Let us remind the definition of covariant derivatives ∇_{μ} acting on tensor fields on $\mathcal{V}$ by the following rules ($\partial_{\mu} = \partial/\partial x^{\mu}$):

1. If $t = t(x)$, $x \in \mathcal{V}$ is a scalar function (invariant), then

$$\nabla_{\mu} t = \partial_{\mu} t.$$

2. If t^{ν} is a vector field on $\mathcal{V}$, then

$$\nabla_{\mu} t^{\nu} \equiv t^{\nu}_{;\mu} = \partial_{\mu} t^{\nu} + \Gamma^{\nu}_{\mu\lambda} t^{\lambda}.$$

3. If t_{ν} is a covector field on $\mathcal{V}$, then

$$\nabla_{\mu} t_{\nu} \equiv t_{\nu;\mu} = \partial_{\mu} t_{\nu} - \Gamma^{\lambda}_{\mu\nu} t_{\lambda}.$$

4. If $u = u^{\nu_1 \ldots \nu_k}_{\lambda_1 \ldots \lambda_l}$, $v = v^{\nu_1 \ldots \nu_r}_{\lambda_1 \ldots \lambda_s}$ are tensor fields on $\mathcal{V}$, then

$$\nabla_{\mu}(u \otimes v) = (\nabla_{\mu} u) \otimes v + u \otimes \nabla_{\mu} v.$$

With the aid of these rules it is easy to calculate covariant derivatives of arbitrary tensor fields. Also, it is easy to check the correctness of the following formulas:

$$\nabla_{\mu} g_{\nu\lambda} = 0, \quad \nabla_{\mu} g^{\nu\lambda} = 0, \quad \nabla_{\mu} \delta^{\nu}_{\lambda} = 0.$$

5. The Clifford derivatives Υ_μ

Let us define the, so called, Clifford derivatives Υ_μ (Upsilon) by the following rules:

1. If $t_{\nu_1...\nu_r}$ is a covariant tensor field on $\mathcal{V}$ of rank $r \geq 0$, then

$$\Upsilon_\mu t_{\nu_1...\nu_r} = \partial_\mu t_{\nu_1...\nu_r}.$$

2. $\Upsilon_\mu dx^\nu = -\Gamma^\nu_{\mu\lambda} dx^\lambda$.

3. If $U, V \in \Lambda(\mathcal{V})$ and UV is the Clifford product of differential forms, then

$$\Upsilon_\mu(UV) = (\Upsilon_\mu U)V + U\Upsilon_\mu V.$$

With the aid of these rules it is easy to calculate how operators Υ_μ act on arbitrary differential forms from $\Lambda_\mathcal{V}$. Namely for $U \in \Lambda^k_\mathcal{V}$, written as (1), we get

$$\Upsilon_\mu U = \frac{1}{k!} u_{\nu_1...\nu_k;\mu} dx^{\nu_1} \wedge \ldots \wedge dx^{\nu_k}. \tag{5}$$

If $U \in \Lambda^k_\mathcal{V}$, then $\Upsilon_\mu U$ is a covector with the value in $\Lambda^k_\mathcal{V}$, that is $\Upsilon_\mu U \in T_1 \Lambda^k_\mathcal{V}$. The formula (5) indicates the connection between operators Υ_μ and ∇_μ.

Consider the change of coordinates $(x) \to (\tilde{x})$

$$p^\mu_\nu = \frac{\partial \tilde{x}^\mu}{\partial x^\nu}, \quad q^\mu_\nu = \frac{\partial x^\mu}{\partial \tilde{x}^\nu}, \quad dx^\mu = q^\mu_\nu d\tilde{x}^\nu,$$

where p^μ_ν, q^μ_ν are functions of $x \in \mathcal{V}$. Then the Clifford derivatives Υ_ν in coordinates x^μ related to the Clifford derivatives $\tilde{\Upsilon}_\nu$ in coordinates $\tilde{x}^\mu$ by the formula

$$\Upsilon_\nu = p^\mu_\nu \tilde{\Upsilon}_\mu \tag{6}$$

exactly the same as formula for partial derivatives $\partial_\nu = p^\mu_\nu \tilde{\partial}_\mu$, where $\partial_\nu = \frac{\partial}{\partial x^\nu}$, $\tilde{\partial}_\nu = \frac{\partial}{\partial \tilde{x}^\nu}$. The proof of this formula is followed from the transformation rule of Christoffel symbols.

The main properties of the operators Υ_μ are listed below.

1) $\Upsilon_\mu I = 0$, where I is the volume form.
2) $\Upsilon_\mu(U^*) = (\Upsilon_\mu U)^*$ for $U \in \Lambda_\mathcal{V}$.
3) $\Upsilon_\mu(\star U) = \star(\Upsilon_\mu U)$ for $U \in \Lambda_\mathcal{V}$.
4) $\Upsilon_\mu(\text{Tr}\, U) = \text{Tr}(\Upsilon_\mu U)$ for $U \in \Lambda_\mathcal{V}$.

6. The main equation

We suppose that the following equation can be considered as a generalization of the Dirac equation for the electron to Riemannian space

$$dx^\mu(\Upsilon_\mu\Psi + \Psi A_\mu + \Psi B_\mu) + m\Psi I = 0, \tag{7}$$

where m is a real constant and

$$\Psi \in \Lambda_\mathcal{V}, \quad A_\mu \in \mathsf{T}_1\Lambda_\mathcal{V}^4, \quad B_\mu \in \mathsf{T}_1\Lambda_\mathcal{V}^2. \tag{8}$$

For the case of Minkowski space this equation was considered in [2],[3]. The equation (7) is a tensor equation. That means all values in it are tensors (differential forms) and all operations in it take tensors to tensors.

Theorem 1. *The equation (7) is invariant under the gauge transformation with the symmetry group* $\mathrm{Spin}_\mathcal{V}$

$$\Psi \to \Psi S, \quad A_\mu \to A_\mu, \quad B_\mu \to S^{-1}B_\mu S - S^{-1}\Upsilon_\mu S \quad \textit{for} \quad S \in \mathrm{Spin}_\mathcal{V}, \tag{9}$$

and also this equation is invariant under the gauge transformation with the symmetry group $\mathrm{U}(1)_\mathcal{V}$

$$\Psi \to \Psi U, \quad A_\mu \to A_\mu - U^{-1}\Upsilon_\mu U, \quad B_\mu \to B_\mu \quad \textit{for} \quad U \in \mathrm{U}(1)_\mathcal{V}. \tag{10}$$

An important role in our model play the 1-form $H \in \Lambda_\mathcal{V}^1$ such that

$$H^2 = 1, \quad \Upsilon_\mu H = [B_\mu, H], \quad \mu = 0, 1, 2, 3. \tag{11}$$

With the aid of this 1-form we define the operation of conjugation

$$\bar{U} = HU^*, \quad U \in \Lambda_\mathcal{V}.$$

Lemma . *Suppose* Ψ, A_μ, B_μ *are chosen as in (8) and*

$$C \equiv \Psi^*(dx^\mu(\Upsilon_\mu\Psi + \Psi A_\mu + \Psi B_\mu) + m\Psi I).$$

Then the conjugated differential form $\bar{C}$ *can be written as*

$$\bar{C} = ((\Upsilon_\mu\bar{\Psi} - A_\mu\bar{\Psi} - B_\mu\bar{\Psi})dx^\mu - mI\bar{\Psi})\Psi.$$

Proof is by direct calculation.

Theorem 2. *Let* Ψ, A_μ, B_μ *satisfy (7),(8) and* $j^\mu \equiv \mathrm{Tr}(\bar{\Psi}dx^\mu\Psi)$. *Then*

$$\partial_\mu(\sqrt{-g}\,j^\mu) = 0. \tag{12}$$

The identity (12) is called a *conservative law* for the equation (7). The vector j^μ is called a *current*.

Proof. It can be checked that

$$\mathrm{Tr}(H(C+C^*)) = \frac{\partial_\mu(\sqrt{-g}\,j^\mu)}{\sqrt{-g}}.$$

For a solution of the equation (7) we have $C = 0$ and so we obtain the conservative law (12). This completes the proof.

Let us define the Lagrangian from which the main equation (7) can be derived

$$\begin{aligned}
\mathcal{L} &= \mathrm{Tr}(H(C - C^*)) \\
&= \mathrm{Tr}(\bar\Psi(dx^\mu(\Upsilon_\mu\Psi + \Psi A_\mu + \Psi B_\mu) + m\Psi I) \\
&\quad -((\Upsilon_\mu\bar\Psi - A_\mu\bar\Psi - B_\mu\bar\Psi)dx^\mu - mI\bar\Psi)\Psi)
\end{aligned}$$

Note that this Lagrangian is invariant under the gauge transformations (9) and (10).

Using the variational principle [4] we suppose that the differential forms Ψ and $\bar\Psi$ are independent in the Lagrangian $\mathcal{L}$ and as variational variables we take 16 functions which are the coefficients of the differential form $\bar\Psi$. The Lagrange-Euler equations with respect to these variables give us the system of equations, which can be written in the form (7).

A gravity field can be included to the model using the same reasoning as in [1].

References

1. Marchuk N.G., A gauge model with spinor group for a description of local interaction of a fermion with electromagnetic and gravitational fields, http://xxx.lanl.gov/abs/math-ph/9912004
2. Marchuk N.G., A tensor form of the Dirac equation, http://xxx.lanl.gov/abs/math-ph/0007025
3. Marchuk N.G., Advances in Applied Clifford Algebras, v.8, N.1, (1998), p.181-225.(http://xxx.lanl.gov/abs/math-ph/9811022)
4. Bogoliubov N.N. and Shirkov D.V., Introduction to the Theory of Quantized Fields. Interscience, New York and London, 1959.

Harmonic Analysis of Dirac Operators on Lipschitz Domains

Andreas Axelsson
Centre for Mathematics and its Applications, Australian National University

René Grognard
PO Box 3635, Marsfield, NSW 2122 Australia

Jeff Hogan
Department of Mathematics, University of Arkansas

Alan McIntosh (alan@wintermute.anu.edu.au)
Centre for Mathematics and its Applications, Australian National University

Abstract. We survey some results concerning Clifford analysis and the L^2 theory of boundary value problems on domains with Lipschitz boundaries. Some novelty is introduced when using Rellich inequalities to invert boundary operators.

Keywords: Clifford analysis, Dirac operator, singular integrals, Rellich inequalities, Maxwell's equations

Math Subject Classifications: 35J55, 35Q60, 45E05

Dedicated to Richard Delanghe on the occasion of his 60th birthday.

1. Introduction

Clifford analysis has a long history and many applications as the book [2] by Brackx, Delanghe and Sommen testifies. It was introduced into the study of the L^2 boundedness of singular integrals on Lipschitz surfaces in the PhD thesis of Murray [23], written under the supervision of Raphy Coifman. She showed how Clifford analysis could be used to prove the L^2 boundedness of the double layer potential operator on surfaces with small Lipschitz constant, a method extended to all Lipschitz constants by McIntosh [15]. More direct proofs and related results were then developed in his joint papers with Li, Qian and Semmes [14], [13] and survey paper [16], as well as in work by Gilbert and Murray [9], David, Journé and Semmes [6], Gaudry, Long and Qian [8], Auscher and Tchamitchian [1], Tao [25] and others. Mitrea has made extensive contributions to this theory, and given a good presentation of the field in his book [20].

231

F. Brackx et al. (eds.), Clifford Analysis and Its Applications, 231–246.

The L^2 boundedness of the double layer potential operator was in fact proved earlier by Calderón for small Lipschitz constants [3], and by Coifman, McIntosh and Meyer [5] for all Lipschitz constants. However, as shown in some of the papers mentioned above, the use of Clifford analysis and Dirac operators gives an increased understanding of the topic. This is particularly true when we turn to other related equations such as Maxwell's equation.

There is a long tradition in applying singular integrals to the study of elliptic and parabolic boundary value problems. As well as proving the L^2 boundedness of singular integrals, one needs to show when they are invertible or at least Fredholm. Classically, invertibility is proved using Fredholm theory, but on Lipschitz domains other techniques are needed. Such tools were developed originally by Rellich, Nečas and others, and were specifically used to invert the L^2 double layer potential operator on the boundary of a Lipschitz domain by Verchota [26]. There is much related work on this topic by Dahlberg, Fabes, Jerison, Kenig, Pipher and others.

Rellich inequalities were adapted to the study of Maxwell's equations by Mitrea, Torres and Welland in [22], [21]. Clifford versions were presented by McIntosh, Mitrea and Mitrea [17], [18].

In this paper we develop a new way of applying Rellich inequalities to invert boundary operators. This method makes essential use of the full Clifford structure.

We also present an outline of the theory of the L^2 boundedness of the singular Cauchy operator on Lipschitz surfaces and of related singular integral operators. In surveying this material we will in general not make specific references to the papers mentioned above.

We have not attempted to survey research on Clifford analysis and Cauchy integrals on domains satisfying stronger smoothness conditions. See the Introduction by Ryan to [24], and the books of Gürlebeck and Sprössig [10] and [11] for more information and further references.

Acknowledgements. We thank Marius Mitrea for sharing with us his insight and knowledge of this topic.

Much of this material was developed while Hogan and McIntosh were at Macquarie University Sydney and Grognard was an Honorary Associate there. While there we received support from the Australian Government through the ARC.

2. Clifford Algebra

The functions we consider in this paper are defined on a subset of $\mathbf{R}^m$, and take their values in the complex Clifford algebra $\mathcal{A} = \mathbf{C}_{(m+1)}$ generated by the basis vectors $e_0, e_1, e_2, \ldots, e_m$ subject to the identification rule

$$e_i e_j + e_j e_i = -2\delta_{ij} , \quad 0 \le i, j \le m .$$

We expand $u \in \mathcal{A}$ as $u = \sum u_S e_S$ where $u_S \in \mathbf{C}$ and $S \subset \{0, 1, \ldots, m\}$. The Euclidean basis element $e_S = e_{j_1} e_{j_2} \cdots e_{j_s}$ if $S = \{j_1, j_2, \ldots, j_s\}$ with $0 \le j_1 < j_2 < \cdots < j_s \le m$. In particular, $e_\emptyset = 1$ and $e_{\{j\}} = e_j$.

There is a natural decomposition

$$\mathcal{A} = \wedge^0 \oplus \wedge^1 \oplus \wedge^2 \oplus \cdots \oplus \wedge^{m+1}$$

into linear subspaces $\wedge^p = \{\sum_{\#S=p} u_S e_S\}$ of elements of degree p. We decompose u into its components of degree p as $u = \sum_{p=0}^{m+1} u^{(p)}$.

The algebra $\mathcal{A}$ naturally supports a *Clifford conjugation*, acting as a complex linear anti-automorphism by

$$\overline{u} := \sum_{p=0}^{m+1} (-1)^{p(p+1)/2} u^{(p)} ,$$

and a *complex conjugation* u^c which just acts on each component of u in the basis generated by the Euclidean basis vectors by ordinary complex conjugation. Furthermore we have the sesquilinear scalar product $(u, v) = (u\overline{v}^c)_\emptyset = \sum u_S v_S^c$, which is used to define the left and right interior product, $\lrcorner$ and $\llcorner$, by

$$(u \lrcorner x, y) := (x, u^c \wedge y) , \quad (x \llcorner u, y) := (x, y \wedge u^c) .$$

When $a \in \wedge^1$ and $v \in \mathcal{A}$, the Clifford product av can be decomposed

$$av = -a \lrcorner v + a \wedge v .$$

Though the Clifford algebra is not commutative, we have the identity $(uv)_\emptyset = \sum u_S v_S e_S^2 = (vu)_\emptyset$. Another important identity is

$$(uv, w) = (v, \overline{u}^c w) = (u, w\overline{v}^c) .$$

3. Clifford Analysis of $\mathbf{D}_k$

With the Clifford algebra comes the Dirac operator $\mathbf{D} = \sum_{j=1}^m e_j \partial_j$. Since our task is to study elliptic boundary value problems arising as time harmonic solutions $f(x) \exp(-i\omega t)$ to the hyperbolic Dirac operator $-\tilde{e}_0 \partial_0 + \mathbf{D}$, $\tilde{e}_0 := -ie_0$ being the forward time direction, we introduce the *k-Dirac operator*

$$\mathbf{D}_k = \mathbf{D} + ke_0 .$$

We assume that the complex parameter k satisfies $\operatorname{Im} k \geq 0$. The Dirac operator does not preserve homogeneity of degree like the exterior derivative $d = \mathbf{D}\wedge$ and the interior derivative $\delta = -\mathbf{D}\lrcorner$, but it maps functions taking values in the even subalgebra $\mathcal{A}^{\text{even}} := \oplus \wedge^{2p}$ to functions taking values in $\mathcal{A}^{\text{odd}} := \oplus \wedge^{2p+1}$ and vice versa.

The relation between the k-Dirac and the Helmholtz operator is

$$\Delta + k^2 = -(\mathbf{D} + ke_0)^2 .$$

The elliptic operator $\mathbf{D}$ has fundamental solution $E(x) := -\frac{x}{\sigma_{m-1}|x|^m}$. Near $x = 0$, the fundamental solution F_k to $\mathbf{D}_k$ (acting either from left or right) behaves like E. When $\operatorname{Im} k > 0$, F_k has exponential decay at ∞ while when $k \in \mathbf{R} \setminus \{0\}$ it satisfies the decay condition

$$F_k(x)e^{-ik|x|} = c_{m,k}\,|x|^{-\frac{m-1}{2}}\left(\tilde{e}_0 + \frac{x}{|x|}\right) + o\left(|x|^{-\frac{m-1}{2}}\right) \quad \text{as } |x| \to \infty$$

with $c_{m,k} \neq 0$. Note that the leading term is directed along the null cone in hyperbolic space. The amplitude of the gradient has decay $o\left(|x|^{-\frac{m-1}{2}}\right)$. See [17].

An explicit expression for F_k is obtained by applying $-\mathbf{D}_k$ to the fundamental solution of the Helmholtz operator, the Bessel potential $B_k(x)$. When $m = 3$, we have $B_k(x) = -\frac{e^{ik|x|}}{4\pi|x|}$ and readily obtain

$$F_k(x) = \left(-\frac{x}{|x|^2} + ik(\tilde{e}_0 + \frac{x}{|x|})\right)\frac{e^{ik|x|}}{4\pi|x|} .$$

In $\mathbf{R}^m$ we will consider Ω^+, being either a bounded domain with strongly Lipschitz boundary Σ, or the region above a Lipschitz graph

$$\Sigma = \{(x', x_m) \; ; \; x_m = \phi(x')\} .$$

Let $\Omega^- := \mathbf{R}^m \setminus (\Omega^+ \cup \Sigma)$. Recall that $\phi : \mathbf{R}^{m-1} \to \mathbf{R}$ being Lipschitz means $|\phi(x') - \phi(y')|/|x' - y'|$ is uniformly bounded, while for a bounded domain, strongly Lipschitz means that for each $y \in \Sigma$ there exists a neighbourhood U_y and a Lipschitz graph Σ_y dividing $\mathbf{R}^m$ into $\Omega_y^{\pm}$ (with respect to some Euclidean coordinate system) such that $\Sigma \cap U_y = \Sigma_y \cap U_y$ and $\Omega^{\pm} \cap U_y = \Omega_y^{\pm} \cap U_y$.

Concerning integral manipulations, the following version of Stokes' theorem, referred to as *the boundary theorem* should be noted.

An integral over Σ, where the integrand contains the outward pointing normal n linearly, equals the integral over Ω^+ with n replaced by $\mathbf{D}$.

In Clifford analysis the most frequently used integrand is gnf, where the boundary theorem tells us that

$$\int_\Sigma g(x)n(x)f(x)\,d\sigma(x) = \int_{\Omega^+} \Big((g\mathbf{D})(x)f(x) + g(x)(\mathbf{D}f)(x)\Big)\,dx \ .$$

Now, consider a k-*monogenic* function f in Ω^+, i.e. a solution to $\mathbf{D}_k f(x) = 0$ there. Applying the identity above with $g(x) = E_k(x) := -F_k(-x)$ (since we need $E_k(\mathbf{D} - ke_0) = \delta_0$), we obtain the reproducing formula

$$f(x) = \int_\Sigma E_k(y - x)n(y)f(y)\,d\sigma(y) \ , \quad x \in \Omega^+. \tag{1}$$

For this to work, we need f to be sufficiently nice up to the boundary.

DEFINITION 3.1. *The Cauchy extension of $f \in L^2(\Sigma)$ is*

$$Cf(x) := \int_\Sigma E_k(y - x)n(y)f(y)\,d\sigma(y) \ , \quad x \in \Omega^+ \cup \Omega^- \ .$$

Let $C^\pm f := Cf|_{\Omega^\pm}$ and in the graph case write $C_\tau f(y) := Cf(y + \tau e_m)$, $\tau \in \mathbf{R} \setminus \{0\}$. The Hardy projections $P^+ f$ and $P^- f$ are the boundary values of $C^+ f$ and of $-C^- f$ respectively in the $L^2(\Sigma)$ sense. The ranges of the projections are called Hardy spaces and will be denoted $P^\pm L^2$. The principal value Cauchy integral is

$$C_\Sigma f(x) = 2\,\text{p.v.} \int_\Sigma E_k(y - x)n(y)f(y)\,d\sigma(y) \ , \quad x \in \Sigma \ .$$

We spend the rest of this and the next section investigating the properties of C, outlining the proof that $P^\pm$ are well defined and bounded complementary projections. Assume here that Σ is a graph and $k = 0$. The case of a bounded domain can be obtained from this, as can the case of a general k since C_Σ^k is a compact perturbation of C_Σ.

PROPOSITION 3.2. *For any $f \in L^2(\Sigma)$*

$$f = \lim_{\tau \to 0^+} (C_\tau f - C_{-\tau} f)$$

both in L^2 and pointwise a.e. Indeed, the difference kernel $K_\tau(x, y) := (E(y - (x + \tau e_m)) - E(y - (x - \tau e_m)))n(y)$ is an approximate unit.

  Proof. Use the estimate $|K_\tau(x, y)| \lesssim \tau/|y - (x + \tau e_m)|^m$ and the fact that $\int_\Sigma K_\tau(x, y)\,d\sigma(y) = 1$ by the boundary theorem.

COROLLARY 3.3. *For any $f \in L^2(\Sigma)$ and $t > 0$, we have*

$$C^-(C_t f) = 0$$
$$C^+(C_t f)(y + \tau e_m) = (C^+ f)(y + (\tau + t)e_m) \ , \quad \tau > 0, \ y \in \Sigma \ .$$

Similar statements holds for $t < 0$.

Proof. Having the estimate

$$|E(y - (x + (t + \tau)e_m)) - E(y - (x + te_m))| \lesssim \frac{\tau}{|y - (x + te_m)|^m} \, ,$$

it follows that $C_t f = \lim_{\tau \to 0^+} C_\tau(C_t f)$. From the proposition we get $\lim_{\tau \to 0^-} C_\tau(C_t f) = 0$. But $C^-(C_t f)$ being monogenic and zero on a hypersurface implies $C^-(C_t f) = 0$ in all Ω^-.

For the second identity, observe that the two sides are equal when $\tau = 0$, which implies equality for all $\tau > 0$ as above.

4. L^2 Boundedness of Hardy Projections

The difficult thing with the Hardy projections is to show that $L^2 = P^+ L^2 \oplus P^- L^2$. By Proposition 3.2 it is enough to show that we have convergence in L^2 for $C_\tau f$ as $\tau \to 0^+$. For this we need uniform L^2 bounds on $C_\tau f$. For $\Sigma = \mathbf{R}^{m-1}$, this follows easily from Fourier theory, since $\hat{E} \in L^\infty$. For a general Lipschitz surface Σ one can proceed as in [14], where the two dimensional case in [5] is generalised to $\mathbf{R}^m$ following [4]. Below we outline very briefly the main steps of the proof in [14] of the needed uniform L^2 estimates. The key ingredient is the following square function estimate, where

$$\|G\|_+^2 := \int_{\Omega^+} |G(x)|^2 \text{dist}(x, \Sigma) \, dx \, .$$

PROPOSITION 4.1. *If G is monogenic in Ω^+ and continuous up to Σ with boundary trace g and satisfies estimates $|G(x)| \leq C_G/(1 + |x|)^{m-1}$ and $|\nabla G(x)| \leq C_G/(1 + |x|)^m$, $x \in \Omega^+$ (e.g. if $g = C_\tau f$ and $f \in L^2(\Sigma)$ has compact support), then*

$$\|g\|_{L^2(\Sigma)} \lesssim \|\nabla G\|_+ \, ,$$

independently of the constant C_G.

With the analogous theorem for right monogenic functions in Ω^-, together with Schur estimates, we prove the following lemma.

LEMMA 4.2. *Let $H \in C_0^\infty(\Omega^+; \mathcal{A})$ and*

$$S_{\tau,j} H(y) := \int_{\Omega^+} \overline{H(x)} \partial_j E(y - (x + \tau e_m)) \text{dist}(x, \Sigma) \, dx, \quad y \in \Sigma \, .$$

Then $\|S_{\tau,j} H\|_{L^2(\Sigma)} \lesssim \|H\|_+$, with constant independent of $\tau > 0$.

Now, combining the proposition and the lemma, using duality and Fubini's theorem, we obtain the desired estimate

$$\|C_\tau f\|_{L^2(\Sigma)} \lesssim \|\nabla C(C_\tau f)\|_+$$

$$\lesssim \sum_{j=1}^m \sup_{\|H_j\|_+ \leq 1} \int_{\Omega^+} \left(\frac{\partial F}{\partial x_j}(x+\tau e_m), H_j(x)\right) \text{dist}(x, \Sigma)\, dx$$

$$\lesssim \sum_{j=1}^m \sup_{\|H_j\|_+ \leq 1} \int_\Sigma \left(S_{\tau,j} H_j(y), \overline{f}(y)\overline{n}(y)\right) d\sigma(y)$$

$$\lesssim \|f\|_{L^2(\Sigma)} \tag{2}$$

for all compactly supported $f \in L^2(\Sigma)$, with a constant independent of f and τ. By Fatou's lemma this can be extended to all $f \in L^2(\Sigma)$.

Recall that *the non-tangential maximal function* of $F : \Omega^+ \to \mathcal{A}$ is

$$\mathcal{N}F(y) := \sup_{x \in y+\Gamma} |F(x)|, \quad y \in \Sigma,$$

where Γ is an open infinite cone with apex at 0 and axis along e_m. The angle of the cone is chosen small enough that $y + \Gamma \subset \Omega^+$ for all $y \in \Sigma$.

PROPOSITION 4.3. *The estimate* $\|\mathcal{N}C^\pm f\|_{L^2(\Sigma)} \lesssim \|f\|_{L^2(\Sigma)}$ *holds for all* $f \in L^2(\Sigma)$.

Proof. Use Corollary 3.3 to obtain

$$\|\mathcal{N}(C^+(C_t f - C_{-t}f))\| = \|\mathcal{N}(C^+(C_t f))\|$$
$$\lesssim \|\mathcal{M}(C_t f)\| \lesssim \|C_t f\| \lesssim \|f\|.$$

The bound of the non-tangential maximal function $\mathcal{N}$ by the Hardy–Littlewood maximal function $\mathcal{M}$ comes from the fact that we can estimate K_τ in Proposition 3.2 by $\tau/|y - (x + \tau e_m)|^m$.

To pass to the limit with t, observe that $\mathcal{N}(C^+(C_t f - C_{-t}f))$ is increasing as $t \to 0^+$ and apply the monotone convergence theorem.

Note that this proposition not only gives control of Cf near Σ but also shows that $C_\tau f \to 0$ in L^2 as $\tau \to \pm\infty$.

THEOREM 4.4. *For any* $f \in L^2(\Sigma)$, *boundary values to* $C^\pm f$ *exist both in* $L^2(\Sigma)$ *and pointwise non-tangentially a.e., so* $\{P^\pm\}$ *are complementary projections in* $L^2(\Sigma)$, *i.e. bounded operators satisfying*

$$(P^\pm)^2 = P^\pm, \quad P^+P^- = 0 = P^-P^+, \quad and \quad P^+ + P^- = I.$$

Thus we have a topological splitting $L^2(\Sigma) = P^+L^2 \oplus P^-L^2$. *We also have* $P^+ - P^- = C_\Sigma$ *and the Plemelj–Sokhotskij jump formulae* $P^\pm =$

$\frac{1}{2}(I \pm C_\Sigma)$.

Proof. Let $f_t = C_t f - C_{-t} f$ and observe that for $C^\pm f_t$ we have boundary values both in L^2 and non-tangentially a.e. Write

$$\|C_\tau f - C_\sigma f\| \leq \|C_\tau(f - f_t)\| + \|C_\tau(f_t) - C_\sigma(f_t)\| + \|C_\sigma(f_t - f)\| \ .$$

Choosing t small and using (2) we can make the first and last term small, and then choosing τ and σ small gives the middle term small. Using the non-tangential maximal function one can similarly prove that pointwise non-tangential boundary values exist a.e.

Finally a remark on the ranges of $C^\pm$. We have seen that $C^+ f$ is a monogenic function in Ω^+ with $\mathcal{N}C^+ f \in L^2(\Sigma)$. The converse is also true, as can be proved with a limiting argument from the reproducing formula (1). A similar result holds in the case of a bounded domain if we use truncated cones for the boundary behaviour and an appropriate radiation condition at infinity depending on k. See [17].

5. Duality

A bilinear pairing $\langle \mathcal{K}, \mathcal{H} \rangle$ between two Hilbert spaces is called a duality if the estimates

$$|\langle k, h \rangle| \lesssim \|k\| \|h\| \ , \quad \|h\| \lesssim \sup_{\|k\|=1} |\langle k, h \rangle| \quad \text{and} \quad \|k\| \lesssim \sup_{\|h\|=1} |\langle k, h \rangle|$$

hold. Denote the adjoint operator of $T : \mathcal{H} \to \mathcal{H}$ relative to this duality by $T' : \mathcal{K} \to \mathcal{K}$.

The relevant duality here is not the $L^2(\Sigma)$ scalar product but rather the weighted duality

$$\langle g, f \rangle := \int_\Sigma (g(y)n(y)f(y))_0 \, d\sigma(y) \ .$$

Here $g \in \overleftarrow{L}^2(\Sigma) = \mathcal{K}$ and $f \in \overrightarrow{L}^2(\Sigma) = \mathcal{H}$, the two spaces being just two identical copies of $L^2(\Sigma)$. Up to now we have made use of $\mathbf{D}_k = \mathbf{D} + k e_0$ acting from the left with Cauchy extensions C, using E_k in the kernel, and Hardy projections $P^\pm$. All this takes place in $\overrightarrow{L}^2$. To emphasise this we sometimes overline these operators, e.g. $\mathbf{D}_k = \overrightarrow{\mathbf{D}}_k$. We may equally well do the same thing in $\overleftarrow{L}^2$. Here we have a Dirac operator $\overleftarrow{\mathbf{D}}_k = \mathbf{D} + k(-e_0)$ acting from the right with Cauchy extension

$$\overleftarrow{C} f(x) = \int_\Sigma f(y)n(y)\overleftarrow{E}_k(y - x) \, d\sigma(y) \ , \quad x \in \Omega^+ \cup \Omega^- \ ,$$

using $\overleftarrow{E}_k(x) := F_k(x) = -\overrightarrow{E}_k(-x)$, The boundary values $I \pm \overleftarrow{C}_\Sigma$ define Hardy projections $\{\overleftarrow{P}^\pm\}$. All the theory in Sections 3 and 4 goes through for these operators in $\overleftarrow{L}^2$, mutatis mutandis.

LEMMA 5.1. *The dualities* $(\overrightarrow{P}^\pm)' = \overleftarrow{P}^\mp$ *hold.*

Proof. Since $P^\pm = \frac{1}{2}(I \pm C_\Sigma)$ it is enough to prove $(\overrightarrow{C}_\Sigma)' = -\overleftarrow{C}_\Sigma$. Formally this follows from the calculation

$$\langle g, \overrightarrow{C}_\Sigma f \rangle = \int_\Sigma \left(g(x)n(x) \int_\Sigma \overrightarrow{E}_k(y-x)n(y)f(y)\, d\sigma(y) \right)_\emptyset d\sigma(x)$$

$$= \int_\Sigma \left(\int_\Sigma g(x)n(x)(-\overleftarrow{E}_k(x-y))d\sigma(x)\ n(y)f(y) \right)_\emptyset d\sigma(y)$$

$$= \langle -\overleftarrow{C}_\Sigma g, f \rangle .$$

6. Splittings of $L^2(\Sigma)$ and Rellich estimates

We now discuss the elliptic boundary value problem for $\mathbf{D}_k$ which was hinted at before. For an elliptic operator of order d one should impose $d/2$ boundary conditions to get a well-posed problem. For $\mathbf{D}_k$ this means half a boundary condition should do.

DEFINITION 6.1. *Let* $Q^\pm : L^2(\Sigma) \to L^2(\Sigma)$ *be the projections*

$$Q^\pm f := \frac{1}{2}(f \pm nfn) .$$

We denote the range of $Q^\pm$ *by* $Q^\pm L^2$ *and set* $\overleftarrow{Q}^\pm = \overrightarrow{Q}^\pm := Q^\pm$.

In case $f \in \mathcal{A}^{\mathrm{odd}}$, then Q^+f is the tangential part of f while Q^-f is the normal part of f, the situation being reversed when $f \in \mathcal{A}^{\mathrm{even}}$. Note that $(\overrightarrow{Q}^\pm)' = \overleftarrow{Q}^\pm$.

The two pairs of complementary projections $\{P^\pm\}$ and $\{Q^\pm\}$ induce the two splittings

$$L^2(\Sigma) = P^+L^2 \oplus P^-L^2 = Q^+L^2 \oplus Q^-L^2 .$$

It is interesting to investigate the geometric relation between them.

DEFINITION 6.2. *Assume* $\{A^\pm\}$ *is a pair of complementary projections on a Hilbert space* $\mathcal{H}$. *We say that a bounded projection* B *is transversal to* $\{A^\pm\}$ *if* $A^+ : B(\mathcal{H}) \to A^+(\mathcal{H})$ *and* $A^- : B(\mathcal{H}) \to A^-(\mathcal{H})$ *are both isomorphisms.*

If the restricted projections are Fredholm with index zero rather than isomorphisms, we say that B *is 0-transversal to* $\{A^\pm\}$.

If B is transversal to $\{A^\pm\}$, then in particular

$$\|A^+x\| \approx \|x\| \approx \|A^-x\| , \quad x \in B(\mathcal{H}) .$$

The case that we have in mind is the following fact concerning the comparability of the normal and tangential part of a k-monogenic function.

THEOREM 6.3. *Let Ω^+ be a bounded open subset of $\mathbf{R}^m$ with strongly Lipschitz boundary. Then there exists a discrete set $S \subset \mathbf{R}$ such that $P^\pm$ is transversal to $\{Q^\pm\}$ for $k \notin S$, and is 0-transversal to $\{Q^\pm\}$ when $k \in S$.*

In case of a Lipschitz graph Σ and $k = 0$, $P^\pm$ is transversal to $\{Q^\pm\}$.

All these statements also hold when the P's and Q's switch roles.

This is a way of stating the well-posedness of the boundary value problem consisting in finding a function F in Ω^+ with boundary trace $f \in L^2(\Sigma)$ such that

$$\begin{cases} \mathbf{D}_k F = 0 & \text{in } \Omega^+ \\ Q^+ f = g \in Q^+ L^2 & \text{on } \Sigma . \end{cases} \tag{3}$$

By Theorem 6.3, if $k \notin S$, there exists a unique solution $f \in P^+L^2$ such that $Q^+f = g$ and thus a unique solution F to (3). If $k \in S$ the result still holds modulo finite dimensions.

Note that the null space of $Q^+|_{P^+L^2}$ is $P^+L^2 \cap Q^-L^2$, it being $\{0\}$ when $k = 0$ and Σ is a graph since Ω^+ then has trivial topology. When $k = 0$ and Ω^+ is bounded, the dimension of this intersection depends on the topology of the domain. See [19] for the classical boundary integral operators.

We begin to sketch the proof of Theorem 6.3 by presenting three abstract lemmata to make the logic more transparent. First we recall the following well-known result relating a priori estimates with the semi–Fredholm property, i.e. having finite dimensional nullspace and closed range.

LEMMA 6.4. *Let $\mathcal{X}$, $\mathcal{Y}$ and $\mathcal{Z}$ be Banach spaces, $T : \mathcal{X} \to \mathcal{Y}$ be bounded, and $K : \mathcal{X} \to \mathcal{Z}$ be compact. Assume the a priori estimate*

$$\|x\|_{\mathcal{X}} \lesssim \|Tx\|_{\mathcal{Y}} + \|Kx\|_{\mathcal{Z}} , \quad x \in \mathcal{X} .$$

Then T is a semi-Fredholm operator.

LEMMA 6.5. *Assume a Hilbert space $\mathcal{H}$ splits as*

$$\mathcal{H} = A^+(\mathcal{H}) \oplus A^-(\mathcal{H}) = B^+(\mathcal{H}) \oplus B^-(\mathcal{H})$$

*with respect to two pairs of complementary projections $\{A^\pm\}$ and $\{B^\pm\}$.
Then a priori estimates for the four restricted projections $A^\pm|_{B^\pm(\mathcal{H})}$
imply estimates for the other four $B^\pm|_{A^\pm(\mathcal{H})}$. If the first four estimates
are strict, i.e. the compact terms are zero, then so are the other four.*

Proof. Assume we have a priori estimates

$$\|u\| \lesssim \|A^\pm u\| + \|K^\pm u\|, \quad u \in B^+(\mathcal{H})$$
$$\|u\| \lesssim \|A^\pm u\| + \|L^\pm u\|, \quad u \in B^-(\mathcal{H})$$

where $K^\pm$ and $L^\pm$ are compact operators. Then by decomposing $A^+(\mathcal{H}) \ni$
$u = B^+u + B^-u$ and observing that $A^-B^+u + A^-B^-u = 0$ we get

$$\|u\| \leq \|B^+u\| + \|B^-u\| \lesssim \|B^+u\| + \|A^-B^-u\| + \|L^-B^-u\|$$
$$= \|B^+u\| + \|A^-B^+u\| + \|L^-B^-u\| \lesssim \|B^+u\| + \|L^-B^-u\|.$$

Last we give a duality lemma for pairs of projections. Here we suppose
that there is a duality $\langle \mathcal{K}, \mathcal{H} \rangle$ between $\mathcal{H}$ and another Hilbert space $\mathcal{K}$.

LEMMA 6.6. *Assume $\mathcal{H}$ splits in two ways as in the previous lemma.
Then $\mathcal{K} = (A^+)'(\mathcal{K}) \oplus (A^-)'(\mathcal{K}) = (B^+)'(\mathcal{K}) \oplus (B^-)'(\mathcal{K})$. If*

$$A^+ : B^+(\mathcal{H}) \longrightarrow A^+(\mathcal{H})$$
$$(A^-)' : (B^-)'(\mathcal{K}) \longrightarrow (A^-)'(\mathcal{K})$$

*both satisfy a priori estimates, then they are Fredholm operators.
    If the estimates are strict, then they are both isomorphisms.
    More generally, if there exists an isomorphism $j : A^-(\mathcal{H}) \cap B^+(\mathcal{H}) \to (A^+)'(\mathcal{K}) \cap (B^-)'(\mathcal{K})$, then A^+ and $(A^-)'$ both have index 0.*

Proof. Observe that $\langle \mathcal{K}, \mathcal{H} \rangle$ restricts to a (non-degenerate) duality

$$\langle (A^+)'(\mathcal{K}) \cap (B^-)'(\mathcal{K}), \ A^+(\mathcal{H}) \ominus A^+B^+(\mathcal{H}) \rangle.$$

This shows that the dimension of the cokernel of $A^+|_{B^+(\mathcal{H})}$ equals that
of the kernel of $(A^-)'|_{(B^-)'(\mathcal{K})}$. An isomorphism j shows that the latter
dimension equals the dimension of the kernel of $A^+|_{B^+(\mathcal{H})}$. Thus it has
index zero. Similarly for $(A^-)'|_{(B^-)'(\mathcal{K})}$.

The application we have in mind is $A^\pm = Q^\pm$ and $B^\pm = P^\pm$ and
the duality is $\langle \overleftarrow{L}^2, \overrightarrow{L}^2 \rangle$ as above with $j(f) = \overline{f}e_0$. Note that j maps
$\overrightarrow{P}^\pm L^2$ to $\overleftarrow{P}^\pm L^2$ and $\overrightarrow{Q}^\pm L^2$ to $\overleftarrow{Q}^\mp L^2$ and thus satisfies the condition
of Lemma 6.6.
    We can now apply the three lemmata if we prove the Rellich-type a
priori estimates for $Q^\pm|_{P^\pm L^2}$.

PROPOSITION 6.7. *Let Ω^+ be bounded. If $f \in P^+L^2$, then*

$$\|f\| \lesssim \|Q^\pm f\| + (|k| + 1)\|Cf\|_{L^2(U \cap \Omega^+)} ,$$

where U denotes a neighbourhood of Σ with compact closure. The operator C is compact from $L^2(\Sigma)$ to $L^2(U \cap \Omega^+)$.

The same estimate holds if $f \in P^-L^2$ and $U \cap \Omega^-$ replaces $U \cap \Omega^+$.

Proof. Following [17] and [18], the proof uses the commutation properties of $\mathcal{A}$ and the boundary theorem. Take $\theta \in C_0^1(U; \wedge_{\mathbf{R}}^1)$ with the property that $(n, \theta) \geq c > 0$ on Σ. We get for $f \in P^\pm L^2$

$$|f|^2(n, \theta) = \tfrac{1}{2}(f, f(n\theta + \theta n)) = -\tfrac{1}{2}\big((f\theta, fn) + (fn, f\theta)\big)$$

$$= -\operatorname{Re}(fn, f\theta) = -\operatorname{Re}(2(Q^\pm f)n, f\theta) \mp \operatorname{Re}(nf, f\theta) .$$

Integrating over Σ, using the boundary theorem on the expression $(nf, f\theta)$ in the second term and writing $F = Cf$, we obtain

$$\|f\|^2 \lesssim \|Q^\pm f\|\|f\|$$

$$+ \left| \int_{\Omega^\pm} \big((-ke_0F, F\theta) - (F, -ke_0F\theta) - (F, \mathbf{D}F\dot\theta)\big) \, dx \right| ,$$

where the dot indicates that $\mathbf{D}$ only acts there. Using the inequality $ab \leq \frac{1}{2\epsilon}a^2 + \frac{\epsilon}{2}b^2$ on the first term with suitable ϵ we obtain the estimate.

That the perturbation term is compact can be shown by using Schur's test with suitable exponents.

Proof of Theorem 6.3:

Bounded Σ, $\operatorname{Im} k \geq 0$: Applying Proposition 6.7, Lemmata 6.5 and 6.6 gives that $P^\pm$ is 0-transversal to $\{Q^\pm\}$ and vice versa.

Bounded Σ, $\operatorname{Im} k > 0$: Here we can use the boundary theorem on (nf, e_0f) to eliminate the compact term. Calculating, we get

$$\int_\Sigma (nf, e_0f) = \int_{\Omega^+} \big((-ke_0F, e_0F) - (F, (-e_0)(-ke_0F))\big)$$

$$= (-2i)\operatorname{Im} k \int_{\Omega^+} |F|^2$$

and $(nf, e_0f) = 2i\operatorname{Im}(Q^\pm f, e_0nf)$. After some algebra this yields the strict a priori estimate

$$\|f\| \lesssim \frac{|k| + 1}{\operatorname{Im} k}\|Q^\pm f\| ,$$

which shows that $P^\pm$ is transversal to $\{Q^\pm\}$ and vice versa via Lemmata 6.5 and 6.6.

Bounded Σ, discreteness of S: This follows from analytic Fredholm theory applied to $Q^{\pm}P^{\pm} : Q^{\pm} \to Q^{\pm}$.

Graph Σ, $k = 0$: The proof of Proposition 6.7 here works with $\theta = e_m$ and gives a strict a priori estimate directly. This implies the transversality as above.

7. Harmonic functions

Rellich's original estimate was formulated as the comparable size in $L^2(\Sigma)$ of the normal and tangential derivatives of a harmonic function $u : \Omega^+ \to \mathbf{R}$. This result follows from the integral identity arising in the proof of Proposition 6.7 by regarding u as a map into $\wedge^0$, letting

$$F = \mathbf{D}u : \Omega^+ \to \wedge^1 ,$$

and noting that F is monogenic, that $Q^+ f$ is the tangential derivative of u, and that $Q^- f$ is the normal derivative of u.

Many integral identities under the name of Rellich in the literature can be derived from the results of Section 6 in a similar way. Applications also include the known estimates [26] for the acoustic double-layer potential operator

$$K_{\Sigma}\phi(x) := 2\,\mathrm{p.v.} \int_{\Sigma} (E_k(y-x), n(y))\phi(y)\,d\sigma(y) = (C_{\Sigma}\phi)_0(x) , \quad x \in \Sigma ,$$

where $\phi \in L^2(\Sigma; \mathbf{C})$.

PROPOSITION 7.1. *The operators $I \pm K_{\Sigma}$ are bounded Fredholm maps on $L^2(\Sigma; \mathbf{C})$ with index 0. When $\mathrm{Im}\, k > 0$ they are isomorphisms.*

Proof. Identify $\mathbf{C}$ with $\wedge^0 \subset \mathcal{A}$. Straightforward calculations give

$$(I \pm K_{\Sigma}^*)\psi = -2nQ^- P_{(-k^c)}^{\pm}(n\psi) , \quad \psi \in L^2(\Sigma; \wedge^0) .$$

Thus Theorem 6.3 gives a priori estimates for $I \pm K_{\Sigma}^*$. Furthermore, as in [7], it can be shown that $\lambda I \pm K_{\Sigma}^*$ is also semi-Fredholm for $|\lambda| \geq 1$, $\lambda \in \mathbf{R}$. It is an isomorphism when $|\lambda| > \|K_{\Sigma}^*\|$, so by general perturbation theory the index is 0.

8. Maxwell's Equation

We conclude by briefly describing how the above theory can be applied to Maxwell's equations

$$\mathbf{D} \wedge B = 0$$
$$\partial_0 B + \mathbf{D} \wedge E = 0$$
$$\partial_0 D + \mathbf{D} \lrcorner H = -J$$
$$\mathbf{D} \lrcorner D = \rho \,,$$

where $E = \epsilon^{-1} D = E_1 e_1 + E_2 e_2 + E_3 e_3$ and $B = \mu H = B_1 e_{23} + B_2 e_{31} + B_3 e_{12}$. From bottom to top these are Gauss' law, Ampère–Maxwell's law, Faraday's law and the magnetic Gauss' law and they take scalar, vector, bivector and trivector values respectively. Taking energy as unit, the physically natural quantity to work with is the electromagnetic field

$$f(t, x) = \epsilon^{1/2} \tilde{e}_0 E + \mu^{-1/2} B \,.$$

As in [12] we observe that Maxwell's equations can be written entirely with Clifford algebra (with time and space dependent ϵ and μ) as

$$(-\tfrac{1}{c} \tilde{e}_0 \partial_0 + \mathbf{D}) f(t, x) + Rf(t, x) = j \,,$$

where the speed of propagation is $c(t, x) := (\epsilon(t, x) \mu(t, x))^{-1/2}$. The zero order term is

$$Rf := T(f)(-\tfrac{1}{c} \tilde{e}_0 \partial_0 + \mathbf{D}) \ln(\epsilon^{-1/2}) + S(f)(-\tfrac{1}{c} \tilde{e}_0 \partial_0 + \mathbf{D}) \ln(\mu^{1/2}) \,,$$

where $T(f) := \tfrac{1}{2}(f - \tilde{e}_0 f \tilde{e}_0) = \epsilon^{1/2} \tilde{e}_0 E$ and $S(f) := \tfrac{1}{2}(f + \tilde{e}_0 f \tilde{e}_0) = \mu^{-1/2} B$, and the "four-current" is $j := \tilde{e}_0 \epsilon^{-1/2} \rho + \mu^{1/2} J$.

Here we will just consider the case when the coefficients are constant on Ω^+ or Ω^- and $j = 0$ there. Then Maxwell's equation takes the form

$$(-\tfrac{1}{c} \tilde{e}_0 \partial_0 + \mathbf{D}) f(t, x) = 0 \,,$$

so that time harmonic solutions $f(t, x) = e^{-i\omega t} f(x)$ are k-monogenic, where $k = \omega/c$ is the wave number.

In solving the boundary value problem (3) in $\mathbf{R}^3$ for Maxwell's equation, the Rellich estimates do not completely solve the problem since we have a constraint on f requiring that it should take values only in $\wedge^2$. Let us see what necessary conditions on $g := Q^+ f = n(n \lrcorner f)$ we have when $f \in P^+ L^2$ and $C^+ f$ takes values in $\wedge^2$. Of course $g \in Q^+ L^2(\Sigma; \wedge^2)$, but we also get, after an integration by parts, that

$$0 = (C^+ f)_\emptyset(x) = \int_\Sigma (E_k(y - x) n(y) f(y))_\emptyset \, d\sigma(y)$$

$$= \int_\Sigma B_k(y - x)(-\delta_\partial(n \lrcorner g)(y) + k e_0 \lrcorner (n \lrcorner g)(y)) \, d\sigma(y),$$

where $B_k(x) = -\frac{e^{ik|x|}}{4\pi|x|}$. Here we have defined the operator δ_∂ in $L^2(\Sigma)$ via duality by

$$\int_\Sigma (\delta_\partial(n \lrcorner g), \Phi|_\Sigma) = \int_\Sigma (n \lrcorner g, (d\Phi)|_\Sigma)$$

for all C^1 functions Φ in a neighbourhood of Σ. Note that this operator only is well-defined for tangential functions and as the notation suggests it is closely related to the usual interior derivative δ_Σ. Now, varying x over Ω^+, we conclude that $\delta_\partial(n \lrcorner g) = ke_0 \lrcorner (n \lrcorner g)$, at least if $\operatorname{Im} k > 0$.

THEOREM 8.1. *If* $\operatorname{Im} k > 0$, $g \in L^2(\Sigma, \wedge^2)$, $n \wedge g = 0$ *and* $\delta_\partial(n \lrcorner g) = ke_0 \lrcorner (n \lrcorner g)$, *then the boundary value problem* (3) *in* $\mathbf{R}^3$ *has a unique solution*

$$F : \Omega^+ \longrightarrow \wedge^2 .$$

Moreover, its boundary trace f satisfies $\|f\|_{L^2(\Sigma)} \lesssim \|g\|_{L^2(\Sigma)}$.

Similar results are true modulo finite dimensions when k is real.

Proof. If $\operatorname{Im} k > 0$, Theorem 6.3 gives a unique $F : \Omega^+ \to A$ with normal boundary trace g. To show that F maps into $\wedge^2$, decompose $f = f^{(0)} + f^{(1)} + f^{(2)} + f^{(3)} + f^{(4)}$, where $f^{(p)} \in \wedge^p$. Since $\mathbf{D}_k$ switches $\mathcal{A}^{\text{even}}$ and $\mathcal{A}^{\text{odd}}$ it follows that $f^{\text{odd}} = f^{(1)} + f^{(3)} \in P^+L^2$. Since $Q^+ f^{\text{odd}} = 0$ we get $f^{(1)} = f^{(3)} = 0$.

By the differentiability condition on g, $\frac{1}{2}(I + K_\Sigma)f^{(0)} = f^{(0)} - (P^+g)_\emptyset = f^{(0)}$, and thus an application of Proposition 7.1 gives $f^{(0)} = 0$. Furthermore $f^{(4)} = 0$ simply because Q^+ is injective on $L^2(\Sigma; \wedge^4)$. Thus $f = f^{(2)} \in P^+L^2$, and this implies that $F : \Omega^+ \longrightarrow \wedge^2$.

References

1. P. Auscher, P. Tchamitchian, *Bases d'ondelettes sur les courbes corde-arc, noyau de Cauchy et espaces de Hardy associés*, Rev. Mat. Iber. 5 (1989), 139-170.
2. F. Brackx, R. Delanghe, F. Sommen, *Clifford Analysis*, Pitman Advanced Publ. Program, London, 1982.
3. A.P. Calderón, *Cauchy integrals on Lipschitz curves and related operators*, Proc. Natl. Acad. Sci. USA 74 (1977), 1324-1327.
4. R.R. Coifman, P. Jones, S. Semmes, *Two elementary proofs of the L^2 boundedness of Cauchy integrals on Lipschitz curves*, J. of Amer. Math. Soc. 2 (1989), 553-564.
5. R.R. Coifman, A. McIntosh, Y. Meyer, *L'intégral de Cauchy définit un opérateur borné sur L^2 pour les courbes Lipschitziennes*, Annals of Math., 116 (1982), 361-387.

6.   G. David, J.-L. Journé, S. Semmes, *Opérateurs de Calderón–Zygmund, fonctions para-accrétives et interpolation*, Rev. Mat. Iber. 1 (1985), 1-57.

7.   L. Escauriaza, E.B. Fabes, G. Verchota, *On a regularity theorem for weak solutions to transmission problems with internal Lipschitz boundaries*, Proc. Amer. Math. Soc. 115 (1992), 1069-1076.

8.   G.I. Gaudry, R.L. Long, T. Qian, *A martingale proof of L^2-boundedness of Clifford-valued singular integrals*, Annali Matematica, Pura Appl. 165 (1993), 369-394.

9.   J. Gilbert, M. A. Murray, *Clifford Algebras and Dirac Operators in Harmonic Analysis*, Cambridge Studies in Advanced Mathematics, 1991.

10.   K. Gürlebeck, W. Sprössig, *Quaternionic Analysis and Elliptic Boundary Value Problems*, Birkhäuser Verlag, Basel, 1990.

11.   K. Gürlebeck, W. Sprössig, *Quaternionic and Clifford Calculus for Physicists and Engineers*, Wiley, Chichester, 1997.

12.   B. Jancewicz, *Multivectors and Clifford Algebra in Electrodynamics*, World Sci. Publ., Singapore, 1988.

13.   C. Li, A. McIntosh, T. Qian, *Clifford algebras, Fourier transforms and singular convolution operators on Lipschitz surfaces*, Rev. Mat. Iber., 10 (1994), 665-721.

14.   C. Li, A. McIntosh, S. Semmes, *Convolution singular integrals on Lipschitz surfaces*, J. Amer. Math. Soc. 5 (1992), no. 3, 455-481.

15.   A. McIntosh, *Clifford algebras and the higher dimensional Cauchy integral*, Approximation and Function Spaces, Banach Center Publ., Warsaw, vol. 22 (1989), 253-267.

16.   A. McIntosh, *Clifford algebras, Fourier theory, singular integrals and harmonic functions on Lipschitz domains*, in: Clifford Algebras in Analysis and Related Topics (J. Ryan ed.), 33-88, CRC Press, Boca Raton, FL, 1996.

17.   A. McIntosh, M. Mitrea, *Clifford algebras and Maxwell's equations in Lipschitz domains*, Math. Meth. Appl. Sci., 22 (1999), 1599-1620.

18.   A. McIntosh, D. Mitrea, M. Mitrea, *Rellich type estimates for one-sided monogenic functions in Lipschitz domains and applications*, in: Analytical and Numerical Methods in Quaternionic and Clifford Algebras, K. Gürlebeck and W. Sprössig eds, the Proceedings of the Seiffen Conference, Germany, 1996, pp. 135-143.

19.   D. Mitrea, *The method of layer potentials for non-smooth domains with arbitrary topology*, Integral Equations Operator Theory 29 (1997), 320-338.

20.   M. Mitrea, *Clifford Wavelets, Singular Integrals and Hardy Spaces*, Lecture Notes in Mathematics, no. 1575, Springer, Berlin, 1994.

21.   M. Mitrea, *The method of layer potentials in electromagnetic scattering theory on nonsmooth domains*, Duke Math. J. 77 (1995), no. 1, 111-133.

22.   M. Mitrea, R.H. Torres, G.V. Welland, *Regularity and approximation results for the Maxwell problem on C^1 and Lipschitz domains*, Clifford Algebras in Analysis and Related Topics (J. Ryan ed.), 297-308, CRC Press, Boca Raton, FL, 1996.

23.   M. Murray, *The Cauchy integral, Calderón commutators and conjugations of singular integrals in $\mathbb{R}^n$*, Trans. of the Amer. Math. Soc. 289 (1985), 497-518.

24.   J. Ryan (editor), *Clifford Algebras in Analysis and Related Topics*, CRC Press, Boca Raton, FL, 1996.

25.   T. Tao, *Harmonic convolution operators on Lipschitz graphs*, Adv. Appl. Clifford Alg. 6 (1996), 207-218.

26.   G. Verchota, *Layer potentials and boundary value problems for Laplace's equation in Lipschitz domains*, J. Funct. Anal. 59 (1984), 572-611.

Weight Problems for Higher Dimensional Singular Integrals via Clifford Analysis

V. Kokilashvili and A. Meskhi

A. Razmadze Mathematical Institute of the Georgian Academy of Sciences

Abstract. In this note we derive two-weight inequalities for higher dimensional singular integrals in the framework of Clifford analysis.

Keywords: Clifford algebra, singular integrals, singular Cauchy integral operator, double-layer potential, two-weight inequalities, Muckenhoupt class

2000 Mathematics Subject Classification: 42B20, 47B47

Dedicated to Richard Delanghe on the occasion of his 60th birthday.

1. Introduction

In the present note we discuss two-weight inequalities for higher dimensional singular integral operators defined on Lipschitz surfaces in Clifford algebras.

The study of the L^2- boundedness for singular integrals on Lipschitz surfaces was introduced by Murray in [16] for small Lipschitz constant. An analogous problem for all Lipschitz constants was considered by McIntosh in [14] (see also [12], [13], [15], [17]).

The L^2- boundedness problem for singular double- layer potential operator was solved by Calderón [1] for small Lipschitz constant and by Coifman, McIntosh and Meyer [2] for arbitrary Lipschitz constants.

2. Preliminaries

Let $\mathcal{A}_m$ denote real or complex Clifford algebra, i.e. $\mathcal{A}_m$ is a 2^m- dimensional real or complex algebra (with identity e_0) generated by the basis vectors $e_0, e_1, \cdots, e_m$ satisfying the conditions:

$$e_j^2 = -e_0,$$

$$e_j e_k = -e_k e_j$$

when $1 \leq j \leq m$, $1 \leq k \leq m$ and $j \neq k$. It is assumed that $R^{m+1} = R \oplus R^m$ is embedded in $\mathcal{A}_m$. We denote by $|x|$ an Euclidean norm

247

F. Brackx et al. (eds.), Clifford Analysis and Its Applications, 247–253.

of $x = x_0e_0 + X \in R^{m+1}$, where $X \in R^m$. The conjugate $\bar{x}$ of $x = x_0e_0 + X \in R^{m+1}$ is $\bar{x} = x_0e_0 - X \in R^{m+1}$.

In the sequel we denote by Σ the Lipschitz surface consisting of all points

$$x = g(X)e_0 + X \in R^{m+1},$$

where $X \in R^m$ and g is a real-valued Lipschitz function which satisfies the condition

$$\|Dg\|_\infty \equiv \operatorname*{ess\,sup}_{X \in R^m} \left\{ \sum_{j=1}^m \left| \frac{\partial g}{\partial x_j}(X) \right|^2 \right\}^{1/2} \leq \tan \omega < \infty$$

where $0 \leq \omega < \pi/2$.

We denote by C_Σ the singular Cauchy integral operator C_Σ, defined for almost all $x \in \Sigma$ by

$$\left(C_\Sigma f \right)(x) \equiv \lim_{\epsilon \to 0} \frac{2}{\sigma_m} \int\limits_{\{y \in \Sigma : |x-y| > \epsilon\}} \frac{\overline{x-y}}{|x-y|^{m+1}} n(y) f(y) dS_y,$$

where dS_y is the surface measure $(dS_y = (1 + |Dg(Y)|^2)^{1/2} dY$, $y = g(Y)e_0 + Y$, $Y \in R^m)$, $n(y)$ is the unit normal,

$$n(y) = \frac{e_0 - Dg(Y)}{(1 + |Dg(Y)|^2)^{1/2}}$$

and $f : \Sigma \to \mathcal{A}_m$ is an $\mathcal{A}_m$ algebra-valued function (we assume that $\| \cdot \|$ is a norm in $\mathcal{A}_m$).

A measurable, locally integrable, a.e. positive function $w : \Sigma \to R$ is called a weight.

We shall need the Muckenhoupt class $A_p(\Sigma)$ $(1 < p < \infty)$ which is the set of all weights w such that

$$\sup_B \left(\frac{1}{|B|} \int\limits_B \overline{w}(X) dX \right) \left(\frac{1}{|B|} \int\limits_B \overline{w}^{1-p'}(X) dX \right)^{p-1} < \infty \quad \left(p' = \frac{p}{p-1} \right),$$

where the supremum is taken over all balls B in R^m and $\overline{w}(x) = w(g(X), X)$, $X \in R^m$. The Muckenhoupt class $A_1(\Sigma)$ is the set of all weights w such that

$$\sup_B \left(\frac{1}{|B|} \int\limits_B \overline{w}(X) dX \right) \operatorname*{ess\,sup}_{X \in B} \frac{1}{\overline{w}(X)} < \infty,$$

where again the supremum is taken over all balls $B \subset R^m$.

We denote by $L_w^p(\Sigma)$, $1 \leq p < \infty$, the space of all $\mathcal{A}_m$ algebra-valued functions f defined on Σ with finite norm

$$\|f\|_{L_w^p(\Sigma)} = \left(\int_{\Sigma} \|f(x)\|^p w(x) dS_x \right)^{1/p},$$

w being an appropriate weight function.

The following statements hold (see e.g. [15], [5], [7], Chapter 5):

THEOREM A. *Let $1 < p < \infty$. Then the integral operator C_Σ is bounded in $L_w^p(\Sigma)$ for $w \in A_p(\Sigma)$.*

THEOREM B. *Let $w \in A_1(\Sigma)$. Then there exists a positive constant c such that for all $\lambda > 0$ and $f \in L_w^1(\Sigma)$ the following inequality holds:*

$$\int_{\{x \in \Sigma : \|C_\Sigma f(x)\| > \lambda\}} w(x) dS_x \leq \frac{c}{\lambda} \|f\|_{L_w^1(\Sigma)}.$$

The analogous results for the Hilbert transform,

$$H\varphi(x) = p.v. \int_{-\infty}^{+\infty} \varphi(y)(x-y)^{-1} dy$$

(φ is a real-valued function) were obtained in [6] (for the Cauchy integrals on regular curves see, e.g., [8]).

3. Main results

Now we pass to two-weight inequalities for the operator C_Σ. We shall denote by $\nu(E)$ a surface measure of the measurable set $E \subset \Sigma$.

THEOREM 1. *Let $1 < p < \infty$, suppose that $\eta_0 \in \Sigma$ and that $\nu(\Sigma) = \infty$. Let σ and u be positive increasing functions on $(0, \infty)$ and let $\rho \in A_p(\Sigma)$. We put $v(x) = \sigma(|x - \eta_0|)\rho(x)$, $w(x) = u(|x - \eta_0|)\rho(x)$. If*

$$\sup_{t>0} \left(\int_{\Sigma \setminus D(\eta_0, t)} v(x)|x|^{-mp} dS_x \right) \left(\int_{D(\eta_0, t)} w^{1-p'}(x) dS_x \right)^{p-1} < \infty, \quad (1)$$

where $D(\eta_0, t) \equiv \Sigma \cap B(\eta_0, t)$, $B(\eta_0, t) \equiv \{x \in \Sigma : |x - \eta_0| < t\}$, then the operator C_Σ is bounded from $L_w^p(\Sigma)$ to $L_v^p(\Sigma)$.

THEOREM 2. *Let $1 < p < \infty$ and let $\eta_0 \in \Sigma$. Suppose that $\nu(\Sigma) = \infty$. Let σ and u be positive decreasing functions on $(0, \infty)$ and let $\rho \in A_p(\Sigma)$. We put $v(x) = \sigma(|x - \eta_0|)\rho(x)$, $w(x) = u(|x - \eta_0|)\rho(x)$. If*

$$\sup_{t>0} \left(\int_{D(\eta_0,t)} v(x)dS_x \right) \left(\int_{\Sigma \backslash D(\eta_0,t)} w^{1-p'}(x)|x|^{-p'm}dS_x \right)^{p-1} < \infty, \quad (2)$$

then the operator C_Σ is bounded from $L_w^p(\Sigma)$ to $L_v^p(\Sigma)$.

Conditions (1) and (2) are necessary in the case of the Hilbert transform H.

Analogous results for Calderón- Zygmund singular integrals in Euclidean spaces were obtained in [3] for radial weights and generalized for homogeneous groups and on spaces of homogeneous type (SHT) in [9], [10], [4] (see also [7], Chapter 9).

For two-weight inequalities of weak type we have the following statements:

THEOREM 3. *Let $\eta_0 \in \Sigma$ and let $\nu(\Sigma) = \infty$. Let σ and u be positive increasing functions on $(0, \infty)$ and suppose that $\rho \in A_1(\Sigma)$. We put $v(x) = \sigma(|x - \eta_0|)\rho(x)$, $w(x) = u(|x - \eta_0|)\rho(x)$. Suppose that*

$$\sup_{\substack{\tau,t \\ \tau>t>0}} \left(\frac{1}{\tau^m} \int_{D(\eta_0,\tau)\backslash D(\eta_0,t)} v(x)dS_x \right) \operatorname*{ess\,sup}_{x \in D(\eta_0,t)} \frac{1}{w(x)} < \infty,$$

then there exists a positive constant c such that for all $\lambda > 0$ and $f \in L_w^1(\Sigma)$, the following inequality holds:

$$\int_{\{x\in\Sigma: \|C_\Sigma f(x)\| > \lambda\}} v(x)dS_x \leq \frac{c}{\lambda}\|f\|_{L_w^1(\Sigma)}. \quad (3)$$

THEOREM 4. *Let $1 < p < \infty, \eta_0 \in \Sigma$ and let $\nu(\Sigma) = \infty$. Suppose that σ and u are positive increasing functions on $(0, \infty)$ and that $\rho \in A_p(\Sigma)$. We put $v(x) = \sigma(|x - \eta_0|)\rho(x)$, $w(x) = u(|x - \eta_0|)\rho(x)$. If*

$$\sup_{\substack{\tau,t \\ \tau>t>0}} \left(\frac{1}{\tau^{mp}} \int_{D(\eta_0,\tau)\backslash D(\eta_0,t)} v(x)dS_x \right) \left(\int_{D(\eta_0,t)} w^{1-p'}(x)dS_x \right)^{p-1} < \infty,$$

then the following inequality holds:

$$\int_{\{x\in\Sigma: \|C_\Sigma f(x)\| > \lambda\}} v(x)dS_x \leq \frac{c}{\lambda}\|f\|_{L_w^p(\Sigma)}^p,$$

where the positive constant c does not depend on $\lambda > 0$ and f, $f \in L_w^p(R)$.

For decreasing weights we have the following result:

THEOREM 5. *Let $\eta_0 \in \Sigma$ and let $\nu(\Sigma) = \infty$. Let σ and u be positive decreasing functions on $(0, \infty)$. Suppose that $\rho \in A_1(\Sigma)$. We put $v(x) = \sigma(|x - \eta_0|)\rho(x)$, $w(x) = u(|x - \eta_0|)\rho(x)$. If the condition*

$$\sup_{t>0} \left(\int_{D(\eta_0,t)} v(x)dS_x \right) \operatorname*{ess\,sup}_{x \in \Sigma \setminus D(\eta_0,t)} \frac{1}{|x|^m w(x)} < \infty$$

is satisfied, then there exists a positive constant c such that for all $\lambda > 0$ and $f \in L_w^p(\Sigma)$ the inequality (3) holds.

Similar results for singular integrals defined on SHT were derived in [4] (for Riesz transforms see [11]).

Now we provide several examples of weights guaranteeing the boundedness of the operator C_Σ in weighted spaces (For the Hilbert transform H see [3]. See also [4]).

EXAMPLE 1. Let $1 < p < \infty$, $\eta_0 \in \Sigma$ and let $b \equiv diam \, \Sigma < \infty$. Let $v(x) = |x - \eta_0|^{(p-1)m}$, $w(x) = |x - \eta_0|^{(p-1)m} \ln^p \frac{4b}{|x-\eta_0|}$. Then the operator C_Σ is bounded from $L_w^p(\Sigma)$ to $L_v^p(\Sigma)$.

EXAMPLE 2. Let $1 < p < \infty$ and $\eta_0 \in \Sigma$. Suppose that $a < \infty$. Let $v(x) = |x - \eta_0|^{-m} \ln^{-p} \frac{4b}{|x-\eta_0|}$, $w(x) = |x - \eta_0|^{-m}$. Then the operator C_Σ is bounded from $L_w^p(\Sigma)$ to $L_v^p(\Sigma)$.

Analogous results for the singular integral operator on Σ,

$$Tf(x) = p.v. \int_{\Sigma} f(y)k(x,y)dS_y, \quad x \in \Sigma,$$

hold, where the kernel k satisfies the following conditions:

(i) there exists a positive constant c such that for all x, y with $y - x \notin \Gamma_a$ (Γ_a is the upright circular cone in the upper- half space R_+^{m+1} having ouverture a, $0 < a < \pi/2 - arctg(\|\nabla g\|_\infty)$, and whose vertex is at the origin), the inequality $\|k(x,y)\| \le c|x - y|^{-m}$ holds;

(ii) for any element $\xi \in \mathcal{A}_m$ there exists $\epsilon = \epsilon(\xi) > 0$ such that, for any $x \in R^{m+1}$,

$$D < k(\cdot,x), \xi >= 0 \text{ on } R^{m+1} \setminus (-\Gamma_a - \epsilon + x),$$
$$< k(\cdot,x), \xi > D = 0 \text{ on } R^{m+1} \setminus (\Gamma_a + \epsilon + x).$$

For real- valued functions all these statements follow from the results obtained in [4].

REFERENCES

1. A. P. Calderón, Cauchy integrals on Lipschitz curves and related operators, *Proc. Natl. Acad. Sci. USA* **74(1977)**, 1324-1327.

2. R.R. Coifman, A. McIntosh and Y. Meyer, L'intégral de Cauchy définit un opérateur borné sur L^2 pour les courbes Lipschitziennes, *Annals of Math.* **116**(1982), 361-387.

3. D. E. Edmunds and V. Kokilashvili, Two-weight inequalities for singular integrals. *Canadian Math. Bull.* **38**(1995), 119-125.

4. D. E. Edmunds, V. Kokilashvili and A. Meskhi, Two-weight estimates for singular integrals defined on homogeneous type spaces. *Canad. J. Math.***52**(2000), No. 3, 468-502.

5. S. Hoffman, Weighted norm inequalities and vector-valued inequalities for certain rough operators, *Indiana Univ. Math. J.* **42**(1993), 1-14.

6. R. A. Hunt, B. Muckenhoupt and R. Wheeden, Weighted norm inequalities for the conjugate function and Hilbert transform. *Trans. Amer. Math. Soc.* **176**(1973), 227–251.

7. I. Genebashvili, A. Gogatishvili, V. Kokilashvili and M. Krbec, Weight theory for integral transforms on spaces of homogeneous type, *Pitman Monographs and Surveys in Pure and Applied Mathematics*, **92**, *Longman, Harlow*, 1998.

8. G. Khuskivadze, V. Kokilashvili and V. Paatashvili, Boundary value problems for analytic and harmonic functions in domains with nonsmooth boundaries. Applications to conformal mappings. *Mem. Differential Equations Math. Phys.*, **14** *Tbilisi*, 1998.

9. V. Kokilashvili and A. Meskhi, Two- weight inequalities for singular integrals defined on homogeneous groups. *Proc. A.Razmadze Math. Inst.* **112**(1997), 57-90.

10. V. Kokilashvili and A. Meskhi, Two-weight inequalities for Hardy-type transforms and singular integrals defined on homogeneous type spaces. *Proc. A. Razmadze Math. Inst.* **114**(1997), 119–123.

11. V. Kokilashvili and A. Meskhi, Boundedness and compactness criteria for some classical integral operators. *In: Lecture Notes in Pure and Applied Mathematics*, **213**, *"Function Spaces V"*, *Proceedings of the Conference, Poznań, Poland* (*Ed. H. Hudzik and L. Skrzypczak*), 279-296, *New York, Bazel, Marcel Dekker*, 2000.

12. C. Li, A. McIntosh and T. Qian, Clifford algebras, Fourier transforms and singular convolution operators on Lipschitz surfaces, *Rev. Mat. Iber.* **10**(1994), 665-721.

13. C. Li, A. McIntosh and S. Semmes, Convolution singular integrals on Lipschitz surfaces, *J. Amer. Math. Soc.* **5**(1992), No. 3, 455-481.

14. A. McIntosh, Clifford algebras and the higher dimensional Cauchy integral, Approximation and Function Spaces, *Banach Center Publ., Warsaw,* **22**(1989), 253-267.

15. M. Mitrea, Clifford Wavelets, singular integrals and Hardy spaces, *Lecture Notes in Math.,* **1575**, *Springer, Berlin,*1994.

16. M. Murrey, The Cauchy integral, Calderón commutators and conjugations of singular integrals in R^n. *Trans. Amer. Math. Soc.* **289** (1985), 497-518.

17. J. Ryan (Editor), Clifford algebras in analysis and related topics, *CRC Press, Boca Raton, Florida,* 1996.

The Conformal Laplacian on Spheres and Hyperbolas via Clifford Analysis

Hong Liu
Department of Mathematics and Computer Science, Embry Riddle Aeronautical University, Daytona, FL 32114, USA

John Ryan (jryan@comp.uark.edu)
Department of Mathematics, University of Arkansas

Abstract. Using Clifford analysis we shall present a factorization of the Laplacian acting on functions defined on domains on spheres. This factorization leads us to investigate properties of a first order differential operator on the sphere, and to determine a Cauchy integral formula for functions annihilated by this operator. We determine a conformal covariance for both the operator and the integral over the sphere and we pull back the operator and representaion formula via a Cayley transformation to obtain similar results on R^{n-1}. The differential operator that we use is different from the spherical Dirac operator introduced by Cnops and Malonek. We use the integral formula to construct a Poisson formula markedly like the one for upper half space. We use this Poisson formula to solve the Dirichlet problems on a hemisphere. Furthermore we show that the factorization gives rise to two seperate Green's type formulas for solutions to our conformal Laplace equation.

Keywords: conformal Laplacian, Green' formula, Poisson kernel, Hardy spaces

Mathematics Subject Classification: 30G35, 53A30, 53A50, 58G28

Dedicated to Richard Delanghe on the occasion of his 60th birthday.

1. Introduction

In [8, 9]and elsewhere Cayley transformations are used to show that a valid analogue of Cauchy's integral formula and a corresponding Dirac operator may be constructed over the sphere and over the hyperbola. These in turn are used to develop a suitable version of Clifford analysis in this context. Independently Cnops and Malonek [3] use Stokes' theorem and an extension argument from the sphere to R^n to give an explicit representation for the spherical Dirac operator. Furthermore Van Lanker in [13, 14, 15] uses Gegenbauer functions to also develop a function theory for this operator and related Dirac operators on the sphere.

While this analysis opens the door to many interesting results it is also natural to ask if there is also a suitable analogue to the Laplacian

255

F. Brackx et al. (eds.), Clifford Analysis and Its Applications, 255–266.

on the sphere and on the hyperbola and if one can introduce and study interesting boundary value problems over domains on the sphere and on the hyperbola. This question was answered in the affirmative in [4] and [5]. In those works a suitable analogue of the Laplacian was found together with other higher order operators. It is shown that these operators are conformally equivalent to the operators D^k in euclidean space, where D is the euclidean Dirac operator and k is an arbitrary positive integer. When $k = 2$ this operator becomes the Laplacian in euclidean space. For the spherical Laplacian a Green's formula is introduced in [4, 5] and this formula bears a striking resemblance to the standard Green's formula in euclidean space.

Here we continue this analysis. We show that the factorization of the spherical Laplacian in terms of Dirac operators leads to two different versions of the spherical Green's formula. In turn this idea leads to a number of inequivalent Cauchy type formulas for the spherical and hyperbolic analogue of the operator D^k for k greater than one. We introduce the fundamental solution to a principle Dirac operator in this family and show that this to, regarded as an integral operator, is conformally invariant on the sphere. We also pull back this kernel and Dirac operator via the Cayley transformation and construct further operators and Cauchy type integral formulas on $\mathbf{R}^{n-1}$

Our analysis also leads to a Poisson kernel on the hemisphere. This kernel bears a marked resemblance to the classical Poisson kernel for upper half space. We use the Poisson kernel introduced here to solve the Dirichlet problems with L^p data for $1 < p < \infty$ on the boundaries of the hemisphere.

Acknowledgement: The second author is grateful to Martin Reimann for very useful discussions on conformally invariant differential operators on the sphere.

2. Preliminaries

Here we will introduce the background material that we need to develop our main results. We begin by considering the real Clifford algebra Cl_n generated from $\mathbf{R}^n$ equipped with a negative definite inner product. So we assume that $\mathbf{R}^n \subset Cl_n$. If $e_1, \ldots, e_n$ is an orthonormal basis for $\mathbf{R}^n$ then $e_i e_j + e_j e_i = -2\delta_{ij}$. For most of the algebraic details on this algebra we refer to [7]. An important group lying within Cl_n is the Pin group, which is defined to be $\{a \in Cl_n : a = a_1 \ldots a_p : p \in \mathbf{N}$ and $a_j \in S^{n-1}$ for $1 \leq j \leq p\}$. This group is denoted by $Pin(n)$. There is an antiautomorphism

$$\sim : Cl_n \to Cl_n :\sim (e_{j_1} \ldots e_{j_r}) = e_{j_r} \ldots e_{j_1}.$$

It is usual to write $\tilde{a}$ instead of $\sim a$ for $a \in Cl_n$.

DEFINITION 1. *Suppose U is a domain in $\mathbf{R}^{n-1}$, the span of $e_1, \ldots e_{n-1}$, then a differentiable function $f : U \to Cl_{n-1}$, is said to be left monogenic if $\sum_{j=1}^{n-1} e_j \frac{\partial f(x)}{\partial x_j} = 0$ for each $x \in U$ where Cl_{n-1} is the Clifford subalgebra of Cl_n generated from $\mathbf{R}^{n-1}$,*

Similarly a differentiable function $g : U \to Cl_{n-1}$ is said to be right monogenic if $\sum_{j=1}^{n-1} \frac{\partial g(x)}{\partial x_j} e_j = 0$ for each $x \in U$. We usually denote the differential operator $\sum_{j=1}^{n-1} e_j \frac{\partial}{\partial x_j}$ by D. So the left monogenic function f satisfies the equation $Df = 0$ on U and the right monogenic function g satisfies the equation $gD = 0$ on U. It should be noted that a function f is left monogenic if and only if $\tilde{f}$ is right monogenic. Also the function $G(x) = \frac{x}{\|x\|^{n-1}}$ is an example of a function which is both left and right monogenic. A simple but important point to note is that $D^2 = -\triangle_{n-1}$ which is the negative of the Laplacian in R^{n-1}.

Following [2] and elsewhere one can determine that if f and g are respectively left and right monogenic functions defined on a domain U and M is a compact, piecewise smooth hypersurface lying in U and bounding a subdomain of U then $\int_M g(x)n(x)f(x)d\sigma(x) = 0$ where $n(x)$ is the outward pointing unit vector at $x \in M$ and normal to M at x and σ is the usual Lebesgue measure for M. Of course this integral formula is a generalization of Cauchy's Theorem.

Suppose now that ψ is a Möbius transformation acting over $\mathbf{R}^{n-1} \cup \{\infty\}$ then in [1] and elsewhere it is shown that there are elements a, b, c and d of Cl_{n-1} such that $\psi(x) = (ax + b)(cx + d)^{-1}$. The Clifford numbers a, b, c and d satisfy certain constraints specified in [1].

Using the previously mentioned Cauchy Theorem one can show, see [11], that if $f(y)$ is left monogenic and $y = \psi(x)$ then the function $J(\psi, x)f(\psi(x))$ is left monogenic in the vector variable x, where $J(\psi, x) = \frac{\widetilde{(cx+d)}}{\|cx+d\|^{n-1}}$. In fact this result primarily follows by noting that under a Möbius transformation the volume element $n(y)d\sigma(y)$ transforms to $\tilde{J}(\psi, x)n(x)J(\psi, x)d\sigma(x)$. See for instance [11] for details.

In Clifford analysis the analogue of Cauchy's integral formula is given by

$$f(y) = \frac{1}{\omega_{n-1}} \int_M G(x - y)n(x)f(x)d\sigma(x)$$

where y is a point in the domain bounded by the hypersurface M and ω_{n-1} is the surface area of the unit sphere in $\mathbf{R}^{n-1}$. Now if $\psi(x) = u$ and $\psi(y) = v$ then it may be determined that

$$G(u - v) = J(\psi, y)^{-1}G(x - y)J(\psi, x)^{-1}.$$

It follows that Cauchy's integral formula remains valid under Möbius transformations.

In [8, 9] we noted that instead of considering Möbius transformations acting over $\mathbf{R}^{n-1} \cup \{\infty\}$ we could consider the Cayley transformation $C(x) = (e_n x + 1)(x + e_n)^{-1}$ from $\mathbf{R}^{n-1}$ to the unit sphere, S^{n-1}, lying in $\mathbf{R}^n$. In [8, 9] we are able to use this transformation to carry over much of basic Clifford analysis to the setting of the sphere. We also used another Cayley transformation to carry over the same type of analysis to the $(n-1)$-dimensional hyperbola. We shall say more on that later. Using Stokes' Theorem and the invariance of monogenicity under Möbius transformations, we were able to infer the existence of a Dirac operator D_S with Cauchy kernel $G'(x-y) = \frac{x-y}{\|x-y\|^{n-1}}$. Here x and $y \in S^{n-1}$. Independently Cnops and Malonek [3] used an extension argument from the sphere to $\mathbf{R}^n$ to show that $D_S = x(\Gamma - \frac{n-1}{2})$, where $x \in S^{n-1}$ and Γ is the restriction to S^{n-1} of $x \wedge D_n$. Furthermore $\wedge$ is the usual vector wedge product and D_n is the Dirac operator in $\mathbf{R}^n$.

Using the fact that for each pair x and $y \in S^{n-1}$ then $\|x - y\|^2 = 2 - 2 <x, y>$ where $<,>$, denotes the usual inner product in $\mathbf{R}^n$, and $x\Gamma <x, y> = x \wedge y$, and $\Gamma x f(x) = -x\Gamma f(x) + (n-1)xf(x)$ for each Cl_n valued C^∞ function f defined on a domain on S^{n-1}, one may show directly that $D_S G' = G' D_S = 0$.

The identity $\Gamma x f(x) = -x\Gamma f(x) + (n-1)xf(x)$ may be easily determined via direct calculation but may also be motivated by the fact shown in [12] that if f is a Cl_n valued, real analytic function defined on S^{n-1} then f is the restriction to S^{n-1} of a left monogenic function F. While the restriction of the left monogenic function $\frac{x}{\|x\|^n} F$ is xf, and the difference between eigenvalues for the operator Γ acting on the functions f and xf is $(n-1)$.

3. The Conformal Laplacian

We now turn to introduce a second order differential operator on the sphere that would play the same role in this context as the Laplacian in Euclidean space. If we consider the fundamental solution $\frac{1}{\|u-v\|^{n-3}}$ to the Laplacian in euclidean space it may be observed that the convolution

$$\int_{R^{n-1}} \frac{1}{\|u-v\|^{n-3}} h(u) du^{n-1}$$

conformally transforms to

$$\int_{S^{n-1}} \frac{1}{\|x-y\|^{n-3}} f(x) d\pi(x)$$

where π is the Lebesgue measure on S^{n-1} and $f(x) = \frac{1}{\|cx+d\|^{n+1}} h(C^{-1}(x))$ with c and d coefficients arising from the inverse Cayley transformation. This would suggest that we are looking for a differential operator that annihilates the function $\frac{1}{\|x-y\|^{n-3}}$. As a first attempt we should ask if it is the case that $D_S^2 \frac{1}{\|x-y\|^{n-3}} = 0$. However a simple direct calculation reveals that

$$D_S \frac{1}{\|x-y\|^{n-3}} = G'(x-y) - \frac{x}{\|x-y\|^{n-3}}.$$

It follows that $(D_S + x)\frac{1}{\|x-y\|^{n-3}} = G'(x-y)$. Consequently

$$D_S(D_S + x)\frac{1}{\|x-y\|^{n-3}} = 0.$$

It follows that a suitable analogue of the euclidean Laplacian on the sphere is the differential operator $D_S(D_S + x)$. We shall denote this operator by $\triangle_S$. The operator $\triangle_S$ is in fact a scalar valued differential operator. This follows by computing D_n^2 in terms of spherical co-ordinates. In this case the usual Laplace-Beltrami operator for the sphere factors as $((n-2)I - \Gamma)\Gamma$. Of course the Laplace-Beltrami operator is a scalar. So for each $\alpha \in \mathbf{C}$ the operator $-(\alpha - n + 2 + \Gamma)(\Gamma - \alpha)$ is a scalar. Moreover when $\alpha = \frac{n-1}{2}$ we get $\triangle_S$.

Now $\triangle_S = D_S^2 + D_S x$. Furthermore $D_S = x(\Gamma - \frac{n-1}{2})x = x^2(-\Gamma + (n-1) - \frac{n-1}{2}) = -xD_S$. So $\triangle_S$ is also equal to $(D_S - x)D_S$. We may now use these two alternative forms for factoring $\triangle_S$ to obtain the following representation formula.

THEOREM 1. *[5] Suppose that U is a domain on S^{n-1} and $h : U \to Cl_n$ is a C^2 function. Suppose also that M is a piecewise smooth $(n-2)$-dimensional manifold lying in U and bounding a subdomain V of U. Then for each $y \in V$ we have*

$$h(y) = \frac{1}{\omega_{n-1}} \int_M (G'(x-y)n(x)h(x) - H'(x-y)n(x)D_S h(x))d\mu(x)$$

$$- \frac{1}{\omega_{n-1}} \int_V H'(x-y)\triangle_S h(x)d\pi(x),$$

where $H'(x-y) = \frac{1}{(n-3)\|x-y\|^{n-3}}$, $n(x)$ is the unit vector lying in the tangent space of S^{n-1} at x and orthogonal to the tangent space of M at x. Moreover $n(x)$ is outward pointing from M. Also μ is the Lebesgue measure on M.

Proof: Let $B_S(y, r)$ be a ball lying in S^{n-1}, centered at y and of radius r and let $S(y, r)$ denote its boundary. Applying Stokes' Theorem to the expression

$$\int_{M-S(y,r)} (G'(x-y)n(x)h(x) - H'(x-y)n(x)D_Sh(x))d\mu(x)$$

gives the term

$$\int_{V\backslash B(y,r)} (G'(x-y)(D_Sh(x)) - (H'(x-y)D_S)D_Sh(x)+$$

$$H'(x-y)D_S^2h(x))d\pi(x).$$

But

$$G'(x-y)(D_Sh(x)) - (H'(x-y)D_S)D_Sh(x) + H'(x-y)(D_S^2h(x)))$$

is equal to

$$G'(x-y)(D_Sh(x) - (H'(x-y)(D_S+x))h(x)$$

$$+H'(x-y)xD_Sh(x) - H'(x-y)(D_S-x)D_Sh(x) - H'(x-y)xD_Sh(x).$$

This expression reduces to $H'(x-y)(D_S - x)D_Sh(x)$ or equivalently $H'(x-y)(D_S(D_S+x)h(x)$. The result now follows by allowing r to tend to zero. $\square$

If we further assume that the function h is real valued then the non-scalar parts of our representation formula must be zero. The representation formula then reduces to

$$h(y) = \frac{1}{\omega_{n-1}} \int_M (< G'(x-y), n(x) > h(x)$$

$$-H'(x-y) < n(x), D_Sh(x) >)d\mu(x) - \frac{1}{\omega_{n-1}} \int_V H'(x-y)\Delta_Sh(x)d\pi(x)$$

where $<, >$ denotes the standard inner product on $\mathbf{R}^n$.

Now $< D_S, n(x) > = < x\Gamma - x\frac{n-1}{2}, n(x) >$. As x and $n(x)$ are orthogonal then $< D_S, n(x) > = < x\Gamma, n(x) >$ and the previous representation formula reduces to

$$h(y) = \frac{1}{\omega_{n-1}}(\int_M (< G'(x-y), n(x) > h(x)$$

$$-H'(x-y) < x\Gamma h(x), n(x) >)d\mu(x) - \int_V H'(x-y)\Delta_Sh(x)d\pi(x)).$$

Furthermore as $G'(x - y) = x\Gamma H'(x - y) - x(n - 3)H'(x - y)$ the representation formula further reduces to

$$h(y) = \frac{1}{\omega_{n-1}}(\int_M (< x\Gamma H'(x - y), n(x) > h(x)$$

$$-H'(x - y) < x\Gamma h(x), n(x) >)d\mu(x) - \int_V H'(x - y)\Delta_S h(x)d\pi(x)).$$

If it is also the case that the scalar valued function h further satisfies the equation $\Delta h = 0$ then we obtain the following version of Green's formula on the sphere:

$$h(y) = \frac{1}{\omega_n} \int_M (< x\Gamma H'(x - y), n(x) > h(x)$$

$$-H'(x - y) < n(x), x\Gamma h(x) >)d\mu(x).$$

If $h : U \to Cl_n$ and $\Delta h = 0$ then we automatically get the following version of Green's formula.

$$h(y) = \frac{1}{\omega_{n-1}} \int_M (G'(x - y)n(x)h(x) - H'(x - y)n(x)D_S h(x)d\mu(x).$$

On the other hand if $h : U \to Cl_n$ is C^2 and $h|_M = 0$ then our representation formula becomes

$$h(y) = \frac{1}{\omega_{n-1}} \int_V H'(x - y)\Delta_S h(x)d\pi(x).$$

In this case we can use this representation formula to adapt arguments presented in the euclidean case in [?] and the Cayley transformation to show that Δ_S is conformally equivalent to Δ_{n-1} the Laplacian in $\mathbf{R}^n$. See [5] for details. Specifically

$$J_{-2}(C, x)\Delta_S h(y) = \Delta_{n-1} J_2(C, x)h(C(x))$$

where $C(x) = y$, $J_{-2}(C, x) = \frac{1}{\|cx+d\|^{n-1}}$ and $J_2(C, x) = \frac{1}{\|cx+d\|^{n-3}}$.

4. Other Operators and Representation Formulas

We have seen in the previous section that $D_S H'(x - y) = G'(x - y) - xH'(x - y)$. From this it is a simple matter to deduce the following Cauchy Integral Formula.

THEOREM 2. *Suppose that for U a domain on S^{n-1} the C^1 function $f : U \to Cl_n$ satisfies the equation $(D + x)f(x) = 0$. Suppose also that M and V are as in Theorem 1 and that $y \in V$. Then*

$$f(y) = \frac{1}{\omega_{n-1}} \int_M (G'(x-y) - xH'(x-y))n(x)f(x)d\mu(x).$$

It is also easy to deduce that if f is just a C^1 function then

$$f(y) = \frac{1}{\omega_{n-1}}(\int_M (G'(x-y) - xH'(x-y))n(x)f(x)d\mu(x)$$

$$- \int_V (G'(x-y) - xH'(x-y))(D_S + x)f(x)d\pi(x)).$$

If $f|_M = 0$ then this formula becomes

$$f(y) = \frac{1}{\omega_{n-1}} \int_V (G'(x-y) - xH'(x-y))(D_S + x)f(x)d\pi(x). \quad (1)$$

These formulas follow as $(D_S - x)(G'(x-y) - xH'(x-y)) = 0$.

As we have previously observed if D_n is the euclidean Dirac operator over $\mathbf{R}^n$ with respect to a variable u, then D_n conformally transforms to $aD_n\tilde{a}$ where D_n in this last expression is the Dirac operator with respect to the vector variable v where $u = av\tilde{a}$ and $a \in Pin(n)$. It follows on splitting up D_n into its spherical and radial parts that if $y\Gamma$ is the spherical part with respect to a variable $y \in S^{n-1}$ then this operator is conformally equivalent to the operator $ax\Gamma\tilde{a}$ with respect to the variable $x \in S^{n-1}$. Consequently the operator D_S is conformally equivalent to $aD_S\tilde{a}$ under the same change of variables and the differential operators $D_S \pm y$ are conformally equivalent to $a(D_S \pm x)\tilde{a}$. For that matter the differential operator $D_S + \alpha y$ is conformally equivalent to $a(D_S + \alpha x)\tilde{a}$ for each $\alpha \in \mathbf{C}$.

If now $u = ax\tilde{a} \in S^{n-1}$ and $v = ay\tilde{a} \in S^{n-1}$ and $(D_S + v)f(v) = 0$ then on changing variables from u and v to x and y we get that

$$f(ay\tilde{a}) = \frac{1}{\omega_{n-1}} \int_{a^{-1}M\tilde{a}^{-1}} a(G'(x-y)-xH'(x-y))\tilde{a}an(x)\tilde{a}f(ax\tilde{a})d\mu(x),$$

where $a^{-1}M\tilde{a}^{-1} = \{x \in S^{n-1} : ax\tilde{a} \in M\}$.

DEFINITION 2. *Suppose that f is as in Theorem 2 so $(D_S+x)f(x) = 0$ then we say that f is left x monogenic.*

We can state a similar definition for right x monogenic functions.

The previous calculation tells us that $f(v)$ is left x monogenic if and only if $\tilde{a}f(ax\tilde{a})$ is left x monogenic. We have also established that the

Dirac type operators $D_S + \alpha x$ and $D_S + \alpha u$ are intertwined by a and $\tilde{a}$. Moreover the convolution operator $(G(x - y) - xH(x - y)) \star |_M$ is also intertwined by the elements a and $\tilde{a}$ of $Pin(n)$. In fact the convolution operator $(G'(x - y) - xH'(x - y)) \star |_V$ is also intertwined by a and $\tilde{a}$. This may be noted by making the appropriate change of variable in Equation 1. By expanding V out to encompass all of S^{n-1} it may be determined that the convolution operator $(G(x-y) - xH(x-y)) \star |_{S^{n-1}}$ is also intertwined by the a and $\tilde{a}$.

We already know from [8] that

$$J_{-1}(C, x)D_S = D_{n-1}J(C, x) \tag{2}$$

where $J_{-1}(C, x) = \frac{\widetilde{cx+d}}{\|cx+d\|^{n+1}}$. Using this fact we want to determine what differential operator over R^{n-1} the differential operator $D_S + x$ is conformally equivalent to. First let us observe that the vector $x \in S^{n-1}$ is also the unit outer vector to S^{n-1} at x. Under the Cayley transformation this vector is transformed to $\frac{(cx+d)e_n(\widetilde{cx+d})}{\|cx+d\|^2}$. It follows from Equation 2 that $(D_S + y)f(y)$ is transformed to

$$J_{-1}(C, x)^{-1}D_{n-1} + \frac{(cx + d)e_n(\widetilde{cx + d})}{\|cx + d\|^2}f(C(x))$$

and this term simplifies to

$$J_{-1}(C, x)^{-1}(D_{n-1} + \frac{e_n}{\|cx + d\|^2})J(C, x)f(C(x)).$$

So

$$J_{-1}(C, x)(D_S + x)f(y) = (D_{n-1} + \frac{e_n}{\|cx + d\|^2})J(C, x)f(C(x))$$

where $y = C(x)$.

In slightly greater generality one also readily has

$$J_{-1}(C, x)(D_S + \alpha x)f(y) = (D_{n-1} + \alpha\frac{e_n}{\|cx + d\|^2})J(C, x)f(C(x)).$$

for any $\alpha \in \mathbf{C}$. This last equation bears a striking resemblance to a formula given in [10] describing the conformal covariance of the differential operator $D_{n-1} + m$ where $m \in \mathbf{R}^+$. In this case the operator satisfies the equation

$$J_{-1}(\psi, x)(D_{n-1} + m)f(y) = (D_{n-1} + \frac{m}{\|cx + d\|^2})J(\psi, x)f(\psi(x))$$

where here $y = \psi(x)$ and ψ is a Möbius transformation over $\mathbf{R}^{n-1} \cup \{\infty\}$. We would obtain the operator $D_{n-1} + \frac{1}{\|cx+d\|^2}$ instead of the operator $D_{n-1} + \frac{e_n}{\|cx+d\|^2}$ if we assumed that the unit 1 of Cl_n is the unit normal vector to $\mathbf{R}^{n-1}$ instead of e_n.

Returning to the Cauchy integral formula given in Theorem 2 it would now be natural to ask how this formula transforms under the Cayley transformation. Again we shall interpret the vector $x \in S^{n-1}$ as the unit outer vector to S^{n-1} at x. In this case the integral becomes

$$\frac{1}{\omega_{n-1}} \int_{C^{-1}(M)} (J(C,v)^{-1} G(u-v) J(C,u)^{-1}$$

$$-\frac{(u+e_n)e_n(u+e_n)}{\|u+e_n\|^2} \|v+e_n\|^{n-3} H(u-v)\|u+e_n\|^{n-3})$$

$$J(C,u)n(u)J(C,u)f(C(u))d\sigma(u).$$

This expression simplifies to

$$J(C,v)^{-1} \frac{1}{\omega_{n-1}} \int_{C^{-1}(M)} (G(u-v)$$

$$-(v+e_n)(u+e_n)^{-1}e_n)n(u)J(C,u)f(C(u))d\sigma(u).$$

As $ue_n = -e_n u$ for each $u \in \mathbf{R}^{n-1}$ the term $(v+e_n)(u+e_n)^{-1}e_n$ is equal to $(v+e_n)e_n(e_n-u)^{-1}$. If we place $G((u-v) + (v+e_n)e_n(u-e_n)^{-1}H(u-v) = Q(u,v)$ then we have established the following:

THEOREM 3. *Suppose U is a domain on S^{n-1} then a C^1 function $f : U \to Cl_n$ satisfies the equation $(D_S + x)f(x) = 0$ on U if and only if*

$$(D_{n-1} + \frac{e_n}{\|cx+d\|^2})J(C,u)f(C(u)) = 0$$

on $C^{-1}(U) \subset \mathbf{R}^{n-1}$. Moreover, if $v \in C^{-1}(V)$ and V is as in Theorem 1 then

$$J(C,v)f(C(v)) = \frac{1}{\omega_{n-1}} \int_{C^{-1}(M)} Q(u,v)n(u)J(C,u)f(C(u))d\sigma(u).$$

Let us return momentarily to the kernel $G'(x-y)$ and the operator D_S. One may use the Cauchy integral formula given in [8, 9] together with standard techniques described in [6] and elsewhere to obtain the following:

THEOREM 4. *Suppose that Σ is the boundary of a strongly Lipschitz domain in $\mathbf{R}^{n-1}$ and $\Pi = C(\Sigma)$ then for $1 < p < \infty$*

$$L^p(\Pi) = H^p(\Pi^+) \oplus H^p(\Pi^-)$$

where $L^P(\Pi)$ is the space of Cl_n valued L^p integrable functions defined on Π, $\Pi^{\pm}$ are the domains in S^{n-1} which complement Π and $H^p(\Pi^{\pm})$ are the Hardy spaces of solutions to the equation $D_S f(x) = 0$ on $\Pi^{\pm}$ that have L^p non-tangential limit functions defined on Π.

Suppose now that f belongs to the Hardy space $H^p(\Pi^+)$ then $D_S(D_S - x)f(x) = 0$. so each component of f is annihilated by $\triangle_S$. In particular the identity component of f is annihilated by this operator.

Let us now consider the special case where Π is the equator S^{n-2} of S^{n-1}. So $S^{n-2} = S^{n-1} \cap \mathbf{R}^{n-1}$. Let us also assume that Π^+ is the hemisphere $S^{n-2,+}$ that has boundary S^{n-2} and contains the point e_n. For $\lambda(x) \in L^p(S^{n-2})$, with $1 < p < \infty$, the function

$$\frac{1}{\omega_{n-1}} \int_{S^{n-2}} G'(x-y)n(x)\lambda(x)d\mu(x) \tag{3}$$

defines a member of $H^p(S^{n-2,+})$. Moreover as y tends non-tangentially to a point $z \in S^{n-2}$ this integral approaches the value

$$\frac{1}{2}\lambda(z) + P.V.\frac{1}{\omega_{n-1}} \int_{S^{n-2}} G'(x-z)n(x)\lambda(x)d\mu(x)$$

almost everywhere and the function defined by this integral belongs to $L^p(S^{n-2})$.

As in this case $n(x) = e_n$ we get when λ is scalar valued then the scalar part of the integral 3 is given by

$$\frac{1}{\omega_{n-1}} \int_{S^{n-2}} P(x-y)\lambda(x)d\mu(x)$$

where $P(x,y)$ is the n-th component $\frac{y_n}{\|x-y\|^{n-1}}$ of $G'(x-y)$. So on multiplying this integral by 2 gives a solution to the Dirichlet problem on $S^{n-2,+}$ for $1 < p < \infty$ and the kernel $2P$ is the analogue of the Poisson kernel in upper half space, but now in the context of the hemisphere.

In conclusion let us place $G(x-y) + xH(x-y) = Q'(x,y)$. Then if h is a C^2 function on U and $\triangle_S h = 0$ and V, M and y are as in Theorem 1 one may apply Stokes' Theorem to obtain the following representation formula.

$$h(y) = \frac{1}{\omega_{n-1}} \int_M (Q'(x,y)n(x)h(x) - H'(x-y)n(x)(D+x)h(x))d\mu(x).$$

This formula gives a second and alternative representation for solutions to the equation $\triangle_S h = 0$. However if h is scalar valued then we only really need the scalar part of this integral formula and one may readily determine that this reduces to the scalar version of Green's formula that we presented earlier in this paper.

In [8, 9] we used an alternative Cayley transformation, to set up analogous results on the hyperbola to those obtained on the sphere. It is a simple exercise to see that all the results presented here and in [5] also carry over to the context of the hyperbola.

References

1. L. V. Ahlfors, *Möbius transformations in R^n expressed through 2×2 matrices of Clifford numbers*, Complex Variables, 5, 1986, 215-24.

2. F. Brackx, R. Delanghe and F. Sommen, *Clifford Analysis*, Pitman, London, 1982.

3. J. Cnops and H. Malonek, *An introduction to Clifford analysis*, Univ. Coimbra, Coimbra, 1995.

4. H. Liu, *Clifford Analysis Techniques for Spherical PDE*, PhD Thesis, University of Arkansas, Fayetteville, USA, 2000.

5. H. Liu and J. Ryan, *Clifford analysis techniques for spherical pde*, to appear.

6. M. Mitrea, *Singular Integrals, Hardy Spaces, and Clifford Wavelets*, Lecture Notes in Mathematics, No 1575, Springer Verlag, Heidelberg, 1994.

7. I. Porteous, *Topological Geometry* CUP, Cambridge, 1981. [?] J. Peetre and T. Qian, *Möbius covariance of iterated Dirac operators*, J. Australian Mathematical Society, Series A 56, 1994, 403-414.

8. J. Ryan, *Dirac operators on spheres and hyperbolae*, Bol. Soc. Mat. Mexicana, 3, 1996, 255-270.

9. J. Ryan, *Clifford analysis on spheres and hyperbolae*, Mathematical Methods in the Applied Sciences, 20, 1997, 1617-1624.

10. J. Ryan, *Conformally covariant operators in Clifford analysis*, Zeitschrift für Analysis und ihre Anwendungen, 14, 1995, 677-704.

11. J. Ryan, *Basic Clifford analysis*, Cubo Matematica Educacional, 2, 2000, 226-256.

12. F. Sommen, *Spherical monogenics and analytic functionals on the unit sphere*, Tokyo J. of Math. , 4, 1981, 427-456.

13. P. Van Lanker, *Clifford Analysis on the Unit Sphere*PhD Thesis, Gent State University, Belgium, 1996.

14. P. Van Lanker, *Clifford analysis on the sphere*, Clifford Algebras and their Applications in Mathematical Physics, V. Dietrich et al (eds), Kluwer, Dordrecht, 1998.

15. P. Van Lanker, *Approximation theorems for spherical monogenics of complex degree*, Bull. Belg. Math. Soc. , 6, 1999, 279-293.

Combinatorics and Clifford analysis

Irene Sabadini (sabadini@mate.polimi.it)
Dipartimento di Matematica
Politecnico di Milano

Frank C Sommen (fs@cage.rug.ac.be)[*]
Department of Mathematical Analysis
Ghent University

Abstract. In this paper we introduce some linear first order systems in the framework of Clifford analysis that are originated by some combinatorial structures. We begin their study by providing their resolutions and we show that most of the systems treated are De Rham like. Being a new field of interest, we also give a list of open problems.

Keywords: combinatorial systems, resolutions, syzygies

Mathematics Subject Classification: 15A66, 30G35, 35A27

Dedicated to Richard Delanghe on the occasion of his 60th birthday.

1. Introduction

In our papers [7] and [9] we have introduced several systems that arise in the framework of Clifford analysis and its developments. In fact, the theory of Dirac or Fueter operators, both in one or several variables, can be considered nowadays a classical topic in Clifford analysis, but there are many other possibilities that can be investigated, e.g. the operators whose nullsolutions are the so–called higher spin monogenic, see [9], [11], [12], [13]. In [7] we started the study of some of the systems we introduced, by looking at the resolution of the associated complex. As it is well known, much of algebraic and analytic information is contained in the resolution, so the resolution is a first global information on how complicated a system is before developing its analysis. Among the systems in [7], we consider particularly original and stimulating those of combinatorial type. In fact we have shown that these systems have, in many cases, a peculiar behaviour since they are De Rham like. The systems of combinatorial type are constructed starting from some incidence structures that are "finite geometries". The "finite geometries" consist of a set of points, also called *tops*, $\{p_1, ..., p_m\}$ and a

---

[*] Senior Research Associate, FWO, Belgium

F. Brackx et al. (eds.), Clifford Analysis and Its Applications, 267–282.
© *2001 Kluwer Academic Publishers. Printed in the Netherlands.*

collection of lines or *blocks* $\{b_1, ..., b_n\}$ where every block b_j is a subset of $\{p_1, ..., p_m\}$. The systems we have in mind consist of operators of the type $\sum_{jk} \pm e_j \partial_{x_k}$ where $x_1, ..., x_m$ are a set of m scalar coordinates, $e_1, ..., e_M$ are generators of the real Clifford algebra $\mathcal{C}_M$, according to the following axioms:

- (A_1) Each operator is of the type $\sum_{jk} \pm e_j \partial_{x_k}$;

- (A_2) Every partial derivative ∂_{x_j} occurs at most once in a given operator (within a term $\pm e_j \partial_{x_k}$);

- (A_3) Also every basis element e_k occurs at most once in a given equation;

- (A_4) Every term $e_j \partial_{x_k}$ occurs at most once in the whole system, preceded by either a plus or minus sign;

- (A_5) the number M of basis elements e_k is minimal.

Then we will associate with each point p_k the partial derivative ∂_{x_k}, so that each block corresponds to set of partial derivatives with which one may form an operator of the above type by attaching to each ∂_{x_j} a well chosen basis element e_k and a signature and taking the sum over all j–indices in the block. A system constructed in this way is not unique and differs from other systems associated with the same finite geometry, either for the assignment of a unit e_j to a given partial derivative ∂_{x_k} in a block, or for the signature of each top of that block. Obviously it is always possible to create such a system if one can choose the units in a set of M elements with M large enough. This is the main reason to explicitly require axiom (A_5).

The construction of a system according to the previous (A_1), ..., (A_5) can be translated into the classical problem of colouring the edges of a certain bipartite graph which is obtained as follows (see e.g. [3]):
The points of the graph consist of two disjoint sets: the set of tops of the finite geometry and the set of blocks. The lines of the graph connect a point in the set of tops to a point in the set of blocks if that top belongs to the block. The set of elements $\{e_1, ..., e_M\}$ may be seen as set of colours with which we have to colour the edges of the graph such that all edges issuing from a given point in the graph have different colour and the total number of colours is minimal. This total number is the edge chromatic number for which Konig proved that it equals the highest number of edges per point.
There is another way to form the operators in a system, namely one can consider a finite geometry of incidence structure with tops $\{p_1, ..., p_M\}$ and blocks $\{b_1, ..., b_n\}$ where to each top we assign a basis element p_k

to a unit e_k. Then for each fixed block we have to choose now a certain "colour" ∂_{x_j} to be assigned to each top e_k of the block, and a signature. The axiom (A_5) has to be replaced by

 − (A_5') the number m of partial derivatives is minimal.

The systems obtained in this second way are called "super-dual systems", in order to distinguish them by the systems obtained by interchanging the words "point" and "line" in a given finite geometry.

We insert in this introduction also a short overview of the algebraic treatment of systems of partial differential equations with the basic tools we need in this treatment. For more details we refer the reader to the fundamental book [5] while, for the applications to Clifford analysis we refer to the papers [1], [2], [6], [7], [8], where some of the matrices we will use in the sequel are directly computed.

Let us note that a system of equations of the form

$$(1) \qquad \left(\sum_{jk} \pm e_j \partial_{x_k} \right) f = g_i$$

where $f, g : U \subseteq \mathbb{R}^m \to \mathcal{C}_M$, can be written in a matrix form by setting Let $\vec{f} = (f_1, \ldots, f_r)$, $r = 2^M$, f_ℓ real differentiable functions on an open set $U \subseteq \mathbb{R}^m$, $\forall \ell = 1, \ldots, r$ and let

$$(2) \qquad \sum_{\ell=1}^{r} P_{i\ell}(D) f_\ell = g_i$$

be a $q \times r$ system of linear constant coefficient partial differential equations. Let $P = [P_{i\ell}]$ be a $q \times r$ matrix of complex polynomials in $\mathbb{C}^m$ and $D = (-i\partial_{x_1}, \ldots, -i\partial_{x_m})$. The polynomial matrix P, which is the symbol of the system, can be obtained from $P(D)$ by replacing (formally) ∂_{x_i} by the complex variable z_i for every $i = 1, \ldots, m$. This procedure is equivalent to taking the Fourier transform of $P(D)$. When we consider a system consisting of equations of the type (1), we take the real components of each equation with respect to each unit in the Clifford algebra $\mathcal{C}_M$ to arrive at a system of the type (2). The transpose matrix P^t of P is an R-homomorphism $R^q \to R^r$ whose cokernel is $\mathcal{M} = R^r / P^t R^q = R^r / < P^t >$, where $R = \mathbb{C}[z_1, \ldots, z_m]$ and $< P^t >$ is the submodule of R^r generated by the columns of P^t. By the Hilbert syzygy theorem, there is a finite free resolution

$$0 \longrightarrow R^{a_s} \xrightarrow{P_{a_s}^t} R^{a_s-1} \longrightarrow \ldots \xrightarrow{P_1^t} R^q \xrightarrow{P^t} R^r \longrightarrow \mathcal{M} \longrightarrow 0$$

that together with its transpose

$$0 \longrightarrow R^r \xrightarrow{P} R^q \xrightarrow{P_1} \ldots \longrightarrow R^{a_s-1} \xrightarrow{P_{a_s}} R^{a_s} \longrightarrow 0$$

are key tools for the algebraic analysis of the system (2). Even though there is a lot of information that arises from the resolutions above, we are mainly interested in the point of view of syzygies: every matrix $P_{a_i}^t(D)$ gives the compatibility conditions for the system whose representative polynomial matrix is $P_{a_{i-1}}^t$. In particular, the matrix $P_1(D)$ gives the compatibility conditions that a datum $\vec{g}$ of a inhomogeneous system $P(D)\vec{f} = \vec{g}$ must satisfy to have solutions f. We have computed the resolutions in this paper running CoCoA, version 3.7 on a Digital AlphaServer 4100/600, with 4 CPU and 3 GB RAM.

2. Affine and projective planes

In this section we will study some systems associated with affine or projective spaces over $\mathbf{Z}_p$, p prime.

The affine geometry over $\mathbf{Z}_3$. This geometry has 8 points, 8 lines and 3 points per line, 3 lines through each point such that any two lines have at most one point in common while through every point outside a given line goes exactly one parallel line. In this case, three Clifford basis elements are sufficient to write the following the combinatorial system:

$$(3) \quad \begin{cases} (e_1\partial_{x_1} + e_2\partial_{x_2} + e_3\partial_{x_3})f = g_1 \\ (e_1\partial_{x_4} + e_2\partial_{x_5} + e_3\partial_{x_6})f = g_2 \\ (e_1\partial_{x_8} + e_2\partial_{x_1} + e_3\partial_{x_4})f = g_3 \\ (e_1\partial_{x_6} + e_2\partial_{x_3} + e_3\partial_{x_7})f = g_4 \\ (e_1\partial_{x_5} + e_2\partial_{x_7} + e_3\partial_{x_1})f = g_5 \\ (e_1\partial_{x_3} + e_2\partial_{x_8} + e_3\partial_{x_5})f = g_6 \\ (e_1\partial_{x_7} + e_2\partial_{x_4} + e_3\partial_{x_2})f = g_7 \\ (e_1\partial_{x_2} + e_2\partial_{x_6} + e_3\partial_{x_8})f = g_8. \end{cases}$$

The operators appearing in the system (3) have been applied to functions $f : \mathbb{R}^8 \longrightarrow \mathcal{C}_3$. Each equation appearing in the system can be associated with an 8×8 matrix U^i, $i = 1, \ldots, 7$ that can be obtained from (15) by substituting (formally) the P_k with the variables x_j if x_j is multiplied by the unit e_k. For example, the matrix symbol of the first operator can be obtained by setting $P_1 = x_1$, $P_2 = x_2$, $P_3 = x_3$. The matrix associated with the system is the 64×8 matrix $U = [U^1 \ldots U^8]^t$. The resolution is contained in the following

PROPOSITION 2.1. *The resolution of the system (3) is*

$$0 \longrightarrow R^8(-8) \longrightarrow R^{64}(-7) \longrightarrow R^{224}(-6) \longrightarrow R^{448}(-5) \longrightarrow$$

$$\longrightarrow R^{560}(-4) \longrightarrow R^{448}(-3) \longrightarrow R^{224}(-2) \longrightarrow R^{64}(-1) \longrightarrow \mathcal{M} \longrightarrow 0.$$

Moreover it is invariant with respect to changes of signature of the terms.

Proof. The rows of the 64×8 matrix $U = [U^1 \ldots U^8]^t$ introduced above generate an R–module M, where $R = \mathbb{C}[x_1, \ldots, x_8]$. The resolution was found by CoCoA giving the command Res(M). The invariance up to signature of the terms was checked using CoCoA by putting coefficients in front of the monomials x_j and by giving the command Rand( ) that let the system chose randomly the coefficients in the set $\{-1, +1\}$. ∎

REMARK 2.2. The resolution of the system (3) has the same length and Betti numbers proportional to those ones appearing in the standard De Rham complex (multiplied by 8 i.e. by the dimension of the overall Clifford algebra) $\partial_{x_1}, \ldots, \partial_{x_8}$. We will call a resolution of this type a *De Rham like* resolution.

The Fano plane. The Fano plane is the projective plane over $\mathbf{Z}_2$, so it is the smallest projective plane. It consists of 7 points, 7 lines, 3 points per line and 3 lines per point such that every two lines have exactly one point in common. More precisely, we have the following points: $\{x_1 = (0, 0, 1), x_2 = (1, 0, 1), x_3 = (0, 1, 1), x_4 = (1, 1, 0), x_5 = (0, 1, 0), x_6 = (1, 0, 0), x_7 = (1, 1, 1)\}$ and the lines joining the points x_i, x_j, x_k with

$$(i, j, k) \in \{(1, 2, 6), (2, 3, 4), (1, 3, 5), (2, 5, 7), (3, 6, 7), (1, 4, 7), (4, 5, 6)\}.$$

Also in this case, a minimal combinatorial Dirac system requires only 3 Clifford basis elements, and an example is given by

$$(4) \quad \begin{cases} (e_1 \partial_{x_1} + e_2 \partial_{x_6} + e_3 \partial_{x_2})f = g_1 \\ (e_1 \partial_{x_2} + e_2 \partial_{x_4} + e_3 \partial_{x_3})f = g_2 \\ (e_1 \partial_{x_5} + e_2 \partial_{x_3} + e_3 \partial_{x_1})f = g_3 \\ (e_1 \partial_{x_7} + e_2 \partial_{x_2} + e_3 \partial_{x_5})f = g_4 \\ (e_1 \partial_{x_3} + e_2 \partial_{x_7} + e_3 \partial_{x_6})f = g_5 \\ (e_1 \partial_{x_4} + e_2 \partial_{x_1} + e_3 \partial_{x_7})f = g_6 \\ (e_1 \partial_{x_6} + e_2 \partial_{x_5} + e_3 \partial_{x_4})f = g_7. \end{cases}$$

Each equation appearing in the system can be associated with an 8×8 matrix U^i, $i = 1, \ldots, 7$, which can be obtained from (15) by substituting (formally) the P_k with the variables x_j, if x_j is multiplied by the unit e_k. For example, the matrix symbol of the first operator can be obtained by setting $P_1 = x_1$, $P_2 = x_6$, $P_3 = x_2$. The matrix associated to the system is then a 56×8 matrix in the monomials $x_1, \ldots, x_7$ and its resolution, that was computed with the same technique used in the previous case, is contained in the following

PROPOSITION 2.3. *The resolution of the system (4) is*

$$0 \longrightarrow R^8(-7) \longrightarrow R^{56}(-6) \longrightarrow R^{168}(-5) \longrightarrow R^{280}(-4) \longrightarrow R^{280}(-3)$$

$$\longrightarrow R^{168}(-2) \longrightarrow R^{56}(-1) \longrightarrow M \longrightarrow 0.$$

Moreover the resolution is De Rham like and invariant with respect to changes of signature of the terms in (4).

Projective plane over Z_3. This projective plane has 13 points, 13 lines, 4 lines per each point, 4 points on each line. A possible system one can obtain requires 4 units, and is

(5)
$$\begin{cases}
(e_1\partial_{x_1} + e_2\partial_{x_2} + e_3\partial_{x_3} + e_4\partial_{x_{12}})f = g_1 \\
(e_1\partial_{x_5} + e_2\partial_{x_6} + e_3\partial_{x_{12}} + e_4\partial_{x_4})f = g_2 \\
(e_1\partial_{x_9} + e_2\partial_{x_{12}} + e_3\partial_{x_7} + e_4\partial_{x_8})f = g_3 \\
(e_1\partial_{x_7} + e_2\partial_{x_4} + e_3\partial_{x_1} + e_4\partial_{x_{10}})f = g_4 \\
(e_1\partial_{x_{10}} + e_2\partial_{x_8} + e_3\partial_{x_2} + e_4\partial_{x_5})f = g_5 \\
(e_1\partial_{x_3} + e_2\partial_{x_{10}} + e_3\partial_{x_6} + e_4\partial_{x_9})f = g_6 \\
(e_1\partial_{x_{11}} + e_2\partial_{x_3} + e_3\partial_{x_5} + e_4\partial_{x_7})f = g_7 \\
(e_1\partial_{x_2} + e_2\partial_{x_9} + e_3\partial_{x_4} + e_4\partial_{x_{11}})f = g_8 \\
(e_1\partial_{x_8} + e_2\partial_{x_1} + e_3\partial_{x_{11}} + e_4\partial_{x_6})f = g_9 \\
(e_1\partial_{x_6} + e_2\partial_{x_7} + e_3\partial_{x_{13}} + e_4\partial_{x_2})f = g_{10} \\
(e_1\partial_{x_{13}} + e_2\partial_{x_5} + e_3\partial_{x_9} + e_4\partial_{x_1})f = g_{11} \\
(e_1\partial_{x_4} + e_2\partial_{x_{13}} + e_3\partial_{x_8} + e_4\partial_{x_3})f = g_{12} \\
(e_1\partial_{x_{12}} + e_2\partial_{x_{11}} + e_3\partial_{x_{10}} + e_4\partial_{x_{13}})f = g_{13}
\end{cases}$$

The matrix symbol of the system is composed of 13 blocks of the type

(6)
$$V^i = \begin{bmatrix} A^i & B^i \\ -(B^i)^t & C^i \end{bmatrix}$$

where A^i, B^i, C^i are given in the Appendix and where the P_j must be substituted by the monomials x_k that are multiplied by e_j. The matrix associated to the system $V = [V^1 \dots V^{13}]^t$ gives rise to a module that has 208 independent generators and CoCoA could not compute the whole resolution. We have obtained, at least partially, the resolution by asking CoCoA to compute the first N syzygies by giving the following commands: GB.Start_Res(M); GB.Steps(M,N);
GB.GetBettiNumbers(M): the system will display a table with the Betti numbers and the degree of the syzygies found; what one obtains is

PROPOSITION 2.4. *The system (5) has 78 linear syzygies at the first step, then 286 linear syzygies, and at least 100 linear syzygies at the third step.*

REMARK 2.5. The first two steps suggest that the resolution is De Rham like, so we tried to compute the resolution for the complex $\partial_{x_1}, \ldots, \partial_{x_{13}}$ in the Clifford algebra $\mathcal{C}_4$. Unfortunately, also in this case it was not possible to compute the whole resolution but, with the same procedure illustrated above, we have taken at least some steps. The two complexes both have 78 then 286 linear syzygies, and the second has at least 224 linear syzygies at the third step, so they coincides at least at the first two steps. It is natural to conjecture that the resolutions coincide entirely, so that we can make the following

CONJECTURE 2.6. The resolution of system (5) is De Rham like and coincides with the following

$$0 \longrightarrow R^{16}(-13) \longrightarrow R^{208}(-12) \longrightarrow R^{1248}(-11) \longrightarrow R^{4576}(-10) \longrightarrow$$

$$R^{12480}(-9) \longrightarrow 0 \longrightarrow R^{20592}(-8) \longrightarrow R^{27456}(-7) \longrightarrow R^{27456}(-6) \longrightarrow$$

$$R^{20592}(-5) \longrightarrow R^{12480}(-4) \longrightarrow R^{4576}(-3) \longrightarrow R^{1248}(-2) \longrightarrow$$

$$\longrightarrow R^{208}(-1) \longrightarrow \mathcal{M} \longrightarrow 0.$$

The results we have obtained so far, have produced examples of De Rham like resolutions. We then tried to deal the case of some other designs. For sake of brevity we do not give the details of the construction of the matrices associated to each system, since they can be easily obtained by the matrices in the Appendix, following the procedure used in the cases we have already treated.

Desargues configuration. Consider 10 points and 10 lines such that every 3 points lie on a line and every point is the intersection of 3 lines. A system we obtained, by using 3 units e_1, e_2, e_3, is the following:

$$(7) \quad \begin{cases} (e_1\partial_{x_1} + e_2\partial_{x_2} + e_3\partial_{x_7})f = g_1 \\ (e_1\partial_{x_9} + e_2\partial_{x_1} + e_3\partial_{x_3})f = g_2 \\ (e_1\partial_{x_8} + e_2\partial_{x_3} + e_3\partial_{x_2})f = g_3 \\ (e_1\partial_{x_4} + e_2\partial_{x_7} + e_3\partial_{x_5})f = g_4 \\ (e_1\partial_{x_6} + e_2\partial_{x_9} + e_3\partial_{x_4})f = g_5 \\ (e_1\partial_{x_5} + e_2\partial_{x_6} + e_3\partial_{x_8})f = g_6 \\ (e_1\partial_{x_{10}} + e_2\partial_{x_4} + e_3\partial_{x_1})f = g_7 \\ (e_1\partial_{x_2} + e_2\partial_{x_5} + e_3\partial_{x_{10}})f = g_8 \\ (e_1\partial_{x_3} + e_2\partial_{x_{10}} + e_3\partial_{x_6})f = g_9 \\ (e_1\partial_{x_7} + e_2\partial_{x_8} + e_3\partial_{x_9})f = g_{10} \end{cases}$$

and the matrix associated with each equation can be obtained by (15).

PROPOSITION 2.7. *The resolution of (7) is*

$$0 \longrightarrow R^8(-10) \longrightarrow R^{80}(-9) \longrightarrow R^{360}(-8) \longrightarrow R^{960}(-7)$$

$$\longrightarrow R^{1680}(-6) \longrightarrow R^{2016}(-5) \longrightarrow R^{1680}(-4) \longrightarrow R^{960}(-3)$$

$$\longrightarrow R^{360}(-2) \longrightarrow R^{80}(-1) \longrightarrow \mathcal{M} \longrightarrow 0.$$

It is De Rham like and invariant with respect to changes of signature of the terms.

Another interesting system that can be obtained with 4 points and 6 lines joining two points at a time, is

$$(8) \quad \begin{cases} (\partial_{x_1} e_1 + \partial_{x_2} e_2)f = g_1 \\ (\partial_{x_3} e_1 + \partial_{x_4} e_2)f = g_2 \\ (\partial_{x_1} e_2 + \partial_{x_3} e_3)f = g_3 \\ (\partial_{x_2} e_1 + \partial_{x_4} e_3)f = g_4 \\ (\partial_{x_4} e_1 + \partial_{x_1} e_3)f = g_5 \\ (\partial_{x_3} e_2 + \partial_{x_2} e_3)f = g_6, \end{cases}$$

where we have considered $f : \mathbb{R}^4 \to C_3$. The matrix associated with each of the operators appearing in the system can be obtained by (15) and the resolution we have found is contained in the following:

PROPOSITION 2.8. *The system (8) has the De Rham like resolution*

$$0 \longrightarrow R^8(-4) \longrightarrow R^{32}(-3) \longrightarrow R^{48}(-2) \longrightarrow R^{32}(-1) \longrightarrow \mathcal{M} \longrightarrow 0,$$

which is invariant with respect to changes of signature of the terms.

Note that the resolution has R^{32} at the beginning. This means that only 32 rows of the module $\mathcal{M}$ are independent; that means, in terms of Clifford relations, that only 4 equations of the system are independent.

REMARK 2.9. All the resolutions we have obtained so far are De Rham like because of the particular structure of the finite geometry considered. To prove that, in general, the behaviour of a combinatorial system is not De Rham like, we treat the case of two simple finite geometries. We first consider the system associated with the dual (in the geometric sense) of the system (8): it has 6 points, 4 lines, 3 points per line. A system arising from this design is for example

$$(9) \quad \begin{cases} (e_1\partial_{x_1} + e_2\partial_{x_6} + e_3\partial_{x_2})f = g_1 \\ (e_1\partial_{x_3} + e_2\partial_{x_2} + e_3\partial_{x_4})f = g_2 \\ (e_1\partial_{x_5} + e_2\partial_{x_1} + e_3\partial_{x_3})f = g_3 \\ (e_1\partial_{x_4} + e_2\partial_{x_5} + e_3\partial_{x_6})f = g_4 \end{cases}$$

The resolution is:

$$0 \longrightarrow R^{16}(-7) \longrightarrow R^{88}(-6) \longrightarrow R^{192}(-5) \longrightarrow R^{200}(-4)$$
$$\longrightarrow R^{24}(-2) \oplus R^{80}(-3) \longrightarrow R^{32}(-1) \longrightarrow \mathcal{M} \longrightarrow 0.$$

If we consider a design with 10 points, 5 lines, 4 points per line, 2 lines through a point, we get for example a system of the type

$$\begin{cases} (e_1\partial_{x_1} + e_2\partial_{x_3} + e_3\partial_{x_6} + e_4\partial_{x_9})f = g_1 \\ (e_1\partial_{x_4} + e_2\partial_{x_1} + e_3\partial_{x_7} + e_4\partial_{x_{10}})f = g_2 \\ (e_1\partial_{x_3} + e_2\partial_{x_2} + e_3\partial_{x_4} + e_4\partial_{x_5})f = g_3 \\ (e_1\partial_{x_2} + e_2\partial_{x_6} + e_3\partial_{x_{10}} + e_4\partial_{x_8})f = g_4 \\ (e_1\partial_{x_8} + e_2\partial_{x_9} + e_3\partial_{x_5} + e_4\partial_{x_7})f = g_5 \end{cases}$$

The resolution is very complicated: since the system could not finish the whole resolution, we have computed only the first two steps and we have found that there is 1 linear and 125 quadratic syzygies at the first step, and then 392 quadratic syzygies. We point out that in these two cases, there are many possibility of choosing the indices for the variables and for the units appearing in each equation. We tried several choices of those labels and we found the same resolutions as above. This fact suggests that the resolution depends only on the underlying geometry but not on the choice of the colouring.

3. Platonic bodies.

We first consider a tetrahedron. This Platonic body has 4 vertices, 4 faces. Every vertex belongs to 3 faces and every face has 3 vertices. A system associated to it can be written in the following way:

(10)
$$\begin{cases} (e_1\partial_{x_1} + e_2\partial_{x_2} + e_3\partial_{x_3})f = g_1 \\ (e_1\partial_{x_2} + e_2\partial_{x_1} + e_3\partial_{x_4})f = g_2 \\ (e_1\partial_{x_3} + e_2\partial_{x_4} + e_3\partial_{x_2})f = g_3 \\ (e_1\partial_{x_4} + e_2\partial_{x_3} + e_3\partial_{x_1})f = g_4. \end{cases}$$

Consider now a cube and its 8 vertices and 6 faces. Every vertex belongs to 3 faces and every 3 faces intersect at most in one point. A system describing this solid is:

(11)
$$\begin{cases} (e_1\partial_{x_1} + e_2\partial_{x_2} + e_3\partial_{x_3} + e_4\partial_{x_4})f = g_1 \\ (e_1\partial_{x_7} + e_2\partial_{x_8} + e_3\partial_{x_5} + e_4\partial_{x_6})f = g_2 \\ (e_1\partial_{x_5} + e_2\partial_{x_6} + e_3\partial_{x_1} + e_4\partial_{x_2})f = g_3 \\ (e_1\partial_{x_8} + e_2\partial_{x_7} + e_3\partial_{x_4} + e_4\partial_{x_3})f = g_4 \\ (e_1\partial_{x_3} + e_2\partial_{x_5} + e_3\partial_{x_7} + e_4\partial_{x_1})f = g_5 \\ (e_1\partial_{x_6} + e_2\partial_{x_4} + e_3\partial_{x_2} + e_4\partial_{x_8})f = g_6. \end{cases}$$

The octohedron that has 6 vertices and 8 faces. Every face has 3 vertices and every vertex belongs to 4 faces. A system associated with it is

(12)
$$
\begin{cases}
(e_1\partial_{x_1} + e_2\partial_{x_2} + e_3\partial_{x_5})f = g_1 \\
(e_1\partial_{x_2} + e_2\partial_{x_1} + e_3\partial_{x_3})f = g_2 \\
(e_1\partial_{x_4} + e_3\partial_{x_1} + e_4\partial_{x_3})f = g_3 \\
(e_2\partial_{x_5} + e_3\partial_{x_4} + e_4\partial_{x_1})f = g_4 \\
(e_1\partial_{x_6} + e_3\partial_{x_2} + e_4\partial_{x_5})f = g_5 \\
(e_1\partial_{x_3} + e_2\partial_{x_6} + e_4\partial_{x_2})f = g_6 \\
(e_2\partial_{x_3} + e_3\partial_{x_6} + e_4\partial_{x_4})f = g_7 \\
(e_1\partial_{x_5} + e_2\partial_{x_4} + e_4\partial_{x_6})f = g_8.
\end{cases}
$$

Another Platonic body is the dodecahedron. It has 20 vertices, 12 faces, every face has 5 vertices and every vertex belongs to 3 faces. A system associated with the dodecahedron consists of 12 equations of the form

(13) $(e_1\partial_{x_i} + e_2\partial_{x_j} + e_3\partial_{x_k} + e_4\partial_{x_\ell} + e_5\partial_{x_n})f = g_\nu$

where $\nu = 12$, (i, j, k, ℓ, n) (in this order) belongs to

$\{(1,2,3,4,5),\ (5,1,6,14,15),\ (2,12,1,13,14),\ (3,10,2,11,12),$

$(4,5,7,6,8),\ (8,3,4,9,10),\ (6,7,15,19,20),\ (16,17,18,20,19),$

$(17,11,12,16,13),\ (7,8,19,18,9),\ (9,18,11,10,17),\ (13,14,20,15,16)\}.$

Finally, we consider an icosahedron. It has 12 vertices, 20 faces, every face has 3 vertices and each vertex belongs to 5 faces. A system associated with this solid contains 20 equations of the type (13) where $\nu = 20$, (i, j, k, ℓ, n) (in this order) belongs to

$\{(1,2,3,0,0),\ (0,1,2,7,0),\ (0,0,1,3,5),\ (4,0,0,5,3),\ (0,6,5,1,0),$

$(8,9,0,2,0),\ (6,0,7,0,1),\ (0,7,0,8,2),\ (3,0,9,0,4),\ (0,0,6,11,7),$

$(7,8,11,0,0),\ (0,11,0,10,12),\ (12,0,0,6,11),\ (11,0,8,0,10),$

$(0,10,0,9,8),\ (9,4,10,0,0),\ (10,0,4,12,0),\ (2,3,0,0,9),\ (0,5,12,4,0),$

$(5,12,0,0,6)\}.$

The last two cases have been treated in the Clifford algebra C_5. The matrix associated with each system can be obtained by suitable formal substitutions from the matrix (7) in the paper [8]. We obtained the following results:

PROPOSITION 3.1. *The systems associated with the Platonic bodies (10), (11), (12), (13) for $\nu = 20$ have the following resolutions:*

Tetrahedron:

$$0 \longrightarrow R^8(-4) \longrightarrow R^{32}(-3) \longrightarrow R^{48}(-2) \longrightarrow R^{32}(-1) \longrightarrow M \longrightarrow 0$$

Cube:

$$0 \longrightarrow R^{32}(-9) \longrightarrow R^{240}(-8) \longrightarrow R^{768}(-7) \longrightarrow R^{1344}(-6) \longrightarrow$$

$$\longrightarrow R^{48}(-4) \oplus R^{1344}(-5) \longrightarrow R^{160}(-3) \oplus R^{720}(-4) \longrightarrow$$

$$\longrightarrow R^{192}(-2) \oplus R^{160}(-3) \longrightarrow R^{96}(-1) \longrightarrow M \longrightarrow 0.$$

Octohedron:

$$0 \longrightarrow R^{16}(-6) \longrightarrow R^{96}(-5) \longrightarrow R^{240}(-4) \longrightarrow R^{320}(-3) \longrightarrow$$

$$\longrightarrow R^{240}(-2) \longrightarrow R^{96}(-1) \longrightarrow M \longrightarrow 0$$

Icosahedron:

$$0 \longrightarrow R^8(-12) \longrightarrow R^{96}(-11) \longrightarrow R^{528}(-10) \longrightarrow R^{1760}(-9) \longrightarrow$$

$$\longrightarrow R^{3960}(-8) \longrightarrow R^{6336}(-7) \longrightarrow R^{7392}(-6) \longrightarrow R^{6336}(-5) \longrightarrow$$

$$\longrightarrow R^{3960}(-4) \longrightarrow R^{1760}(-3) \longrightarrow R^{528}(-2) \longrightarrow R^{96}(-1) \longrightarrow M \longrightarrow 0$$

REMARK 3.2. The systems (10), (12), (13) for $\nu = 20$ are De Rham like while the system associated with the cube has a more complicated structure. Note also that, despite the fact that the system (13) contains 20 equations, it behaves like a De Rham system in only 12 operators. This fact proves that not all the equations in the system are independent. The case of the dodecahedron is very complicated and exceeds the capabilities of the system. The only fact we were able to prove is that there are linear relations in the first syzygies. As in the case of the cube–system, we are certain that the dodecahedron–system cannot be of De Rham type since the number of equations is less than the number of points, i.e. the total dimension of the system.

PROPOSITION 3.3. *The system (10) associated with a tetrahedron can be reduced to a De Rham system of the type*

$$\begin{cases} \partial_{x_1} f = h_1 \\ \partial_{x_2} f = h_2 \\ \partial_{x_3} f = h_3 \\ \partial_{x_4} f = h_4 \end{cases}$$

Proof. If we consider the second equation of the system (10) multiplied on the left by e_{12} we get that the first pair of equations in the system become

$$(e_1\partial_{x_1} + e_2\partial_{x_2} + e_3\partial_{x_2})f = g_1$$
$$(e_2\partial_{x_2} - e_1\partial_{x_1} + e_{123}\partial_{x_4})f = e_{12}g_2,$$

so that they can be rewritten in the form

(14) $$\qquad 2e_1\partial_{x_1}f = g_1 - e_{12}g_2 - e_3\partial_{x_3}f + e_{123}\partial_{x_4}f$$
$$2e_2\partial_{x_2}f = g_1 + e_{12}g_2 - e_3\partial_{x_3}f - e_{123}\partial_{x_4}f.$$

Now we can multiply the third equation by $-e_{23}$ so that we arrive to the new equation

$$e_2\partial_{x_2} = -e_{23}g_3 + e_{123}\partial_{x_3}f + e_3\partial_{x_4}f,$$

which can be combined with the second equation in the system (14) to give

$$(e_3 + 2e_{123})\partial_{x_3}f = g_1 + e_{12}g_2 + 2e_{23}g_3 - (2e_3 - e_{123})\partial_{x_4}f.$$

This allows us to write ∂_{x_3} in terms of ∂_{x_4} since the coefficient $e_3 + 2e_{123}$ admits an inverse. By substituting ∂_{x_1} and ∂_{x_3} in the fourth equation of (10), we can get ∂_{x_4} and, by consequence also ∂_{x_3}. This completes the proof.■

REMARK 3.4. We expect that similar reductions may also be established also in the case of the other De Rham systems we have obtained even though the reduction is not obvious, in general.

We finish this section by considering the super–dual of some of the systems we have considered. To that end, it suffices to rewrite the systems already considered by interchanging, formally, ∂_{x_i} with e_i. In the case of dodecahedron and icosahedron, it was not possible to handle the case of super–dual systems since they involve 12 or 20 units respectively, so that the matrices one has to write are too big to be treated. Some of the systems obtained produce an identical resolution, as described in the following propositions.

PROPOSITION 3.5. *The super–dual of systems associated to the Fano plane, the affine geometry over* $\mathbf{Z}_3$, *the Desargues configuration, and the system (8) have the same De Rham like resolution with respect to three operators, so they have 3 linear first syzygies, then other 3 and 1 linear syzygies.*

Proof. Once one has written the super–dual system in the various cases, one has to find the associated matrix. In the cases of the Fano plane and the affine geometry over Z_2, the matrix is composed of 7 and 8 respectively 16×16 blocks coming from the symbol of the Dirac operator in C_8 via suitable substitutions, and the resolution is

$$0 \longrightarrow R^{16}(-3) \longrightarrow R^{48}(-2) \longrightarrow R^{48}(-1) \longrightarrow \mathcal{M} \longrightarrow 0.$$

In the case of system (8) the matrix consists of 4 blocks, again 16×16, coming from suitable substitutions in the matrix associated with the Dirac operator in C_4; the resolution obtained is the same as above. Finally, in the case of the Desargues configuration, the matrix symbol of the system is made by 10 blocks of size 32×32, which can be obtained by applying the spinor formalism to the original equations which are C_{10}-valued. After the reduction, each equation can be written in $C_4 \oplus iC_4$ and once one has the matrix symbol, one can produce with CoCoA the resolution, which is

$$0 \longrightarrow R^{32}(-3) \longrightarrow R^{96}(-2) \longrightarrow R^{96}(-1) \longrightarrow \mathcal{M} \longrightarrow 0.$$

This completes the proof.∎

PROPOSITION 3.6.   *The super-duals of the systems (10), (11), (12) have the following resolutions*

$$0 \longrightarrow R^8(-4) \longrightarrow R^{32}(-3) \longrightarrow R^{48}(-2) \longrightarrow R^{32}(-1) \longrightarrow \mathcal{M} \longrightarrow 0$$

$$0 \longrightarrow R^{16}(-4) \longrightarrow R^{64}(-3) \longrightarrow R^{96}(-2) \longrightarrow R^{64}(-1) \longrightarrow \mathcal{M} \longrightarrow 0$$

$$0 \longrightarrow R^{16}(-6) \longrightarrow R^{96}(-5) \longrightarrow R^{240}(-4) \longrightarrow R^{320}(-3) \longrightarrow R^{240}(-2)$$
$$\longrightarrow R^{96}(-1) \longrightarrow \mathcal{M} \longrightarrow 0$$

so they are De Rham like.

Proof. The first resolution can be obtained by associating with the system a suitable matrix built with blocks of the type (6). The second system is associated to a matrix built with 6 blocks of size 16×16 that can be obtained from the matrix associated with the Dirac operator in C_8, with suitable substitutions (see [8]). The matrix symbol of the system has 96 rows of which only 64 are independent. The last system is associated with a matrix with blocks of size 8×8, coming from the matrix representing the Dirac operator in C_6. All the resolutions were obtained with CoCoA.∎

4. Open problems

The cases we have treated in this paper and in [7] are only a starting point in studying systems of combinatorial type. We think that several problems can be addressed in this field; some of them arise naturally in the combinatorial setting, some others arise from our treatment. Classical problems are for example:

- (P_1) In how many ways can the basis elements e_k be chosen such that all conditions A_i in the introduction are met?

- (P_2) What is the finite group of permutations of the tops leaving the set of blocks invariant which, if combined with a permutation of the elements e_k, leave the set of operators invariant?
  Less classical problems are:

- (P_3) To extend the finite group invariance to a continuous group invariance whereby the continuous extensions of the permutations of the sets $\{\partial_{x_1}, ..., \partial_{x_m}\}$ and $\{e_1, ..., e_M\}$ are subgroups of $SO(m)$ and $SO(M)$ respectively.

- (P_4) In all the cases in which we have tried to choose the signature randomly, we have found that the resolution remains the same. This fact suggests the following question: to what extent is the resolution of the system dependent on the chosen colouring or the choice of the signatures of the terms?

- (P_5) In the same spirit as the previous question we ask: for which finite geometries is the resolution only dependent on the geometry and not on the colouring or choice of signature?

- (P_6) When is the resolution of a system proportional to a De Rham complex? We have found this behaviour in several cases and we conjecture that this happens in the case of all affine and projective geometries over $\mathbf{Z}_p$, p prime, and also in the case of self-dual designs and for designs in which the number of lines is not less than the number of points.

- (P_7) What is the class of geometries that correspond to the same Betti numbers, ignoring the dimension of the Clifford algebra? What geometric information about a finite geometry is included in the sequence of degrees and Betti numbers of a resolution?

5. Appendix I

$$A = \begin{bmatrix} 0 & -P_1 & -P_2 & -P_3 & -P_4 & 0 & 0 & 0 \\ P_1 & 0 & 0 & 0 & 0 & P_2 & P_3 & P_4 \\ P_2 & 0 & 0 & 0 & 0 & -P_1 & 0 & 0 \\ P_3 & 0 & 0 & 0 & 0 & 0 & -P_1 & 0 \\ P_4 & 0 & 0 & 0 & 0 & 0 & 0 & -P_1 \\ 0 & -P_2 & P_1 & 0 & 0 & 0 & 0 & 0 \\ 0 & -P_3 & 0 & P_1 & 0 & 0 & 0 & 0 \\ 0 & -P_4 & 0 & 0 & P_1 & 0 & 0 & 0 \end{bmatrix},$$

$$B = \begin{bmatrix} 0 & 0 & 0 & 0 & 0 & 0 & 0 & 0 \\ 0 & 0 & 0 & 0 & 0 & 0 & 0 & 0 \\ P_3 & P_4 & 0 & 0 & 0 & 0 & 0 & 0 \\ -P_2 & 0 & P_4 & 0 & 0 & 0 & 0 & 0 \\ 0 & -P_2 & -P_3 & 0 & 0 & 0 & 0 & 0 \\ 0 & 0 & 0 & -P_3 & -P_4 & 0 & 0 & 0 \\ 0 & 0 & 0 & P_2 & 0 & -P_4 & 0 & 0 \\ 0 & 0 & 0 & 0 & P_2 & P_3 & 0 & 0 \end{bmatrix},$$

$$C = \begin{bmatrix} 0 & 0 & 0 & -P_1 & 0 & 0 & -P_4 & 0 \\ 0 & 0 & 0 & 0 & -P_1 & 0 & P_3 & 0 \\ 0 & 0 & 0 & 0 & 0 & -P_1 & -P_2 & 0 \\ P_1 & 0 & 0 & 0 & 0 & 0 & 0 & P_4 \\ 0 & P_1 & 0 & 0 & 0 & 0 & 0 & -P_3 \\ 0 & 0 & P_1 & 0 & 0 & 0 & 0 & P_2 \\ P_4 & -P_3 & P_2 & 0 & 0 & 0 & 0 & -P_1 \\ 0 & 0 & 0 & -P_4 & P_3 & -P_2 & P_1 & 0 \end{bmatrix}$$

(15)
$$U^i = \begin{bmatrix} 0 & P_1^i & P_2^i & P_3^i & 0 & 0 & 0 & 0 \\ P_1^i & 0 & 0 & 0 & P_2^i & P_3^i & 0 & 0 \\ P_2^i & 0 & 0 & 0 & -P_1^i & 0 & P_3^i & 0 \\ P_3^i & 0 & 0 & 0 & 0 & -P_1^i & -P_2^i & 0 \\ 0 & -P_2^i & P_1^i & 0 & 0 & 0 & 0 & -P_3^i \\ 0 & -P_3^i & 0 & P_1^i & 0 & 0 & 0 & P_2^i \\ 0 & 0 & -P_3^i & P_2^i & 0 & 0 & 0 & -P_1^i \\ 0 & 0 & 0 & 0 & P_3^i & -P_2^i & P_1^i & 0 \end{bmatrix}$$

References

1. W.W. Adams, C.A. Berenstein, P. Loustaunau, I. Sabadini, D.C. Struppa, *Regular functions of several quaternionic variables and the Cauchy-Fueter complex*, J. Geom. Anal. **9**, (1999), 1–16.

2. W.W. Adams, P. Loustaunau, *Analysis of the module determining the proper-ties of regular functions of several quaternionic variables*, to appear in Pacific J.

3. S. Fiorini, R. J. Wilson *Edge colourings of graphs*, Res. Notes in Math. **16**, Pitman 1977,

4. J. Gilbert, M. Murray, *Clifford algebras and Dirac Operators in Harmonic Analysis*, Cambridge, Cambridge Univ. Press n. 26, 1990.

5. V.P. Palamodov, *Linear Differential Operators with Constant Coefficients*, Springer Verlag, New York 1970.

6. I. Sabadini, M. Shapiro, D.C. Struppa, *Algebraic analysis of the Moisil-Theodorescu system*, Compl. Var. **40**, (2000), 333–357.

7. I. Sabadini, F. Sommen, *Special first order systems and resolutions*, preprint, (2000).

8. I. Sabadini, F. Sommen, D.C. Struppa, P. Van Lancker, *Complexes of Dirac operators in Clifford algebras*, to appear in Math. Z.

9. F. Sommen, *Special first order systems in Clifford Analysis*, Proc. Symp. Anal. Num. Meth. Quat. Clif. Anal., Seiffen 1996, 169–274.

10. F. Sommen, *Monogenic Functions of Higher Spin*, Z. A. A. **15**, (1996), n.2, 279–282.

11. F. Sommen, *Clifford analysis on super-space*, to appear in Proc. Cetraro Conf. 1998, Adv. Clif. Alg and Appl.

12. F. Sommen, G. Bernardes, *Multivariable monogenic functions of higher spin*, to appear in J. Nat. Geom.

13. V. Souček, *Clifford analysis for higher spins*, Proc. third Intern. Conf. Clif. Alg. Appl. Math. Phys., Deinze 1993, Kluwer Acad. Publ. (1993), 223–232.

The Spherical X-Ray Transform of Texture Goniometry

Helmut Schaeben (schaeben@geo.tu-freiberg.de)
Geomathematics and Geoinformatics, Freiberg University of Mining and Technology

Wolfgang Sprößig (sproessig@math.tu-freiberg.de)
Applied Mathematics, Freiberg University of Mining and Technology

Gerald van den Boogaart
Graduate College "Spatial Statistics", Freiberg University of Mining and Technology

Abstract. In crystallography, a probability density function $f(g)$ defined on the group $SO(3)$ of rotations is referred to as orientation density function; the analysis of preferred crystallographic orientation is referred to as texture analysis. In non-destrcutive practical applications it is common practice to measure diffraction pole figures $\tilde{P}(h, r)$ of a few crystallographic forms h with a texture goniometer. A pole figure is the result of sampling a spherical probability density function which is defined as tomographic projection of the orientation density function f.

In terms of quaternions a pole density function is properly defined as average of f along circles and therefore referred to as the spherical X-ray transform of texture goniometry. Properties of this spherical X-ray transform are presented, including a range theorem, and open problems are addressed.

Keywords: spherical Radon transform, quaternionic representations, rotations

Mathematics Subject Classification: 86A22, 44A12, 15A66

Dedicated to Richard Delanghe on the occasion of his 60th birthday.

1. Introduction to a real world problem

The subject of texture analysis is the experimental determination (collecting and processing data) and interpretation of the statistical distribution of orientations of crystals within a specimen of polycristalline materials, which could be metals or rocks (cf. [9]; [3]). The orientation of an individual crystal is given by the rotation $g \in SO(3)$ which brings a coordinate system K_A fixed to the specimen into coincidence with a coordinate system K_B fixed to the crystal, $g : K_A \mapsto K_B$. Thus, in the context of texture analysis, rotation always means passive or "frame" rotation.

The coordinates of a unique direction represented by $h \in S^2 \subset \mathbb{R}^3$ with respect to the crystal coordinate system K_B (referred to a crystal-

F. Brackx et al. (eds.), Clifford Analysis and Its Applications, 283–291.
© *2001 Kluwer Academic Publishers. Printed in the Netherlands.*

lographic direction) and by r with respect to the specimen coordinate system K_A (referred to as specimen direction) are related to each other by

$$h = gr. \tag{1}$$

Applying the parametrization of rotations by Euler angles (α, β, γ) of 3 successive rotations about conventionally fixed axes: the first rotation by $\alpha \in [0, 2\pi)$ about the Z-axis, the second by $\beta \in [0, \pi]$ about the (new) Y'-axis, the third by $\gamma \in [0, 2\pi)$ about the (new) Z''-axis, g in equation (1) has the matrix representation

$$M(g) = \begin{pmatrix} \cos\gamma & \sin\gamma & 0 \\ -\sin\gamma & \cos\gamma & 0 \\ 0 & 0 & 1 \end{pmatrix} \begin{pmatrix} \cos\beta & 0 & -\sin\beta \\ 0 & 1 & 0 \\ \sin\beta & 0 & \cos\beta \end{pmatrix} \begin{pmatrix} \cos\alpha & \sin\alpha & 0 \\ -\sin\alpha & \cos\alpha & 0 \\ 0 & 0 & 1 \end{pmatrix}$$

$$= \begin{pmatrix} \cos\alpha\cos\beta\cos\gamma - \sin\alpha\sin\gamma & \sin\alpha\cos\beta\cos\gamma + \cos\alpha\sin\gamma & -\sin\beta\cos\gamma \\ -\cos\alpha\cos\beta\sin\gamma - \sin\alpha\cos\gamma & -\sin\alpha\cos\beta\sin\gamma + \cos\alpha\cos\gamma & \sin\beta\sin\gamma \\ \cos\alpha\sin\beta & \sin\alpha\sin\beta & \cos\beta \end{pmatrix}$$

(cf. [1],[4]).

Texture analysis with X-ray diffraction data is the analysis of the orientation distribution by volume and asks for a measure of the volume portion $\Delta V/V$ of the polycrystalline specimen of total volume V carrying crystal grains with orientations within a range (volume element) $\Delta G \subset G$ of the subgroup G of all feasible orientations $G \subset SO(3)$.

Assuming that the measure possesses a probability density function $f : G \mapsto \mathbb{R}_+^1$, then

$$\text{prob}(g \in \Delta G) = \int_{\Delta G} f(g) dg$$

and f is referred to as the orientation density function by volume and $dg = \sin\beta d\alpha d\beta d\gamma$ is the invariant measure of $SO(3)$ (cf. [5]).

It is emphasized that size, shape, and spatial location of the contributing grains are not considered. An observed orientation distribution by volume may actually be predominated by a single grain.

Eventually, texture analysis is the determination and interpretation of the orientation density function of a polycrystalline specimen. It is often referred to as analysis of crystallographic preferred orientation, and the objective of the interpretation is to relate an observed pattern of preferred orientation to its generating processes.

In X-ray diffraction experiments the orientation density function f cannot be directly measured but with a texture goniometer only pole density functions $\tilde{P}(h, r)$ can be sampled, which represent the probability that a (fixed) crystal direction h or its antiparallel $-h$ statistically coincide with the specimen direction r. With respect to the experiment the feasible crystal directions are the normals of the crystallographic lattice planes. A pole density function is the tomographic projection of an orientation density function which is basically provided by

$$\tilde{P}(h,r) = \frac{1}{2}\Big(P(h,r) + P(-h,r)\Big) \qquad (2)$$

with

$$P(h,r) = \frac{1}{2\pi} \int_{\{g \in SO(3)\,|\,h=gr\}} f(g)dg = (\mathcal{P}_h f)(r). \qquad (3)$$

Observing that

$$g(\alpha,\beta,\gamma) \begin{pmatrix} \cos\alpha\sin\beta \\ \sin\alpha\sin\beta \\ \cos\beta \end{pmatrix} = \begin{pmatrix} 0 \\ 0 \\ 1 \end{pmatrix}, \quad \gamma \in [0, 2\pi)$$

the path of integration $\{g \in SO(3)\,|\,h = gr\}$ in equation (3) may be represented as the set of two successive (passive) rotations in terms of Euler angles. The first is a fixed rotation $g_1 = g_1(\varphi, \vartheta, 0)$ given in terms of the spherical coordinates (φ, ϑ) of r mapping r onto $e_3 = (0, 0, 1)$, the second is a variable rotation $g_2^{-1} = g_2^{-1}(\alpha, \beta, \omega)$ given in terms of the spherical coordinates (α, β) of h and a variable angle $\omega \in [0, 2\pi)$ mapping e_3 onto h for all $\omega \in [0, 2\pi)$, resulting in

$$\{g \in SO(3) \mid h = gr\} = \{g_2^{-1}(\alpha, \beta, \omega)g_1(\varphi, \vartheta, 0) \mid \omega \in [0, 2\pi)\} \quad (4)$$
$$= \{g_2^{-1}(\alpha, \beta, 0)g_1(\varphi, \vartheta, \omega) \mid \omega \in [0, 2\pi)\}. \quad (5)$$

Thus, the defining equation (3) of the h–pole density function may be explicitly rewritten as

$$P((\cos\alpha\sin\beta, \sin\alpha\sin\beta, \cos\beta)^t, (\cos\varphi\sin\vartheta, \sin\varphi\sin\vartheta, \cos\vartheta)^t)$$
$$= \frac{1}{2\pi} \int_0^{2\pi} f(g_2^{-1}(\alpha, \beta, \omega)g_1(\varphi, \vartheta, 0))d\omega. \qquad (6)$$

2. The spherical X-ray transform of texture analysis in terms of quaternions

2.1. Definitions

Let $\mathbb{H}$ denote the skew-field of real quaternions. An arbitrary quaternion $q \in \mathbb{H}$ has the representation

$$q = q_0 + \underline{q}$$

with $\underline{q} = \sum_{i=1}^{3} q^i e_i = \text{Vec } q$ and $q_0 = \text{Sc } q$, where $\text{Vec } q$ denotes the vector part of q, and $\text{Sc } q$ denotes the scalar part of q. The basis elements

e_i $(i = 1, 2, 3)$ fulfil the relations $e_i e_j + e_j e_i = 0$ $(i = 1, 2, 3)$. S^2 denotes
the unit sphere in Vec $\mathbb{H} \simeq \mathbb{R}^3$ which is generated by all unit vectors,
and S^3 denotes the sphere in $\mathbb{H} \simeq \mathbb{R}^4$ which is generated by all unit
quaternions.

Now let $\underline{h}$ and $\underline{r}$ be unit vectors. We are going to identify

$$\underline{h} = h \quad \text{and} \quad \underline{r} = r$$

which means that the vectors $\underline{h}$ and $\underline{r}$ can also be considered as the
vector parts of the quaternions h and r, respectively.

Any passive ("frame") rotation can be written as

$$R_q(r) = q^{-1} r q$$

with $q^{-1} = \bar{q} |q|^{-2}$, where $\bar{q}$ denotes the conjugate quaternion of q.
$R_q(r)$ provides the coordinate transformation of r corresponding to the
rotation represented by q (cf [1],[2],[4]).

Defining

$$q_1 := h + r \tag{7}$$
$$q_2 := 1 - rh \tag{8}$$

it is easy to show that for $-h \neq r$

$$q_1^{-1} r q_1 = h = R_{q_1}(r) \tag{9}$$
$$q_2^{-1} r q_2 = h = R_{q_2}(r) \tag{10}$$

hold. Since $h = q^{-1} r q$ is equivalent to $qh = rq$ for $q \neq 0$, the proofs
follow from

$$(h + r)h = -|h|^2 + rh = rh - 1 = rh - |r|^2 = r(h + r)$$

$$(1 - rh)h = h - rhh = h + r = -rrh + r = r(1 - rh).$$

q_1 represents the rotation about the normalized axis of rotation

$$n_1 = \frac{h + r}{|h + r|} = \frac{1}{2\cos(\eta/2)}(h + r)$$

by π, which is the largest feasible angle of rotation, and q_2 represents
the rotation about the normalized axis of rotation

$$n_2 = \frac{h \times r}{|h \times r|} = \frac{1}{\sin \eta}(h \times r)$$

by $\eta = \arccos([h, r])$, which is the smallest feasible angle of rotation,
where $[h, r] := \text{Sc}\,(r\bar{h})$ denotes the scalar product in $\mathbb{H}$.

With a real–valued probability density function f defined on S^3 the texture goniometer diffraction experiment is mathematically modeled by an averaging operator $(Af)(h,r)$ which is defined pointwise as the mean of f for all $q \in S^3$ with $q^{-1}rq = h$, and is subsequently introduced step by step. Geometrically this mean may be interpreted as being taken for all passive rotations representing the unique unit vector r with respect to the coordinate system K_A as $h = q^{-1}rq$ with respect to the rotated coordinate system K_B.

Since only rotations about h leave h fixed, the mean $(Af)(h,h)$ is given as the mean of f over all rotations with axis h for any given unit vector $h \in S^2$:

$$(Af)(h,h) := \frac{1}{2\pi} \int_0^{2\pi} f(\cos t + h \sin t)dt. \tag{11}$$

Due to the invariance of the X–ray experiment with respect to any rotation of the crystal and specimen coordinate system Af has to be defined such that

$$(Af^{(q_1,q_2)})(q_1^{-1}hq_1, q_2^{-1}rq_2) = (Af)(h,r)$$

holds with $f^{(q_1,q_2)}(q) := f(q_2qq_1^{-1})$, $q_1, q_2 \in S^3$.

Choosing $q \in \mathbb{H}$ such that $q^{-1}rq = h$, the mean for some arbitrary given unit vectors $h, r \in \text{Vec } \mathbb{H}$ is thus defined as

$$(Af)(h,r) := (Af^{(1,q)})(h,h)$$

$$= \frac{1}{2\pi} \int_0^{2\pi} f(q(\cos t + h \sin t))dt$$

$$= \frac{1}{2\pi} \int_0^{2\pi} f(q\cos t + qh \sin t)dt. \tag{12}$$

It should be noted that

$$R_q(r) = q^{-1}rq = h$$
$$R_{qh}(r) = \overline{qh}rqh = (\overline{rq})r(rq) = q^{-1}\overline{r}r(rq) = q^{-1}rq = h$$
$$R_{\alpha q + \beta qh}(r) = h, \quad \alpha, \beta \in \mathbb{R}^1$$
$$q(qh) = -|q|^2h$$
$$[q, qh] = 0.$$

Thus, q and qh are seen to be orthogonal with respect to the scalar product in $\mathbb{H}$.

The dependence of $(Af)(h,r)$ on r becomes more obvious when $q = (h+r)/|h+r| \in S^3$ in particular is chosen. Then $qh = (rh-1)/|h+r| \in S^3$, and

$$(Af)(h,r) = \frac{1}{2\pi} \int_0^{2\pi} f\left(\frac{h+r}{|h+r|}\cos t + \frac{rh-1}{|h+r|}\sin t\right)dt. \qquad (13)$$

The set

$$Q(h,r) := \{(h+r)\cos t+(rh-1)\sin t, \ t \in [0,2\pi]\} = \{q \in \mathbb{H} \mid q^{-1}rq = h\}$$

may generally be interpreted geometrically as an ellipse, which degenerates to a circle if $h+r$ and $rh-1$ are normalized. In this case, $(Af)(h,r)$ is the average of $f(q)$ along the circle $\{q \in \mathbb{H} \mid q^{-1}rq = h\}$. While q varies within $Q(h,r)$, the angle of the corresponding rotations varies in $[\eta,\pi]$ and the axis of rotation along the circle of S^2 spanned by n_1 and n_2, because the rotation axes Vec $(q\cos t + qh\sin t)/\|$Vec $(q\cos t + qh\sin t)\|$ are the normalized vector parts of quaternions constituting an ellipse.

The case $-h = r$ must be considered separately. The rotations q with $q^{-1}rq = -r$ are provided by any rotation with its axis in the orthogonal complement of Vec r, which again represents a circle if normalized, and its angle constantly equal to π.

Summarizing, $(Af)(h,r)$ is well defined and may be referred to as spherical X-ray transform; it actually provides the proper mathematical model for the tomographic pole figure projection of texture goniometry.

2.2. PROPERTIES

In the following some properties including invariance properties are listed.

$$(Af)(-h,r) = (Af)(h,-r) \qquad (14)$$

$$\left(Af(\cdot)\right)(h,r) = \left(Af(\bar{\cdot})\right)(r,h) \qquad (15)$$

$$(Af)(qhq^{-1},r) = (Af^{(q,q)})(h,q^{-1}rq) \qquad (16)$$

$$(Af)(qhq^{-1},r) = (Af)(h,q^{-1}rq) \qquad (17)$$

where the last equation holds if f is independent of Vec q, in which case $f(q) = f(q_0 q q_0^{-1})$ for any $q_0 \in S^3$.

$(Af)(h,r)$ satisfies a spherical variant of Asgeirsson's mean value theorem which may be stated as follows. Let $\mu(h,r;\rho)$, $\nu(h,r;\rho)$, and

$w(h, r; \rho, \tau)$ be defined as

$$\mu(h, r; \rho) := \int_{[h, h']=\rho} (Af)(h', r) dt(h') \tag{18}$$

$$\nu(h, r; \rho) := \int_{[r, r']=\rho} (Af)(h, r') dt(r') \tag{19}$$

$$w(h, r; \rho, \tau) := \int_{[r, r']=\tau} \int_{[h, h']=\rho} (Af)(h', r') dt(h') dt(r') \tag{20}$$

where dt denotes the one-dimensional boundary element of the small circle $S^2 \cap S_{h, \sqrt{2(1-\rho)}} = C_{h, \rho} = \{h' \in S^2 \,|\, [h', h] = \rho\}$ with center $h \in S^2_+$ and radius $\sin(\arccos \rho)$ normalized in such a way that the integral of the function 1 is 1. Then

$$\mu(h, r; \rho) = w(h, r; \rho, 1) \tag{21}$$
$$\nu(h, r; \rho) = w(h, r; 1, \rho) \tag{22}$$

and

$$\mu(h, r, \rho) = \nu(h, r, \rho). \tag{23}$$

More generally

$$w(h, r; \rho, \tau) = w(h, r; \tau, \rho) \tag{24}$$

the double mean value is symmetric in the radii ρ and τ.

Since equation (23) holds for all $\rho \in (-1, 1)$ it provides at once a mean value theorem for the spherical surface element $\Omega_{h, \rho} = \{h' \in S^2 \,|\, \rho \leq [h', h]\}$ with center $h \in S^2_+$ by integration with respect to ρ

$$\int_{[1, \delta]} \mu(h, r, \rho) d\rho = \int_{[1, \delta]} \nu(h, r, \rho) d\rho \tag{25}$$

or more explicitly

$$\int_{[1, \delta]} \int_{[h, h']=\rho} (Af)(h', r) dt(h') d\rho = \int_{[1, \delta]} \int_{[r, r']=\rho} (Af)(h, r') dt(r') d\rho. \tag{26}$$

Eventually, $(Af)(h, r)$ satisfies the differential equation

$$(\Delta_h - \Delta_r)(Af)(h, r) = 0. \tag{27}$$

A range theorem characterizing the image of the X-ray transform in terms of the differential equation states that any function $u \in C^\infty(S^2 \times S^2)$ satisfying $(\Delta_h - \Delta_r)u(h, r) = 0$ is the X-ray transform of some function $f \in C^\infty(S^3)$ (Nikolayev and Schaeben, 1999).

3. Conclusions and Outlook

A spherical X-ray transform has been well defined in terms of quaternions which provides a mathematical model for the pole figure projection of texture goniometry. The quaternion model provides more instructive insight in the structure of the involved mathematics than any other model. Results in [8],[9] may prove helpful to find satisfying answers to the such open problems as:

— What is the kernel of the operator

$$(Xf)(h,r) := \frac{1}{2}((Af)(h,r) + (Af)(-h,r))$$

in terms of quaternions?

— Is there a nice inversion formula?

— Is there a criterion for $(Af)(h,r)$ to be the image of a non–negative f?

— Is it possible to reconstruct a non–negative f from experimental images?

— Is it possible to expand the functions into series of simple quaternionic functions?

References

1. Altmann, S.L., 1986, Rotations, Quaternions, and Double Groups: Oxford University Press, Oxford
2. Gürlebeck, K., Sprößig, W., 1997, Quaternionic Calculus for Physicists and Engineers: Wiley
3. Kocks, U.F., Tomé, C.N., Wenk, H.R., 1998, Texture and Anisotropy - Preferred Orientations in Polycrystals and their Effect on Materials Properties: Cambridge University Press, Cambridge
4. Kuipers, J.B., 1999, Quaternions and Rotation Sequences û A Primer with Applications to Orbits, Aerospace, and Virtual Reality: Princeton University Press
5. Moran, P.A.P., 1975, Quaternions, Haar measure, and the estimation of a paleomagnetic rotation, in Gani, J., (ed.), Perspectives in Probability and Statistics, Applied Probability Trust, 295-301
6. Nikolayev, D.I, Schaeben, H., 1999, Characteristics of the ultrahyperbolic differential equation governing pole density functions: Inverse Problems 15, 1603-1619
7. Sommen, F., 1987, Spingroups and spherical means II: Supplemento ai Rendiconti del Circolo Matematico di Palermo, ser II, 14, 157-177

8. Sommen, F., 1989, Spingroups and spherical means III: Supplemento ai
    Rendiconti del Circolo Matematico di Palermo, ser II, 21, 295-323
9. Wenk, H.R., (ed.), 1985, Preferred Orientation in Deformed Metals and Rocks
    - An Introduction to Modern Texture Analysis: Academic Press, Orlando

Quaternionic Complexes in Clifford Analysis

Petr Somberg (somberg@karlin.mff.cuni.cz)*
Mathematical Institute of Charles University Prague

Abstract. The series of complexes, studied in [3], are interpreted as quaternionic complexes of invariant differential operators discussed for the first time in [1]. The starting point is the discovery of the full automorphism groups of the operators appearing in the Hilbert syzygies of [3]; then the result follows from the comparison of dimensions of places and homogeneities of the symbols of morphisms in both complexes.

Keywords: Hilbert syzygies, invariant differential operators, BGG resolutions

Mathematics Subject Classification 2000: 18G40,18G35,13D02,35R20

Dedicated to Richard Delanghe on the occasion of his 60th birthday.

1. Introduction

In [3], the authors interested in Clifford analysis study Hilbert syzygies associated with "n-Dirac operators" as morphism between the 1st and 0th places of Hilbert syzygies. In particular, let R be a finitely generated graded ring. A free resolution of graded R-modules

$$\ldots \to F_n \overset{\varphi_n}{\to} \ldots \to F_1 \overset{\varphi_1}{\to} F_0 \qquad (1)$$

is called a Hilbert syzygy (φ_1=symbol of "n-Dirac operators"). The Hilbert syzygy theorem ensures that there exists a resolution of finite length, bounded from the top by the number of primitive elements of the ring R. The methods used in [3] cannot take into account the invariance properties of the structure considered.

The fact that the algebro-geometric counterpart in projective geometry of a finitely generated free module over a graded ring is a coherent sheaf, together with a module structure on coherent cohomology, leads to an expectation of the interpretation of the series of complexes in [3] inside parabolic invariant theory. As we shall see, the singular BGG-resolutions, with a reductive subgroup of the parabolic group equal to the automorphism group of the quaternionic projective space, naturally fits into this concept.

---

* This work was supported by grant GAČR No. 201/00/P070.

F. Brackx et al. (eds.), Clifford Analysis and Its Applications, 293–301.

In fact, we shall show that these complexes correspond in the sense of the dimensions of all places and the homogeneity of all homomorphisms in the resolutions, to quaternionic complexes, introduced for the first time in [1]. In other words, the main observation leading to this interpretation is the discovery for the full group of automorphisms of "n-Dirac operators" not to be n-copies of the conformal group of the ordinary Dirac operator, but rather the full group of automorphisms of quaternionic projective space. This assertion has been discussed in [4].

In conclusion, in the case of (basic) quaternionic geometry, we use following vocabulary in the respective languages of [3] and [1]:

(i) Hilbert syzygy ⟿ singular BGG resolution

(ii) linear syzygy ⟿ (image of the symbol of) standard singular invariant differential operator

(iii) quadratic syzygy ⟿ (image of the symbol of) non-standard singular invariant differential operator

(iv) free module in the resolution ⟿ Lie algebra (co)homology, valuation module of sections.

We would like to warn the reader. In differential geometry, people usually study the dual of BGG sequences. In order to compare the results with [3], we use (in their language) the transpose of their complexes. Of course, the second possibility would be to compare directly the Hilbert syzygies with the resolution by generalized Verma modules.

2. Penrose transform and its relation to non-standard morphisms

Let X, Y, Z be complex manifolds, $\eta : Y \longrightarrow Z$, $\tau : Y \longrightarrow X$ fibrations, i.e. continuous epimorphisms which need not be holomorphic, such that the couple $(\eta, \tau) : Y \hookrightarrow Z \times X$ embeds Y in Cartesian product of Z and X. The diagram

is called the **correspondence** of (X, Y, Z). We regard the points $z \in Z$ $(x \in X)$ as parametrizing submanifolds of X (Z) via

$$z \longrightarrow \tau(\eta^{-1}(z)) \subset X \ (x \longrightarrow \eta(\tau^{-1}(x)) \subset Z). \tag{2}$$

Moreover, we impose two technical conditions: η should be a topologically trivial fibration in the sense of fiberwise homotopy, and τ should have compact fibers.

The heart of the Penrose transform is the following principle. Let E be a holomorphic bundle on Z ($E \longrightarrow Z$), and $\mathcal{O}(E)$ be its sheaf of sections. Let us choose an analytic cohomology class $\omega \in H^p(Z, \mathcal{O}(E))$ on Z. Then the Penrose transform is a map

$$\omega \longrightarrow P\omega$$
$$H^p(Z, \mathcal{O}(E))|_{\eta(\tau^{-1}(-))} \longrightarrow \Gamma(X, H^p(\eta(\tau^{-1}(-)), \mathcal{O}(E))), \quad (3)$$

such that $P\omega(x) \in H^p(\eta(\tau^{-1}(x)), \mathcal{O}(E)) \equiv H^p(Z, \mathcal{O}(E))|_{\eta(\tau^{-1}(x))}$. However, as we shall see, the sections in the image $P\omega$ will be suitably restricted to lie in the kernels of invariant differential operators (the system of PDE's). In conclusion, we have the correspondence:

(analytic) cohomology class of Z $\leadsto$

$\qquad\qquad$ kernels of invariant differentialoperators on X .

2.1. HOMOGENEOUS GENERALIZED PENROSE TRANSFORM

Let G be a semisimple complex Lie group (with Lie algebra $\mathfrak{g}$), and P, R standard parabolic subgroups (with Lie algebras $\mathfrak{p}$, $\mathfrak{r}$); then $Q := P \cap R$ is also a parabolic subgroup (with Lie algebra $\mathfrak{q}$).

AIM 2.1. *We would like to systematically generate non-standard singular invariant morphisms w.r.t. a given structure carried by a parabolic system (geometry) via unfolding (relative) regular resolutions by standard invariant operators (regular Bernstein-Gelfand-Gelfand (B-G-G) sequences) into singular resolutions by (NON-)standard invariant operators (singular B-G-G sequences). The image of the symbol of any of these non-standard operators will correspond to quadratic syzygies of complexes.*

The homogeneous case of the Penrose transform is described by the correspondence

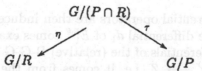

where the fibers of η resp. τ are generalized flag manifolds $R/(P \cap R)$ resp. $P/(P \cap R)$. The correspondence (2) is recovered by choosing an affine subspace ("big cell", or open orbit of maximal compact subgroup of G etc.) $X \subset G/P$, and then $Y := \tau^{-1}(X), Z := \eta(Y)$.

The construction of (2.1) can be described in two steps:

1. pull-back stage

The fiber-wise homotopy equivalence implies, for all p,

$$H^p(Z, \mathcal{O}(E)) \xrightarrow{\sim} H^p(Y, \eta^{-1}\mathcal{O}(E)), \tag{4}$$

where $\eta^{-1}\mathcal{O}(E)$ denotes the topological inverse image sheaf. In the homogeneous case, $\mathcal{O}(E) := \mathcal{O}_{\mathfrak{r}}(\lambda)$ is the holomorphic homogeneous vector bundle associated with the $\mathfrak{r}$-module with dominant weight λ and restricted to Z. On Y, there is a fiber-wise relative B-G-G resolution of $\eta^{-1}\mathcal{O}_{\mathfrak{r}}(\lambda)$ by $\mathfrak{q}$-modules,

$$0 \longrightarrow \eta^{-1}(\mathcal{O}_{\mathfrak{r}}(\lambda)) \longrightarrow \Delta_{\mathfrak{q}}^{\star}(\lambda)$$
$$\Delta_{\mathfrak{q}}^p(\lambda) := \oplus_{w \in W_{\mathfrak{r}}^{\mathfrak{q}}, l(w)=p} \mathcal{O}_{\mathfrak{q}}(w\lambda), \tag{5}$$

where $W_{\mathfrak{r}}^{\mathfrak{q}}$ is the relative Hasse diagram with affine action on weights, and $l(w)$ is the length of minimal representant $w \in W_{\mathfrak{r}}^{\mathfrak{q}}$.

2. push-forward stage

This step is technically more difficult. Let $\tau_{\star}^q$ denote the q-th (derived) direct image of τ in the sense of sheaf theory. In the homogeneous case, $\tau_{\star}^q$ can be computed using (a relative version of) the Bott-Borel-Weil (B-B-W) theorem. The B-B-W theorem identifies cohomology groups of homogeneous sheaves on homogeneous spaces with finite dimensional representations, via the module structure on cohomology. There is (see [2]) a hyper-cohomology spectral sequence $E_1^{p,q}$,

$$E_1^{p,q} := H^p(Y, \Delta_{\mathfrak{q}}^q(\lambda)) \Longrightarrow H^{p+q}(Y, \eta^{-1}\mathcal{O}_{\mathfrak{r}}(\lambda)).$$

Using the B-B-W theorem, we have $H^q(Y, \Delta_{\mathfrak{q}}^q(\lambda)) \xrightarrow{\sim} \Gamma(X, \tau_{\star}^q \Delta_{\mathfrak{q}}^q(\lambda))$, which finally gives the desired convergence of the spectral sequence $(E_m^{p,q}, d_m)$:

$$E_1^{p,q} = \Gamma(X, \tau_{\star}^q \Delta_{\mathfrak{q}}^q(\lambda)) \Longrightarrow H^{p+q}(Z, \mathcal{O}_{\mathfrak{r}}(\lambda)). \tag{6}$$

Invariant differential operators are then induced from differentials d_m of $E_m^{p,q}$; the differential d_1 of $E_1^{p,q}$ comes exactly from the push-forward of differentials of the (relative) B-G-G sequence (5) on the fibers of $\eta : Y \longrightarrow Z$, i.e. it comes from the standard invariant homomorphisms.

3. Example - non-standard morphisms for quaternionic structure on A_4

As an example, we consider the complex Lie group

$$G = A_7 \equiv \overset{1\ \ 2\ \ 3\ \ 4\ \ 5\ \ 6\ \ 7}{\bullet\!-\!\bullet\!-\!\bullet\!-\!\bullet\!-\!\bullet\!-\!\bullet\!-\!\bullet}\,,$$

here the numbers over the nodes correspond to the numbering of the simple roots. The fibrations

$$\eta : \overset{1\ \ 2\ \ 3\ \ 4\ \ 5\ \ 6\ \ 7}{\times\!-\!\times\!-\!\bullet\!-\!\bullet\!-\!\bullet\!-\!\bullet\!-\!\bullet} \longrightarrow \overset{1\ \ 2\ \ 3\ \ 4\ \ 5\ \ 6\ \ 7}{\times\!-\!\bullet\!-\!\bullet\!-\!\bullet\!-\!\bullet\!-\!\bullet\!-\!\bullet}\,,$$

$$\tau : \overset{1\ \ 2\ \ 3\ \ 4\ \ 5\ \ 6\ \ 7}{\times\!-\!\times\!-\!\bullet\!-\!\bullet\!-\!\bullet\!-\!\bullet\!-\!\bullet} \longrightarrow \overset{1\ \ 2\ \ 3\ \ 4\ \ 5\ \ 6\ \ 7}{\bullet\!-\!\times\!-\!\bullet\!-\!\bullet\!-\!\bullet\!-\!\bullet\!-\!\bullet}\,,$$

have kernels

$$Ker(\eta) \simeq \overset{2\ \ 3\ \ 4\ \ 5\ \ 6\ \ 7}{\times\!-\!\bullet\!-\!\bullet\!-\!\bullet\!-\!\bullet\!-\!\bullet}\,,$$

and

$$Ker(\tau) \simeq \overset{1}{\times}$$

respectively.

This case will correspond to the resolution in [3] starting with "3-Dirac operators".

The regular Hasse diagram for the quaternionic structure on A_7 is

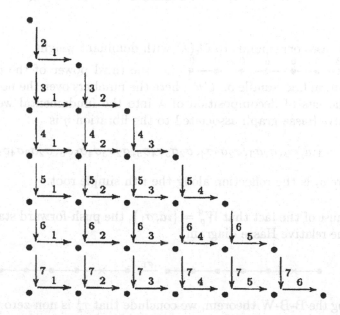

where the numbers correspond to the reflections in the weight lattice along (one of the seven) simple roots.

The singular cases of interest correspond to dominant weights lying in the closure of the Weyl chamber (but not in the intersection of planes perpendicular to simple roots). There are seven possibilities for singular Hasse diagrams, but we shall study only one of them:

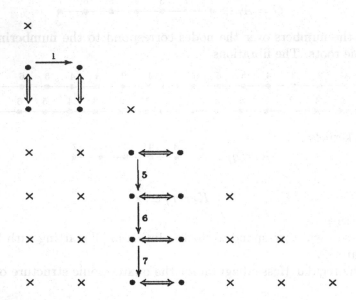

This case corresponds to $\mathcal{O}_{\mathbf{r}}(\lambda)$ with dominant weight

$$\underset{\times}{\overset{-3}{\bullet}}\ \underset{}{\overset{0}{\bullet}}\ \underset{}{\overset{0}{\bullet}}\ \underset{}{\overset{0}{\bullet}}\ \underset{}{\overset{0}{\bullet}}\ \underset{}{\overset{0}{\bullet}}\ \underset{}{\overset{0}{\bullet}}$$

i.e. the third power of the dual of the canonical line bundle on CP^7; here the numbers over the nodes are the coefficients of decomposition of λ into the fundamental weights! The relative Hasse graph associated to the fibration η is

$$W_{\mathbf{r}}^{\mathbf{q}} = \{id, \sigma_2, \sigma_2\sigma_3, \sigma_2\sigma_3\sigma_4, \sigma_2\sigma_3\sigma_4\sigma_5, \sigma_2\sigma_3\sigma_4\sigma_5\sigma_6, \sigma_2\sigma_3\sigma_4\sigma_5\sigma_6\sigma_7\},$$

where σ_i is the reflection along the i-th simple root.

Because of the fact that $W_{\mathbf{p}}^{\mathbf{q}} = \{id, \sigma_1\}$, the push-forward stage consists of the relative Hasse diagram

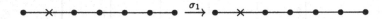

Using the B-B-W theorem, we conclude that τ_*^1 is non-zero on the first two terms only, and τ_*^0 is zero on the first three terms (and the identity on the remaining places). Thus the first list of hypercohomology spectral sequence reads

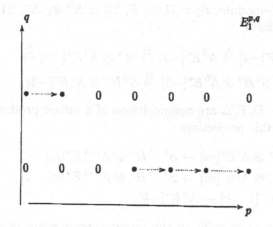

and its first derived list is

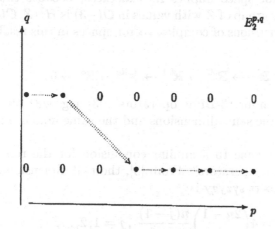

In this case there appears in the second list of spectral sequences the operator $\Delta := d_2 : E_1^{1,1} \longrightarrow E_1^{3,0}$, and so it must inevitably be a non-standard singular invariant differential operator; the convergence is then manifestly $E_3^{p,q} = E_\infty^{p,q}$.

Let us finally rewrite these sequences in a slightly more differential geometric form. In the standard notation used for quaternionic structures,

$$E \equiv \overset{0}{\bullet} \overset{0}{\times} \overset{1}{\bullet} \overset{0}{\bullet} \overset{0}{\bullet} \overset{0}{\bullet} \overset{0}{\bullet}, E^\star \equiv \overset{0}{\bullet} \overset{-1}{\times} \overset{1}{\bullet} \overset{0}{\bullet} \overset{0}{\bullet} \overset{0}{\bullet} \overset{0}{\bullet},$$

$$H \equiv \overset{1}{\bullet} \overset{0}{\times} \overset{0}{\bullet} \overset{0}{\bullet} \overset{0}{\bullet} \overset{0}{\bullet} \overset{0}{\bullet}, H^\star \equiv \overset{1}{\bullet} \overset{-1}{\times} \overset{0}{\bullet} \overset{0}{\bullet} \overset{0}{\bullet} \overset{0}{\bullet} \overset{0}{\bullet},$$

and as an $(A_1 \times A_5)$-module, $T_{\mathbb{C}} \simeq H \otimes_{\mathbb{C}} E$, $T_{\mathbb{C}}^{\star} \simeq H^{\star} \otimes_{\mathbb{C}} E^{\star}$. Then the sequence takes the form

$$H^{\star}[-1] \xrightarrow{F} E^{\star}[-2] \xrightarrow{\Delta} \Lambda^3 E^{\star}[-3] \xrightarrow{D} H^{\star} \otimes \Lambda^4 E^{\star}[-3] \xrightarrow{D}$$

$$\xrightarrow{D} S^2 H^{\star} \otimes \Lambda^5 E^{\star}[-3] \xrightarrow{D} S^3 H^{\star} \otimes \Lambda^6 E^{\star}[-3] , \qquad (7)$$

where the operators D, F, Δ are compositions of a tensor product with $T_{\mathbb{C}}^{\star}$ followed by suitable projection:

$$D : \quad S^k H^{\star} \otimes \Lambda^l E^{\star}[m] \to S^{k+1} H^{\star} \otimes \Lambda^{l+1} E^{\star}[m],$$

$$F : \quad S^k H^{\star} \otimes \Lambda^l E^{\star}[m] \to S^{k-1} H^{\star} \otimes \Lambda^{l+1} E^{\star}[m-1],$$

$$\Delta : \quad \Lambda^{k-2} E^{\star}[1-k] \to \Lambda^k E^{\star}[-k].$$

We now consider an open orbit in the homogeneous space of quaternionic subgroups $GL(1, \mathbb{H}) \subset GL(2, \mathbb{C})$ etc., regarded as real forms of complex Lie groups. The convergence of the sequence to the analytic cohomology of twistor space implies its exactness, because the only non-zero cohomology group of Z with values in $\mathcal{O}(-3)$ is $H^1(Z, \mathcal{O}(-3))$.

Counting real dimensions of complex vector spaces in this resolution reveals

$$0 \to \mathbb{R}^4 \to \mathbb{R}^{12} \to \mathbb{R}^{40} \to \mathbb{R}^{60} \to \mathbb{R}^{36} \to \mathbb{R}^8 \to 0, \qquad (8)$$

which is the resolution for "3-Dirac operators", [3], e.g. with the same number of places of the same dimensions and the same homogeneity of symbols (syzygies).

It is easy now to come to a similar conclusion for the resolution starting with "n-Dirac operators". From [3], the real dimension of the j-th place of the Hilbert syzygy is

$$dim_{\mathbb{R}} F_j = 4 \binom{2n-1}{j} \frac{n(j-1)}{j+1} , j = 1, 2, ... \qquad (9)$$

On the other hand, one easily computes the complex dimensions of places appearing in quaternionic complexes [1]:

$$dim_{\mathbb{C}} V_j = (j+1) \binom{2n}{j+3} , j = 3, 4, ... \qquad (10)$$

It is easy to see that, after substitution $j \to j-2$, these combinatorial formulas agree.

THEOREM 3.1. *For any number n of Dirac operators, such that "n-Dirac operators" over $\mathbb{R}^4$ starts the Hilbert syzygy, the places of the Hilbert syzygies in [3] have the same dimensions and the same homogeneities of morphisms as the quaternionic complexes in [1]. Thus* *it is natural to interpret the "n-Dirac operators"-operator to have as the automorphism group, the full automorphism group of quaternionic projective space, and not just the product of n-copies of 4-dimensional conformal group.*

Furthermore appeared in [3] other interesting Hilbert syzygies related to a suitable number of "Dirac operators" over $\mathbb{R}^5$ or $\mathbb{R}^6$. I think that they should also have an interpretation inside parabolic invariant theory, but at present it is not clear yet which interpretation it should be.

References

1. Baston, R.J. - Quaternionic complexes, Journal of Geometry and Physics 8, (1992), 29-52 .
2. Baston R.J., Eastwood M.G. - The Penrose transform - its interaction with representation theory, Oxford University Press, New York, 1989.
3. I.Sabadini, F.Sommen, D.C.Struppa, P.Van Lancker - Complexes of Dirac operators, To appear in Complex Variables, 2000.
4. V.Souček - Clifford analysis as a study of invariant operators, in: Proceedings of the NATO ARW "Clifford Analysis and its Applications", F Brackx, JSR Chisholm & V Souček (eds.), Kluwer Academic Publishers, Dordrecht, 2001

Clifford Analysis on the Level of Abstract Vector Variables

F. Sommen* (fs@cage.rug.ac.be)
Department of Mathematical Analysis, Ghent University

Abstract. In this paper we present a survey of results as well as new ideas in connection with analysis on the level of abstract vector variables. In particular we define abstract Dirac operators and monogenic objects. We also present various Fischer decompositions.

Keywords: Clifford analysis, abstract vector variables

Mathematics Subject Classification: 30G35

Dedicated to Richard Delanghe on the occasion of his 60th birthday.

Introduction

The most basic algebra of vector variables is the free associative algebra $R(S)$ generated by a set S of so-called abstract vector variables together with the defining relations

(A)
$$z(xy + yx) = (xy + yx)z.$$

These so-called radial algebras $R(S)$ are quite similar to algebras of vector variables in a Clifford algebra R_m generated by a basis $\{e_1, \ldots, e_m\}$ with defining relations $e_i e_j + e_j e_i = -2\delta_{ij}$. Indeed for each $x \in S$ one may produce the vector variable $\underline{x} = \sum e_j x_j$ depending on scalar variables x_j. Hence if $P(m, S)$ denotes the algebra of "Clifford polynomials" generated by the set $\{e_1, \ldots, e_m\} \cup \{x_1, \ldots, x_m : x \in S\}$ one obtains the representation

$$\underline{\ } : R(S) \to P(m, S) : x \to \underline{x}$$

which becomes isomorphic in case Card $S \leq m$.

Whatever is defined as Clifford vector variables (in a coordinate free way) may be reintroduced on the level of abstract vector variables and in that way one would obtain a version of Clifford analysis which is defined independent of coordinates, quadratic forms and dimension. This was one of the goals of the book [HS] and our motivation to consider radial algebras in [So1], [So2] was first of all to provide a better axiomatic foundation for these ideas. In particular one may

* Senior Research Associate, FWO, Belgium

F. Brackx et al. (eds.), Clifford Analysis and Its Applications, 303–322.
© 2001 *Kluwer Academic Publishers. Printed in the Netherlands.*

redefine the Dirac operator ∂_x (or vector derivative $\partial_{\underline{x}}$) abstractly as an endomorphism on $R(S)$ acting from the left and the right by imposing the axioms:

(D1) $\qquad \partial_x[fF] = \partial_x[f]F + f\partial_x[F], \quad [fF]\partial_x = F\partial_x[f] + f[F]\partial_x,$

whereby $F \in R(S)$ and $f \in Z(R(S))$ is central (scalar),

(D2) $\qquad \partial_x[FG] = \partial_x[F]G, \quad [GF]\partial_x = G[F]\partial_x,$

$\qquad \qquad$ for $\quad G \in R(S\backslash\{x\})$

(D3) $\qquad [\partial_x F]\partial_y = \partial_x[F\partial_y],$

(D4) $\qquad \partial_x\, x^2 = x^2\, \partial_x = 2x, \quad \partial_x\{x,y\} = \{x,y\}\partial_x = 2y, \quad x \neq y.$

At first sight this work of reformulating Clifford analysis for abstract vector variables seems quite elementary. But differences arise from the fact that Clifford analysis is defined on a space R^m of fixed dimension m provided with a fixed quadratic form. For example any wedge product of more than m vectors vanishes, which leads to the fact that e.g. the Fischer decomposition as formulated in [So3] is not always valid. Counter examples to this fundamental theorem play an important role in connection with the Dirac complex (see also [ABLSS] and [SSSVL]). In radial algebras there are no limitations having to do with the choice of a dimension or quadratic form and theorems from Clifford analysis must hence be reconsidered. In this paper it will be shown that the abstract Fischer decomposition holds in general. In fact, using the above axioms one may prove that the object $M = \partial_x[x]$ is an abstract scalar parameter which replaces the dimension $M = m$ of the underlying space of vectors R^m in Clifford analysis. This means that the radial algebra $R(S)$ is generated over the field of rational functions in the scalar variable M and when restricting M to a specific value $M = m \in \mathbb{N}$, one may have to deal with zeros and poles, causing restriction on the validity of theorems after making the above Clifford vector representation $x \to \underline{x}$.

The idea of "abstract dimension M" is not new; in special function theory the dimension of space usually corresponds to a complex parameter. But defining vector variables and derivatives using an abstract dimension M is already much more. Moreover, in our papers [So4, So5, So6] we defined a "super-space representation" of $R(S)$ of the form

$$x \to \underline{\dot{x}} + \underline{x}$$

where as before, $\underline{x} = \sum e_j x_j$ and whereby $\underline{\dot{x}} = \sum e_j\dot{x_j}$, $\dot{x_1}, \ldots, \dot{x}_{2n}$ being anti-commuting variables $\dot{x_i}\dot{x_j} = -\dot{x_j}\dot{x_i}$ and $\dot{e_1}, \ldots, \dot{e}_{2n}$ being a set of generators for the Crumeyrolle Clifford algebra (see [Cr]) with symplectic defining relations

$$e_i \dot{e_j} - e_j \dot{e_i} = f_{ij} = -f_{ji}.$$

This leads to a natural way to introduce Dirac-type operators, monogenic functions and other Clifford analysis objects on super-spaces by simply using the above projection $x \to \underline{\dot{x}} + \underline{x}$. Remarkable hereby is also the fact that on super-space the abstract dimension M assumes the integer value

$$M = m - 2n.$$

One also obtains a synthesis of an orthogonal and a symplectic structure.

A Dirac operator for symplectic forms was previously introduced and studied in [Ha] also using the Crumeyrolle Clifford algebra (or Weyl algebra), but expressed in commuting rather than anti-commuting variables. It is possible to link the operator from [Ha] to the radial algebra only by extending the radial algebra $R(S)$, including also abstract vector differentials

$$dx \to d\underline{x} = \sum_{j=1}^{m} e_j \, dx_j$$

and introducing abstract contraction operators

$$\partial_x| \to \partial_{\underline{x}}| = \sum_{j=1}^{m} e_j \partial_{x_j}|$$

as endomorphisms (see [So4]). The Dirac operator introduced in [Ha] then corresponds to a fermionic representation of the operator $\partial_x|$. The vector differentials $d\underline{x}$ and contraction operators $\partial_{\underline{x}}|$ play an important role in theories of monogenic differential forms (see [So7], [SS]).

We may thus conclude that the extended radial algebra provides a kind of unifying abstract background for both orthogonal and symplectic Clifford analysis in commuting and anti-commuting variables.

Without claiming completeness we also refer to related results obtained in [Be], [CRS], [Pa], [VV] while for Clifford analysis we refer to [DSS], [GM]. In section one of this paper we repeat the axiomatic formulation of radial algebra in its general form, including differential forms. We also show that there exists an alternative formulation inspired by [Ha], in which the the anti-commutator of vector variables is replaced by a commutator and the standard Clifford algebra by the Crumeyrolle algebra [Cr]. But it turns out that this "alternative radial algebra" is transformable to the original one so that there is only one unified axiomatic system.

In section two we give a definition of monogenicity on the level of abstract vector variables and prove several Fischer decompositions. An independent proof for the Fischer decomposition in Clifford analysis was obtained by P. Van Lancker, using Helgason's version of the Fischer decomposition for scalar systems (see [He]) and the representation theory of Spin(m) (see [VLSC]). Explicit computer verifications were made by D. Constales.

1. Radial algebras with differential forms

We start from a general set S of which the elements $x \in S$ are called abstract vector variables and define the radial algebra $R(S)$ as the free associative algebra generated by S, using the above relations (A). This may be done over any field, in particular over the field of rational functions of the scalar variable $M = \partial_x[x]$, ∂_x being the above defined abstract vector derivative.

To extend this radial algebra $R(S)$ with differential forms, we first produce an exact disjoint copy dS of the set S assuming the "unpronounced axioms"

(U1) $d : S \to dS$ is a bijection,

(U2) $S \cap dS = \varnothing$.

The axioms which define the extended radial algebra $R(S \cup dS)$ were established in our paper [So4] and were also inspired by the following representation in terms of Clifford vector variables and their differentials

$$x \to \underline{x} = \sum_{j=1}^{m} e_j x_j, \qquad dx \to d\underline{x} = \sum_{j=1}^{m} e_j dx_j,$$

whereby $x_1, \ldots, x_m$ are commuting coordinates, $dx_1, \ldots, dx_m$ their differentials and $e_1, \ldots, e_m$ is the standard Clifford basis $e_i e_j + e_j e_i = -2\delta_{ij}$. The symbol "$d$" hereby stands for the classical exterior derivative which is determined by axioms of the form

(i) $dd = 0$

(ii) $d(fF) = df F + f dF$, for $f \in \mathrm{Alg}\{x_1, \ldots, x_m; \ldots\}$

which also leads to the property

(iii) $\quad d(dx_j F) = -dx_j \, dF.$

When working with abstract vector variables, it seems natural to define the algebra $R(S \cup dS)$ as the associative algebra which is freely generated by the set $S \cup dS$ over the field of rational functions in "M" taking the following rules or axioms into account:

(i) the bijective operator "d" extends to an endomorphism on $R(S \cup dS)$ with the properties:

(R1) $\quad d(xF) = dxF + xdF, \quad x \in S, \quad F \in R(S \cup dS),$

(R2) $\quad d(dxF) = -dxdF, \quad x \in S, \quad F \in R(S \cup dS),$

which in particular imply that $dd = 0$, as well as the relations

$$d\{x, y\} = \{dx, y\} + \{x, dy\}, \quad d\{dx, y\} = -[dx, dy],$$

(ii) the product in $R(S \cup dS)$ satisfies the extra relations for any triple $x, y, z \in S,$

(R3) $\quad [z, \{x, y\}] = 0,$

(R4) $\quad [dz, \{x, y\}] = 0,$

(R5) $\quad [z, \{dx, y\}] = 0,$

(R6) $\quad \{dz, \{dx, y\}\} = 0,$

(R7) $\quad [z, [dx, dy]] = 0,$

and as property one has the extra identity

(R8) $\quad [dz, [dx, dy]] = 0.$

These simple rules indeed define the computational system for working with vector variables and their differentials on an abstract basis and in [So4] we proved that the assignment

$$\underline{} : x \in S \to \underline{x} = \sum_{j=1}^{m} e_j x_j, \quad dx \in dS \to d\underline{x} = \sum_{j=1}^{m} e_j dx_j,$$

leads non-isomorphic representations

$$\underline{} : R(S \cup dS) \to P(m, S, dS)$$

whereby $P(m, S, dS)$ is simply the algebra generated by $\{e_1, \ldots, e_m\}$ together with the sets $\bigcup_{x \in S} \{x_1, \ldots, x_m\}$ and $\bigcup_{x \in S} \{dx_1, \ldots, dx_m\}$. Important is the fact that all these representations together leads to an injective mapping; we say that the totality of all these representations forms an isomorphism.

Our next task is study the endomorphisms acting on $R(S \cup dS)$ and in particular to extend the definition of the vector derivative "∂_x" to the algebra $R(S \cup dS)$ as well as to introduce an abstract version "$\partial_x|$" of the contraction operator $-\partial_{\underline{x}}| = -\sum e_j \partial_{x_j}|$ (see also [So7]), thus extending the above representations to

$$\partial_x \to -\partial_{\underline{x}} = -\sum_{j=1}^m e_j\, \partial_{x_j}, \quad \partial_x| \to -\partial_{\underline{x}}| = -\sum_{j=1}^m e_j\, \partial_{xj}|.$$

The correct identities needed for this are in fact determined by these representations and they were worked out also in [So4]. In particular we have to assume that the abstract dimension M is given by

$$\partial_x|dx = \partial_x x = M$$

and corresponds to the dimensions m of the underlying space of vectors R^m.

An important identity satisfied by the operators $\partial_x = \partial_x.$ and $\partial_x = \partial_x|.$ acting from the left is the following (see also [So7])

$$\partial_x = d\partial_x| + \partial_x|d,$$

meaning that $\partial_x.$ is a kind of Lie derivative acting on differential forms. A similar identity may be obtained for the operators acting from the right. Moreover, when assuming these identities as axioms, it is sufficient to define the contraction operator $\partial_x|.$

Another option is to define (dual to the operator d) the "primitivation operator p" via the axioms

(i) $p[x] = 0,$ $p[dx] = x,$ i.e. $p : dS \to S$ inverts "d",

(ii) $p[xF] = x\, p[F],$

(iii) $p[dx\ F] = x\, F - dx\, p[F].$

Note that while in coordinates the operator "d" corresponds to the sum of the operators $\sum dx_j \partial_{x_j}$ over all variables $x \in S$, the operator "p" would rather correspond to the sum of the operators $\sum x_j \partial_{x_j}|.$

Hence for the left operators $\partial_x = \partial_x.; \partial_x| = \partial_x|.$ one may establish the dual to the above identity:

$$\partial_x| = -p\, \partial_x + \partial_x\, p$$

which makes it possible to define $\partial_x|$ in terms of ∂_x and so it is sufficient to define either ∂_x, or $\partial_x|$ as endomorphisms on $R(S \cup dS)$. Finally, to extend the operator ∂_x to $R(S \cup dS)$ we first introduce

DEFINITION 1. *The "scalar subalgebra" $r(S \cup dS)$ is the subalgebra of $R(S \cup dS)$ generated by the elements $\{x, y\}, \{dx, y\}, [dx, dy], x, y \in S$.*

Note that $r(S \cup dS)$ is not the center of $R(S \cup dS)$ but rather corresponds to the set of scalar-valued objects in the Clifford algebra representation $P(m, S, dS)$ of $R(S \cup dS)$ i.e. those elements that are mapped on the subalgebra generated by the set $\{x_j, dx_j : j = 1, \ldots, m; x \in S\}$. Also the operators "$d$" and "$p$" are in this sense scalar-valued.
To define now the operator $\partial_x, x \in S$ we note that every $F \in R(S \cup dS)$ may be written into the form

$$F = \sum fG + x \sum fG, \quad f \in r(S \cup dS), \quad G \in R((S \backslash \{x\}) \cup dS)$$

and, sticking to this notation, we assume the modified axioms

(V1) $\partial_x[fG] = \partial_x[f]G$

(V2) $\partial_x[xfG] = \partial_x[f]xG + \partial_x[x]fG, \quad \partial_x[x] = M,$

and similar rules for the action of ∂_x from the right. Hence it suffices to define the action of ∂_x on the scalar subalgebra $r(S \cup dS)$ by

(V3) $\partial_x[fg] = \partial_x[f]g + f\partial_x[g], \quad \partial_x[f] = [f]\partial_x,$

(V4) $\partial_x\{x, y\} = 2y, \quad \text{for} \quad x \neq y, \partial_x x^2 = 2x$

(V5) $\partial_x\{x, dy\} = 2dy, \quad \partial_x[dy, dz] = 0.$

This completes the axiomatic definition of vector variables, vector derivatives, vector differentials and vector contractors on an abstract level. There are of course many more interesting endomorphisms that may be considered as we did for vector variables in [So1]. This however requires extensive treatment which is beyond the scope of this paper.
Next let us see what becomes of the algebra of vector variables if we replace the vector variables $\underline{x} = \sum_{j=1}^{m} e_j x_j$ by the "symplectic alternative"

$\underline{X} = \sum_{j=1}^{2n} \grave{e_j} X_j$ whereby the standard Clifford algebra defining relations are replaced by those of the Crumeyrolle Clifford algebra (see [Cr]):

$$\grave{e}_{2j-1} \grave{e}_{2j} - \grave{e}_{2j} \grave{e}_{2j-1} = 1, \qquad \grave{e}_j \grave{e}_k = \grave{e}_k \grave{e}_j \qquad \text{otherwise,}$$

which may be realized when making an identification with the Weyl algebra

$$\grave{e}_{2j-1} = \partial_{a_j}, \qquad \grave{e}_{2j} = a_j$$

as was chosen in [Ha] and [So4].

The vector variables $\underline{X}, \underline{Y}$ then satisfy

$$[\underline{X}, \underline{Y}] = \sum \{X_{2j-1} Y_{2j} - X_{2j} Y_{2j-1}\}$$

which is a commutative scalar.

In this setting it is also natural to define a Dirac-type operator by

$$\partial_{\underline{X}} = 2 \sum_{j=1}^{m} \{\dot{e}_{2j} \, \partial_{X_{2j-1}} - \dot{e}_{2j-1} \, \partial_{X_{2j}}\}$$

(compare with [Ha]) for which one has the special relations

$$\partial_{\underline{X}}[\underline{X}, \underline{Y}] = 2\underline{Y} = [\underline{X}, \underline{Y}]\partial_{\underline{X}},$$

$$\partial_{\underline{X}} \underline{X} = -2n = -\underline{X}\partial_{\underline{X}}$$

and more general relations of the form

$$\partial_{\underline{x}} \underline{X}^l = C(l, n)\underline{X}^{l-1}$$

for suitable constants $C(l, n)$. Apart from these there are differentiation rules similar to the rules (D1),(D2), (D3).

This suggests defining an "alternative radial algebra" $R(S`)$ starting from a set $S`$ of abstract vector variables $X \in S`$ using the axiom

(A`)     $[Z, [X, Y]] = 0, X, Y, Z \in S`$

and then defining abstract vector derivatives and exterior differentiation following ideas similar to the above ones for abstract Clifford vectors. In particular as abstract laws for differentiation we assume laws similar to (D1), (D2), (D3) together with evaluations of the form

$$\partial_X[X, Y] = [X, Y]\partial_X = 2Y,$$

$$\partial_X[X] = -[X]\partial_X = -2N, \qquad \partial_X[X^l] = C(l, M)X^{l-1}.$$

In this way it suffices to make the assignments

$$X \to \underline{X}, \qquad \partial_X \to \partial_{\underline{X}}, \qquad N \to n$$

in order to arrive at a representation carrying over all identities correctly. Next to define differential forms one would introduce a version "D" of the exterior derivative via laws of the form

(R1`) $D(X \, F) = DX \, F + X \, DF$

(R2`) $D(DX \, F) = -DX \, DF$

while this time $[X, Y], [DX, Y]$ and $\{DX, DY\}$ are expected to play the role of generalized scalars. This would lead to identities of the form

(R3') $[Z, [X, Y]] = 0,$

(R4') $[DZ, [X, Y]] = 0,$

(R5') $[Z, [DX, Y]] = 0,$

(R6') $\{DZ, [DX, Y]\} = 0,$

(R7') $[Z, \{DX, DY\}] = 0,$

from which one may derive the identity

(R8') $[DZ, \{DX, DY\}] = 0.$

But this formalism is in fact fully equivalent with the formalism first introduced which was based on Clifford vectors. In this we make the following identifications:

$$X \in S \to e\, dx, dx \in dS,$$

$$D \to fep,$$

$$DX \in DS \to f\, pdx = f\, x, \quad x \in S$$

whereby e, f are Clifford algebra generators satisfying $e^2 = f^2 = 1$ and $e f = -f e$ and whereby p is the primitivation operator. Note that we also have that in this correspondence

$$[X, Y] \to [dx, dy], \quad [DX, Y] = fe\{x, dy\}, \quad \{DX, DY\} \to \{x, y\}$$

and the axioms (R1'), (R2') transform into the laws (ii), (iii) for the primitivation operator p, while the remaining laws (R3')-(R8') transform each into one of the laws (R3)-(R8). Now these laws (R3)-(R8) together with the primitivation operator also lead back to the reintroduction of the operator d via laws (R1), (R2). On the symplectic level this would correspond to the introduction of a primitivation operator P via laws of the similar form

(i) $P\, DX = X, \quad P\, X = 0,$

(ii) $P(X\, F) = X\, PF,$

(iii) $P(DX\, F) = X\, F - DX\, PF$

and, when making the identification $P \to e\ f\ d$, the identity (i) holds in the S, dS setting and the identities (ii),(iii) lead back to (R1),(R2).

Finally for the vector derivative ∂_x inspired by [Ha] we make the identification

$$\partial_X[F] \to e\partial_x|[F], \quad [F]\partial_X \to [F](-e\partial_x|), \quad -2N \to M$$

to arrive at the correct identities for $\partial_x|$:

$$\partial_X[X,Y] = e\partial_x|[dx,dy] = 2e\ dy = 2Y = [X,Y]\partial_X = -e[dx,dy]\partial_x|,$$

$$\partial_X X = e\partial_x|e\ dx = M = -2N = -X\partial_X = e\ dx\ e\ \partial_x|,$$

$$\partial_X[X,DY] = 2DY = [X,DY]\partial_X$$
$$= e\ \partial_x|e\ f\{dx,y\} = 2fy = e\ f\{dx,y\}(-\partial_x|).$$

There is hence just one axiomatic formalism that can be written in two styles namely "orthogonal style" and "symplectic style". For technical simplicity of calculus we advise keeping the use of both styles in spite of their equivalence, especially in coordinate representations.

2. Monogenic objects and Fischer decompositions in radial algebras

In this section we restrict ourselves to the "orthogonal style" of formulating the radial algebra. The results obtained are however also valid in the symplectic setting i.e. for the Dirac operators considered in [Ha] and also for differential forms and also in the super-space setting.

To define "monogenic radial objects" we start off with a set of vector variables of the special form $S \cup T$, whereby we have in mind the disjoint union of a set S of so-called "true" vector variables $x, y, \ldots \in S$ and a set T of so-called "parameter vectors" $u, v, \ldots \in T$.

DEFINITION 2. *An element $F \in R(S \cup T)$ is called monogenic with respect to S if for every $x \in S$ we have the equation $\partial_x F = 0$. More in general, an element $F \in R(S \cup T \cup dS \cup dT)$ is called monogenic with respect to dS if for every $x \in S$ we have the equation $\partial_x|F = 0$.*

In most practical cases the set S of true vector variables will be a finite set of the form $S = \{x_1, \ldots, x_k\}$ and in particular for $S = \{x\}$ we have to do with monogenic objects in the vector variable x (or in dx or in both x and dx). The so-called parameter vectors $u \in T$ behave in much the same way as general Clifford vectors provided one makes no use of the vector derivatives with respect to these variables and also no differentials du. One may even assume the use of vector variables like e_1, e_2,... with constraints $e_j\, e_k + e_k\, e_j = -2\delta_{jk}$ thus reintroducing Clifford algebra within a radial algebra or one could consider nullvectors $t_1, \ldots, t_k$ using constraints of the form $t_j\, t_k + t_k\, t_j = 0$.

But all these cases are restrictions of the general case whereby $u \in T$ are just abstract vector variables, which allows one to use ∂_u, du and $\partial_u|$. Radial algebra is hence to be developed under the assumption that also the parameter vectors are variables.

This distinction between fixed and variable vectors becomes very important when introducing the so-called Fischer inner product.

In the standard Clifford analysis setting whereby one uses a Clifford basis $e_1, \ldots, e_m$ with $e_j\, e_k + e_k\, e_j = -2\delta_{jk}$ one first defines the conjugate a^+ of $a \in C_m$ by:

$$(a + i\, b)^+ = \bar{a} - i\, \bar{b} \qquad \text{with} \qquad a,\, b \in R_m$$

whereby $a \to \bar{a}$ is the main anti-involution given by $\bar{e}_j = -e_j$.

This means that on C_m one has a Hermitian inner product given by

$$(a, b) = [a^+ b]_0.$$

Next for C_m-valued polynomials $R(\underline{x}_1, \ldots, \underline{x}_k)$ one defines the Fischer inner product by

$$(R,\, S) = \left[R^+(\partial_{\underline{x}_1}, \ldots, \partial_{\underline{x}_k}) S(\underline{x}_1, \ldots, \underline{x}_k) \right]_0 \big|_{\underline{x}_j = 0},$$

i.e. one replaces the variables x_{jk} by the corresponding partial derivatives $\partial_{x_{jk}}$ which one can write as $(x_{jk})^+$ so that in fact

$$(R(\underline{x}_1, \ldots, \underline{x}_k))^+ = R^+(\partial_{\underline{x}_1}, \ldots, \partial_{\underline{x}_k}).$$

It is easy to extend this inner product to a Hermitian positive definite inner product on the space of all polynomial differential forms in k variables $R(\underline{x}_1, \ldots, \underline{x}_k;\, d\underline{x}_1, \ldots, d\underline{x}_k)$ by defining the Hermitian conjugate of a coordinate form dx_{jk} by $(dx_{jk})^+ = \partial_{x_{jk}}|$ and by stating that the Hermitian conjugation is an anti-involution.

One may now deduce from this an inner product on a general radial algebra $R(S \cup dS)$ by first defining the Hermitian conjugate $F \to F^+$ by the axioms

$$(F\, G)^+ = G^+ F^+,$$

$$x^+ = \partial_x, \quad dx^+ = \partial_x| \qquad \text{for any} \qquad x \in S$$

and for any F, $G \in R(S \cup dS)$ the Fischer inner product could hence be defined by (see also [Sol])

$$(F, G) = [F^+[G]]_0,$$

where this time the "scalar part" refers to the part which is homogeneous of degree zero in all variables $x \in S$ and $dx \in dS$. But this inner product takes values in the field of rational functions in M, so it is not positive definite except under the Clifford algebra representation

$$x \to \underline{x}, \quad dx \to d\underline{x}, \quad \partial_x \to -\partial_{\underline{x}}, \quad \partial_x| \to -\partial_{\underline{x}}|, \quad M \to m$$

where it coincides with the Fischer inner product for Clifford objects and this ONLY in case the above representation is an ISOMORPHISM. Now this is only the case for radial algebras $R(S)$ when $\geq$ Card(S). Hence in the case of vector variables one could say that the abstract Fischer inner product is positive definite if one assumes that M is real and very large. In case one also uses differentials dx, one never obtains a positive definite inner product for the simple reason that dx^l never vanishes.

However, if one restricts the degree of homogeneity of objects in the differential form variables dx as well as the number of such variables considered, the above application becomes an isomorphic representation provided that m is large enough.

The above way to introduce an inner product also works for the algebra $R(S \cup T \cup dS \cup dT)$ but that means that the parameter $u \in T$ is considered to be a vector variable rather than a fixed real Clifford vector. In the case of elements $u \in T$ considered as fixed vectors one would have to assume that $u^+ = -u$ and $-u^2$ is positive. But that excludes the use of the basic vector differentials du, which are needed in analysis also as "parameter objects". Moreover, the assumption $-u^2 > 0$ also excludes the consideration of super-vectors and even of complex vectors, in particular nullvectors $u^2 = 0$.

Next suppose that we are in a position to use a positive Fischer inner product on some suitable restriction of $R(S \cup T \cup dS \cup dT)$, assuming that Card(S) and Card(T) are finite, and that one considers subspaces of homogeneous objects. Then from the positive definiteness one obtains

THEOREM 1. *(Simple Fischer decomposition)*
Every $F \in R(S \cup T), S = \{x_1, \ldots, x_k\}$ may be written in a unique way as

$$F = M(F) + \sum_{j=1}^{k} x_j F_j, \quad \partial_{x_j} M(F) = 0, \qquad x_j \in S,$$

Similarly every $F \in R(S \cup T \cup dS \cup dT)$ may be written in a unique way as

$$F = M'(F) + \sum_{j=1}^{k} dx_j F_j, \quad \partial_{x_j}| M'(F) = 0, \qquad x_j \in S,$$

$$F = M''(F) + \sum_{j=1}^{k} x_j F_j + \sum_{j=1}^{k} dx_j G_j,$$

$$\partial_{x_j} M''(F) = \partial_{x_j} | M''(F) = 0, x_j \in S.$$

<u>Proof</u>. In fact when considering single elements $F \in R(S \cup T)$ one always works on the assumption of finitely many vector variables and bounded degrees of homogeneity, so that the above decomposition holds for large positive values of the dimension M (see also [So3]). However, in the Fischer decomposition process the only numbers that arise are expressible as rational functions of M. Hence the above Fischer decomposition remains true over the field of rational functions in M. ∎

Note that the above also implies that the Fischer decomposition holds in cases where M is negative and close to ∞. This gives another proof of the Fischer decomposition in the fermionic case (see [So5]) but also in the case of the symplectic Dirac operators (see [Ha]), where $M = -2N < 0$. Next, by recursive application of the above result one may in the case of $F \in R(S \cup T)$ arrive at a decomposition of the form

$$F = M_0(F) + M_1(F) + M_2(F) + \ldots$$

whereby $M_s(F)$ is expressible in the form

$$M_s(F) = \sum x_{j_1} \ldots x_{j_s} F_{j_1 \ldots j_s} \quad \text{with} \quad \partial_{x_j} F_{j_1 \ldots j_s} = 0, \quad x_j \in S.$$

But it is not clear whether the pieces $F_{j_1} \ldots j_s$ are unique and even not whether the total terms $M_s(F)$ are unique. In the Clifford analysis setting we proved in [So3] that the terms $M_s(F)$ are orthogonal for the Fischer product. We first proved this under the assumption that $F(\underline{x}_1, \ldots, \underline{x}_k)$ is a Clifford tensor i.e. it is linear in each of the variables $\underline{x}_j$. This is no true restriction because a general polynomial can be obtained from a Clifford tensor by making identifications between vector variables of the form $\underline{x}_1 = \ldots = \underline{x}_{l_1}, \underline{x}_{l_1+1} = \ldots \underline{x}_{l_1+l_2}, \ldots$ and the monogenic decomposition remains a monogenic decomposition under this projection. The orthogonality result then follows from the fact that in the multilinear case

$$\partial_{x_{j_1}} \ldots \partial_{x_{j_l}} \underline{x}_1 \ldots \underline{x}_s G(x_{s+1}, \ldots, \underline{x}_k) = 0$$

whenever $l > s$ and G is multilinear monogenic, hereby making permutations of this result w.r.t. the indices of the variables.
It is hence not hard to prove that a similar orthogonality result remains

valid in the radial algebra setting $F \in R(S \cup T)$ and also there one may make use of objects F which are multilinear in the abstract vector variables $x_j \in S$.

On the level of differential forms one may for any $F \in R(S \cup T \cup dS \cup dT)$ produce a decomposition of the form

$$F = M_0'(F) + M_1'(F) + M_2'(F) + \ldots,$$

whereby $M_s'(F)$ has the special form

$$M_s'(F) = \sum dx_{j_1} \ldots dx_{j_s} F_{j_1 \ldots j_s} \quad \text{with} \quad \partial_{x_j}|F_{j_1 \ldots j_s} = 0, \quad x_j \in S.$$

Similarly to the above case one may prove that the terms $M_0'(F)$, $M_1'(F), \ldots$ are orthogonal with respect to the Fischer inner product. Indeed once again by making identification of vector differentials dx_j it is sufficient to prove the above orthogonality in the case where F is multilinear in the vector differentials $dx_1, \ldots, dx_k$, and in that case the result follows from the identities for monogenic forms G:

$$\partial_{x_{j_1}}| \ldots \partial_{x_{j_l}}|dx_1 \ldots dx_s G(dx_{s+1}, \ldots, dx_k) = 0 \quad \text{for} \quad l > s.$$

Note that this result also leads to an orthogonal Fischer decomposition in the case of the symplectic Dirac operator in [Ha].

More complicated is the decomposition of the form

$$F = M_0''(F) + M_{1,0}''(F) + M_{0,1}''(F) + \ldots + M_{s,t}''(F) + \ldots$$

whereby

$$M_{s,t}''(F) = \sum \text{perm}(x_{j_1}, \ldots, x_{j_s}, dx_{l_1}, \ldots, dx_{l_t}) F_{\text{perm},j_i,k_i},$$

$$\partial_{x_j} F_{\text{perm},j_i,k_i} = \partial_{x_j}|F_{\text{perm},j_i,k_i} = 0, x_j \in S,$$

and in which "perm" refers to "taking permutations and multiply". This decomposition is in general not orthogonal although it turns out to be a direct sum. Moreover, if one groups terms together as follows:

$$M_1''(F) = M_{1,0}''(F) + M_{0,1}''(F), M_2''(F) = M_{2,0}''(F) + M_{1,1}''(F) + M_{0,2}''(F) \text{ etc.}$$

then one may prove the orthogonality of the decomposition

$$F = M_0''(F) + M_1''(F) + M_2''(F) + \ldots$$

similar to [So3] by once again proving things in the multilinear case and using identities for monogenic forms $G(x_{s+1}, \ldots; dx_{t+1}, \ldots)$ of the form:

$$\text{perm}(\partial_{x_{j_1}}, \ldots, \partial_{x_{j_l}}, \partial_{x_{k_1}}|, \ldots, \partial_{x_{k_n}}|) \text{perm}(x_1, \ldots, x_s, dx_1, \ldots, dx_t) F$$

$$= 0,$$

which hold in case $l + n > s + t$ and permutations of this identity.

We may summarise these results in:

THEOREM 2. *(Orthogonal Fischer decomposition)*
Any $F \in R(S \cup T)$ resp. $F \in R(S \cup T \cup dS \cup dT)$ admits orthogonal
decompositions of the above forms:

$$F = M_0(F) + M_1(F) + \ldots,$$
$$F = M_0'(F) + M_1'(F) + \ldots, \qquad F = M_0''(F) + M_1''(F) + \ldots.$$

Much more difficult is to examine the uniqueness of the terms in a
decomposition of the form

$$M_s(F) = \sum x_{j_1} \ldots x_{j_s} F_{j_1 \ldots j_s}, \quad \partial_{x_j} F_{j_1 \ldots j_s} = 0, \quad x_j \in S,$$

and this even in the case of multilinear objects $F \in R(S \cup T)$ with
respect to the variables $x_j \in S$. First of all, except for the cases $s = 0, 1, 2$, it is easy to see that the monogenic elements $F_{j_1 \ldots j_s}$ in the above
decomposition can't be unique; the defining relations of radial algebra
$x_1(x_2 x_3 + x_3 x_2) = (x_2 x_3 + x_3 x_2)x_1$ serve as counter-examples of this.
We hence first have to express all products $x_{j_1} \ldots x_{j_s}$ in terms of a
basis for the radial algebra $R(\{x_{j_1}, \ldots, x_{j_s}\})$. There are many choices
for such bases and any choice leads to an equivalent result. For example
one may make use of the "lexicographic basis":

$$x_1^{s_1} \ldots x_k^{s_k} \prod_{j<k} \{x_j, x_k\}^{s_{jk}}, \quad s_j, s_{jk} \in \mathbb{N},$$

with $s_1 + \ldots + s_k + 2 \sum s_{jk} = s$. These basis-elements are linearly
independent over the rational functions of M, but one has to show
that they are also linearly independent after right multiplication with
monogenic objects.

In the multilinear case there is a simple proof for $s = 1$, i.e. we have

LEMMA 1. *Any equation of the form*

$$\sum_{j=1}^{k} x_j F_j(x_1, \ldots, x_{j-1}, x_{j+1}, \ldots, x_k) = 0,$$

with $F_j \in R(S \cup T)$ multilinear monogenic only has the nullsolution.

<u>Proof.</u> Letting the operator $\partial_{x_1}, \partial_{x_2}, \ldots$ act on this equation leads to

$$M F_1(x_2, \ldots, x_k) + 2F_2(x_2, \ldots, x_k) + \ldots + 2F_k(x_k, x_2, \ldots, x_{k-1}) = 0,$$

$$2F_1(x_1, x_3, \ldots, x_k) + MF_2(x_1, x_3, \ldots, x_k) + \ldots +$$
$$2F_k(x_k, x_1, x_3, \ldots x_{k-1}) = 0,$$

etc., which may be solved for high values of M simply by means of a Gauss procedure for solving linear systems. The fact that this is possible is not obvious because when performing the Gauss elimination procedure, several changes and permutations of the variables x_j take place; yet this causes no extra problems. Depending on the size of the system (which is only a function of the number k of vector variables) this Gauss method works from a certain high enough value of M on. Hence the result also remains true over the field of rational functions of M.   ∎

The above suggests once again concentrating on the multilinear case and we have the following

LEMMA 2. *Any equation of the form*

$$F = \sum x_1^{s_1} \ldots x_k^{s_k} \prod \{x_j, x_k\}^{s_{jk}} F_{s_j, s_{jk}} = 0$$

with $\sum s_j + 2 \sum s_{jk} = s$, *$F$ multilinear and $F_{s_j, s_{jk}}$ monogenic in the remaining vector variables, only has the nullsolution $F_{s_j, s_{jk}} = 0$.*

Proof. In the same way as above, we consider the action of the set of differential operators

$$\partial_{x_k}^{s_k} \ldots \partial_{x_1}^{s_1} \prod \{\partial_{x_j}, \partial_{x_k}\}^{s_{jk}}$$

whereby $s_j, s_{jk} \in \{0, 1\}$ are chosen such that the above is multilinear and $\sum s_j + 2 \sum s_{jk} = s$. These actions transform the equation in the statement into a linear system consisting of equations of the form

$$\sum \text{coefficients } F_{s_j, s_{jk}}(\text{variables}) = 0,$$

whereby the coefficients have the form: constant $\times M^t$ for some $t \le s$. It is now a matter of arranging the system in a suitable way and using a Gauss elimination procedure whereby one starts with the highest powers of M and tries to eliminate the corresponding terms in the other equations.

This time however one has to be aware of the fact that the terms $F_{s_j, s_{jk}}$ together with some of their permutations occur in the same equation; one cannot just apply the Gauss procedure as for linear systems. Yet any equation can be permuted w.r.t. the x_j-variables and one can consider the Gauss procedure for the system together with its permutations taking into account that the number of permutations only depends on

k. It speaks for itself that the system as a whole must first be divided into subsystems $S_0, S_1, \ldots, S_l, \ldots$ corresponding to the cases $\sum s_{jk} = l$ and to carry out the Gauss procedure first in S_0, then in S_1 etc.. For example, system S_0 is obtained by letting act $\partial_{x_k}^{s_k} \ldots \partial_{x_1}^{s_1}$ and leads to equations of the form: $M^s \, F_{s_1 \ldots s_k, 0} + l.o.t = 0$, whereby $l.o.t.$ refers to terms with lower order power coefficients in M. It speaks for itself that one tries to eliminate $F_{s_1 \ldots s_k, 0}$ in all the equations together with their permutations except in the equation where it has the highest power coefficient in M. At any step there is always a unique highest power term in the variable M and at the end of the procedure one obtains corrections of the form $F_{s_1 \ldots s_k, 0} = \frac{1}{M} \times$. Something , which depend on the total size of the system. As this is a function of k, the procedure can be carried out for high enough values of M within S_0. But of course one also has to eliminate $F_{s_1 \ldots s_k, 0}$ in the next systems $S_1, S_2, \ldots$ and one can solve the system only after the whole procedure. The main problem is that e.g. when letting act an S_1-operator like

$$\partial_{x_k}^{s_k} \ldots \partial_{x_3}^{s_3} \langle \partial_{x_2}, \, \partial_{x_1} \rangle$$

there are several terms with highest power in M, namely in the above case the term $C_1 M^{s-1} F_{1,1,s_3 \ldots, s_k}; 0 \ldots$ and $C_2 M^{s-1} F_{s_3, \ldots, s_k; 1, 0 \ldots}$. But when eliminating the first term using the system S_0, the power of M in the first term decreases with 1 and again we are left with one leading power in M. Also in the remaining systems S_l , after action of an operator like

$$\partial_{x_k}^{s_k} \ldots \partial_{x_1}^{s_1} \prod \langle \partial_{x_j}, \partial_{x_k} \rangle^{s_{jk}}, \quad \text{with} \quad \sum s_{jk} = l,$$

one may prove by induction on l that there is only one term with highest power M^{s-l} that has the form $F_{s_j, s_{jk}}, \sum s_{jk} = l$, while all other leading terms have the same power M_{s-l} and come from previous systems. After eliminating these terms coming from previous systems, the power of M drops so that only one leading power term remains. This is enough to explain the solvability of the system for high values of M.   ∎

In the previous proof one might be tempted to simply taking the limit fo $M \to \infty$, which allows one to simply cancel all the lower order terms. It would then follow immediately from system S_0 that all terms $F_{s_j, 0}$ vanish and then one could prove the same for the other terms by using induction on l in system S_l. This would be easy but one forgets then that the monogenic terms $F_{s_j, s_{jk}}$ themselves also depend on the variable M. Hence one can only consider the system for fixed M and prove the solvability for high values of M. Once this is done, the result also remains valid over the field of rational functions of M since there are never more than finitely many of them involved in the Fischer decomposition process.

Note that we have established the following:

THEOREM 3. *(Complete Fischer decomposition)*
Any $F \in R(S \cup T)$ may be written in a unique way as a sum of the form

$$F = \sum x_1^{s_1} \ldots x_k^{s_k} \prod \{x_j, x_k\}^{s_{jk}} F_{s_j, s_{jk}},$$

with $\partial_{x_k} F_{s_j, s_{jk}} = 0, x_k \in S$.

Note that the above holds over the field of rational functions of M. When assigning fixed valued to M, in order to apply the result to Clifford analysis one must take care of singularities in M. In [So5] we have shown that in the super-space setting these may indeed arise. But in standard Clifford analysis in R^m there always is a Fischer decomposition, which coincides with the above one when the representation

$$x \to \underline{x}, x \in S, \qquad u \to \underline{u}, u \in T$$

is an isomorphism, i.e. in case $M = m \geq \mathrm{Card}(S) + \mathrm{Card}(T)$. In that case there is a unique Fischer decomposition for those polynomials that are images of elements $F \in R(S \cup T)$, following from the isomorphism. Also in case of non-isomorphic representations , e.g. in case where $S = \{x_1, \ldots, x_k\}$ and $T = \{u_1, \ldots, u_l\}$ with $m < k + l$, the Fischer decomposition on the level of abstract vector variables may project down to a Fischer decomposition in the Clifford algebra setting. But the uniqueness result is certainly lost because it suffices to consider the Fischer decomposition of

$$F = x_1 \wedge \ldots \wedge x_k \wedge u_1 \wedge \ldots \wedge u_l = \sum x_1^{s_1} \ldots x_k^{s_k} \prod \{x_j, x_k\}^{s_{jk}} F_{s_j, s_{jk}}$$

which for $m < k + l$ projects down to an identity of the form

$$0 = \sum \underline{x}_1^{s_1} \ldots \underline{x}_k^{s_k} \prod \{\underline{x}_j, \underline{x}_k\}^{s_{jk}} F_{s_j, s_{jk}}$$

which may serve as a counter-example.

Also in the more general case $F \in R(S \cup T \cup dS \cup dT)$ one may prove direct sum Fischer decompositions over the field of rational functions in M (i.e. for high positive or negative real values of M). Hereby as basis for $R(S \cup dS)$ one may make use of the lexicographic basis (use axioms (R3)-(R7))

$$x_1^{s_1} \ldots x_k^{s_k} dx_1^{t_1} \ldots dx_k^{t_k} \times \prod \{x_j, x_k\}^{s_{jk}} \prod \{dx_j, x_k\}^{r_{jk}} \prod [dx_j, dx_k]^{t_{jk}}$$

and consider linear combinations with monogenic coefficients. Again the multilinear case will lead to linear systems which may be dealt with using a Gauss procedure.

REFERENCES

[ABLSS] W.W. Adams, C.A. Berenstein, P. Loustaunau, I. Sabadini, D. Struppa, *Regular functions of several quaternion variables and the Cauchy-Fueter complex*, J. Geom. Anal. **9** (1999) 1-16.

[Be] F.A. Berezin, *Introduction to algebra and analysis with anti-commuting variables*, Moscow University, Moscow 1983; English Transl.: Introduction to super-analysis, (Reidel, Dordrecht, 1987).

[Cr] A. Crumeyrolle, *Orthogonal and symplectic Clifford algebras*, Math. and Its Appl. **57**, (Kluwer Acad. Publ., Dordrecht, 1990).

[CRS] W. Chan, G.C. Rota, J.A. Stein, *The power of positive thinking*, in *Invariant methods in Discrete and Computational Geometry* (Curacao, 1994), Kluwer Acad. Publ., Dordrecht, (1995) 1-36.

[DSS] R. Delanghe, F. Sommen, V. Souček, *Clifford algebra and spinor valued functions: a function theory for the Dirac operator*, Math. and Its Appl. **53**, (Kluwer Acad. Publ, Dordrecht, 1992).

[GM] J. Gilbert, M.A.M. Murray, *Clifford algebras and Dirac operators in harmonic analysis*, Cambridge studies in advanced math. **26**, (Cambridge Univ. Press, Cambridge, 1991).

[Ha] K. Habermann, *The Dirac operator on symplectic spinors*, Annals of Global Analysis and Geometry **13**, (1995), 155-168.

[He] S. Helgason, *Groups and Geometric Analysis*, Pure and Applied Math., (Acad. Press, Orlando, London, 1984).

[HS] D. Hestenes, G. Sobczyk, *Clifford algebra to geometric calculus*, (D. Reidel, Dordrecht, 1985)

[Pa] V. Palamodov, *Cogitions over Berezin's integral*, Preprint series No.14, Univ. of Oslo, 1994.

[SS] F. Sommen, V. Souček, *Monogenic differential forms*, Complex Variables: Theory Appl. **19** (1992) 81-90.

[SSSVL] I. Sabadini, F. Sommen, D. Struppa, P. Van Lancker, *Complexes of Dirac operators in Clifford algebras*, to appear in Math. Z.

[So1] F. Sommen, *An algebra of abstract vector variables*, Portugaliae Math. Vol. **54**, Fasc. 3 (1997) 287-310.

[So2] F. Sommen, *The problem of defining abstract bivectors*, Results in Math., **31** (1997) 148-160.

[So3] F. Sommen, *Clifford tensor calculus*, Proc. XXIIth Conf. on D.G.M. Theor. Phys. (Eds. J. Keller, Z. Oziewicz), Ixtapa 1993, A.A.C.A., **4** (S1), (1994), 423-436.

[So4] F. Sommen, *An extension of Clifford analysis towards supersymmetry*, Proc. Conf. Clifford algebras and their applications in Math. Phys., Ixtapa 1999, Progress in Physics Vol.**19**, Birkhauser (2000),199-224.

[So5] F. Sommen, *Clifford analysis on super-space*, to appear in Proc. Conf. Dirac operators, Cetraro, 1998.

[So6] F. Sommen, *Clifford analysis on super-space II*, submitted to A.I.M..

[So7] F. Sommen, *Monogenic differential calculus*, Trans. A.M.S. Vol. 326 Nr. **2**, (1991), 613-632.

[VLSC] P. Van Lancker, F. Sommen, D. Constales, *Models for irreducible representations of Spin(m)*, to appear in Prof. Cetraro Conf. 1998.

[VV] V.S. Vladimirov, I.V. Volovich, *Superanalysis*, Transl. from Teoreticheskaya i Matematicheskaya Fizika, Vol. 59, No.1 (1984) 3-27.

Clifford Analysis as a Study of Invariant Operators

Vladimír Souček (soucek@karlin.mff.cuni.cz)
Mathematical Institute of Charles University Prague

Abstract. The conformal invariance of the Dirac equation and its consequences for monogenic functions in Clifford analysis are carefully discussed. The paper is trying to attract the attention of people interested in Clifford analysis to other types of symmetries useful for research in Clifford analysis. First, the general framework of parabolic geometries is recalled. Then two special examples (quaternionic geometry and its symmetry group and projective contact geometry and related symplectic geometry) are discussed in more detail, for possible future applications and as an inspiration for a search for other useful symmetries.

Keywords: Invariant differential operators, exact complexes, BGG sequences, Dirac operators

Mathematics Subject Classification: 53C15, 53A40, 53A30, 53A55, 53C05

Dedicated to Richard Delanghe on the occasion of his 60th birthday.

1. Introduction

The main stream of Clifford analysis has been devoted to a study of properties of the Dirac operator. The question, what is special about the Dirac operator, was discussed many times. One of the systematic answers to the question is the fact that the Dirac operator is quite special and distinguished operator due to its invariance properties. Its simplest definition is that the Dirac operator is the unique (up to a multiple) first order elliptic conformally invariant differential operator acting on spinor fields. This definition remains true even in the more general situation of spinor fields on a general manifold with a conformal structure and a chosen compatible spinor structure.

The conformal invariance of an operator is, as we have just seen, quite a strong requirement. At present, a complete classification of all conformally invariant differential operators on a conformally flat manifold is known and many important classes of operators have been extended to a general manifold with a conformal structure (see e.g. [36]). There are other elliptic first order conformally invariant operators acting on fields with values in more complicated (irreducible) Spin-modules. Study of such operators (e.g. the so called Rarita-Schwinger operator) created recently a new branch in Clifford analysis ([10, 11]).

323

F. Brackx et al. (eds.), Clifford Analysis and Its Applications, 323–339.

Conformal invariance of solutions of the Dirac equation was recently used very systematically in Clifford analysis ([32, 33]).

The aim of this paper is to show that the role of operators invariant with respect to groups of transformations different from the conformal one could increase in importance. The first very interesting example is the case of monogenic functions of several variables. This is a topic which already has a good tradition in Clifford analysis. The simplest case is that of several quaternionic variables. It will be shown in this paper that the group of transformations leaving invariant solutions of the Fueter operator in several quaternionic variables is much bigger than the product of conformal maps in individual variables. It is shown below that this system of differential equations is preserved by the whole group of automorphisms of the quaternionic projective space under the condition that (a generalized) conformal weight is chosen appropriately. The quaternionic projective space is the homogeneous model for the so called quaternionic geometry and the case of several quaternionic variables should be treated from that point of view.

Both conformal and quaternionic geometries are examples of so called parabolic geometries. Such geometries can be divided into families characterized by the so called $|k|$-grading. Both of the geometries mentioned belong to the simplest $|1|$-graded case. But it is very inspiring to look also to more complicated examples of parabolic geometries. A very interesting one (from the point of view of (generalized) Clifford analysis) is the so called projective contact geometry. It leads to a super-version of the Dirac equation. The orthogonal Clifford algebras are substituted by symplectic ones, finite dimensional spinor representation has to be substituted by a very interesting infinite dimensional representation of the metaplectic group (the so called Segal-Shale-Weil representation) , but the analogy with classical Clifford analysis is very strong and nice. There is a broad framework of parabolic geometries where a lot of information on invariant differential operators is already at our disposal. The two examples mentioned are just an indication of such possibilities (e.g. the symmetry behind the case of several Clifford variables is still to be found).

In the Sect. 2, a summary of facts needed for a study of invariant operators on homogeneous models for parabolic geometries is presented. In the next section, the case of several quaternionic variables is discussed. The last section is devoted to the projective contact geometry and its relation to symplectic Clifford analysis.

2. Parabolic geometries

2.1. THE GENERAL SCHEME

To begin, let us recall briefly the general scheme for the so called *invariant* (or *homogeneous*) operators on homogeneous bundles.

The basic object to be used is a homogeneous space $M = G/P$, where G is a Lie group (usually a real one but complex cases are also very much studied, leading to complex geometry and analysis) and P is a Lie subgroup of G. The natural left action of G on M is clearly transitive.

In an opposite way, if a smooth transitive action of G on a manifold M is given, then P can be defined as a stabilizer of a chosen point and M can be identified with the quotient G/P.

If $\mathbb{E}$ is any P-module (the action of P on elements of $\mathbb{E}$ is usually denoted by a dot), then we can consider the associated vector bundle $E = G \times_P \mathbb{E}$ and the space $\Gamma(E)$ of its global smooth sections. In most situations, only finite dimensional P-modules are considered but we shall see also an interesting and distinguished exception later on. There is a natural left action of G induced on the space $\Gamma(E)$ of all global sections. This is already an infinite dimensional module standardly called the induced module (or more precisely, the module induced by the representation $\mathbb{E}$ of P).

We would like now to define the action of G on $\Gamma(E)$. Let us recall first that the space $\Gamma(E)$ of global sections is automatically identified with the space $C^\infty(G, \mathbb{E})^P$ of smooth P-equivariant maps on G. By definition, an element $f \in C^\infty(G, \mathbb{E})$ is p-equivariant iff $p \cdot f(gp) = f(g)$ holds for all $g \in G, p \in P$.

THEOREM 2.1. *Let $M = G/P$ be a homogeneous space, let $\mathbb{E}$ be a representation of P and let $E = G \times_P \mathbb{E}$ be the associated homogeneous vector bundle. Then there is a 1-1 correspondence between the space $\Gamma(M, E)$ of all its sections and the space*

$$C^\infty(G, \mathbb{E})^P = \{f \in C^\infty(G, \mathbb{E}) | p \cdot f(gp) = f(g); g \in G; p \in P\}$$

of equivariant maps from G to the representation space $\mathbb{E}$.

Proof of the theorem is a standard and easy consequence of the definitions.

Now, the action of $g \in G$ on $f \in C^\infty(G, \mathbb{E})$ is defined by the left regular action, i.e.

$$[g' \cdot f](g) := f((g')^{-1}g), g \in G.$$

Due to the fact that the left action commutes with right action, p-equivariant maps are transformed again into P-equivariant maps.

2.2. THE PARABOLIC CASE

The parabolic case means the case of generalized flag manifolds G/P with G simple, P parabolic. It is well known that on the level of the Lie algebras, the choice of such a pair $(\mathfrak{g}, \mathfrak{p})$ is equivalent to a choice of the so called $|k|$-grading of a semisimple $\mathfrak{g}$:

Let us consider Lie algebras over the field $\mathbb{K} = \mathbb{R}, \mathbb{C}$. A $|k|$-graded Lie algebra over $\mathbb{K}$, $k \in \mathbb{N}$ is a Lie algebra $\mathfrak{g}$ over $\mathbb{K}$ together with a decomposition $\mathfrak{g} = \mathfrak{g}_- \oplus \mathfrak{g}_0 \oplus \mathfrak{g}_+$, where

$$\mathfrak{g}_- = \mathfrak{g}_{-k} \oplus \ldots \oplus \mathfrak{g}_{-1}, \quad \mathfrak{g}_+ = \mathfrak{g}_1 \oplus \ldots \oplus \mathfrak{g}_k$$

with the property that $[\mathfrak{g}_i, \mathfrak{g}_j] \subset \mathfrak{g}_{i+j}$ and such that the subalgebra $\mathfrak{g}_-$ is generated by $\mathfrak{g}_{-1}$. By $\mathfrak{p}$ we will denote the subalgebra $\mathfrak{g}_0 \oplus \mathfrak{g}_+$ of $\mathfrak{g}$.

There is always a unique element $E \in \mathfrak{g}$ whose adjoint action is given by $[E, X] = j X$ for $X \in \mathfrak{g}_j$. The element E is contained in the center of the subalgebra $\mathfrak{g}_0$, which is always reductive. For each $i = 1, \ldots, k$, the Killing form of $\mathfrak{g}$ induces an isomorphism $\mathfrak{g}_i \cong \mathfrak{g}_{-i}^*$ of $\mathfrak{g}_0$-modules. See e.g. [39, Section 3] for details.

In the complex case, giving a $|k|$-grading on $\mathfrak{g}$ is the same thing as giving a parabolic subalgebra $\mathfrak{p}$ of $\mathfrak{g}$ (a parabolic subalgebra is one containing the Borel subalgebra). It is well-known that such parabolic subalgebras are (up to conjugation) in one-to-one correspondence to subsets of simple roots of $\mathfrak{g}$. Therefore, complex $|k|$-graded semisimple Lie algebras can be conveniently denoted by Dynkin diagrams with crossed nodes. That is, given a $|k|$-graded semisimple complex Lie algebra, we may assume that $\mathfrak{p}$ is the standard parabolic subalgebra corresponding to a set Σ of simple roots. Then we denote the $|k|$-graded Lie algebra $\mathfrak{g}$ by crossing out the nodes corresponding to the simple roots contained in Σ in the Dynkin diagram of $\mathfrak{g}$. See the book [5] for a detailed discussion of the Dynkin diagram notation for parabolic subalgebras.

Finally note that for a $|k|$-graded Lie algebra $\mathfrak{g}$ over $\mathbb{R}$ the complexification $\mathfrak{g}^{\mathbb{C}}$ of $\mathfrak{g}$ is $|k|$-graded, too. So in general we deal with certain real forms of pairs $(\mathfrak{g}, \mathfrak{p})$, where $\mathfrak{g}$ is complex and semisimple and $\mathfrak{p}$ is a parabolic in $\mathfrak{g}$. The classification of all these real forms is provided in [39, Section 4].

For a Lie group G with $|k|$-graded semisimple Lie algebra $\mathfrak{g}$ and the subgroup P corresponding to the Lie algebra $\mathfrak{p}$, we shall consider the homogeneous space G/P.

The tangent and cotangent bundles of G/P are homogeneous vector bundles. One easily verifies that they are vector bundles associated with the representations of P on $\mathfrak{g}_- \cong \mathfrak{g}/\mathfrak{p}$ and $\mathfrak{p}_+$ induced by the adjoint action, respectively.

2.3. EXAMPLES

The simplest and best known situation occurs for $|1|$–graded algebras, i.e. $\mathfrak{g} = \mathfrak{g}_{-1} \oplus \mathfrak{g}_0 \oplus \mathfrak{g}_1$. The examples include the conformal, almost Grassmannian, and almost quaternionic structures. The projective structures correspond to $\mathfrak{g} = \mathfrak{sl}(m+1, \mathbb{R})$, $\mathfrak{g}_0 = \mathfrak{gl}(m, \mathbb{R})$. The series of papers [19–21] is devoted to all these geometries.

Next, the $|2|$–graded examples include the so called projective contact geometries, which will be treated in detail below and CR-structures. See e.g. [39] for more detailed discussions.

2.4. INVARIANT OPERATORS ON HOMOGENEOUS MODELS

DEFINITION 2.2. *Let D be a linear operator acting between the spaces $\Gamma(E)$ and $\Gamma(F)$ of two homogeneous bundles E, resp. F in M. Then D is called invariant, resp. homogeneous operator, if it commutes with the action of G on the spaces of sections. So for all $g \in G$ and for all $f \in \Gamma(E)$,*

$$D(g \cdot f) = g \cdot (Df).$$

In particular, if f is a solution of the equation $Df = 0$ on a domain Ω, then $g \cdot f$ is again a solution on the image of Ω by the map induced by the element $g \in G$.

The definition of invariant operators in this homogenous setting is hence very simple. Specific and detailed information on their complete classification is already worked out in many important cases. The most important and the best studied case is the case of conformally invariant operators (see e.g. a review paper [36]). A quite general scheme is that of parabolic geometries, where G is semisimple Lie group and P its parabolic subgroup. This scheme is broad enough to include quite a lot of interesting and important structures. In the lecture, we would like to show some less known consequences, which can be deduced from the fact that a certain operator is invariant and to relate then to questions and problems studied recently in Clifford analysis.

2.5. The construction of first order invariant operators

If D is an invariant first order differential operator $D : \Gamma(E) \to \Gamma(F)$ on a homogeneous space $M = G/P$, then D is induced by a vector bundle homomorphism $\tilde{D} : J^1(E) \to F$, where $J^1(E)$ is the first jet prolongation of E. Now $J^1(E)$ is again a homogeneous vector bundle, and the invariance of D is equivalent to the fact that $\tilde{D}$ is equivariant for the G-actions on $J^1(E)$ and F. Since G acts transitively on G/P, the homomorphism $\tilde{D}$ is actually determined by its restriction $\tilde{D} : J^1(E)_o \to F_o$ to the fiber over $o \in G/P$, and by invariance of D, this map is P-equivariant.

Conversely, a P-homomorphism $J^1(E)_o \to F_o$ extends uniquely to a G-homomorphism $J^1(E) \to F$ and thus gives rise to an invariant differential operator. Thus, invariant differential operators $\Gamma(E) \to \Gamma(F)$ of order 1 are in bijective correspondence with P-homomorphisms $J^1(E)_o \to F_o$.

As G_0-module, $J^1(E)_0 \simeq \mathbb{E} \oplus (\mathfrak{g}_+ \otimes \mathbb{E})$. Hence a standard method of construction of invariant first order differential operators is to take a G_0-invariant projection from $\mathfrak{g}_+ \otimes \mathbb{E}$ to its irreducible component and to choose the (generalized) conformal weight (see below) of the irreducible representation $\mathbb{E}$ in such a way that the projection is a P-homomorphism. There is a linear relation using Casimir expressions to compute the appropriate value w of the (generalized) conformal weight. For example, in the conformal case, this is a question of a choice of appropriate conformal weight.

2.6. Invariant properties of solutions

Every element $g \in G$ induces a point transformation ϕ_g on the homogeneous space $M = G/P$ by the left action. The space of solutions of any invariant (homogeneous) equation discussed above is preserved by the action of the transformation group G. In the conformal case, it is just a property of conformal invariance of solutions of the Dirac (or other invariant) equation. Note that a proper choice of the conformal weight was significant for these invariant properties.

In the general case, we need to recall a few facts about irreducible representations of reductive Lie groups. Such a group H is a product $H_1 \times H_2$ of a commutative subgroup H_1 and a semisimple Lie group H_2. Any irreducible representation $\mathbb{V}$ of H is a tensor product $\mathbb{V} = \mathbb{V}_1 \otimes \mathbb{V}_2$ of an irreducible representation $\mathbb{V}_1$ of H_1 and an irreducible representation $\mathbb{V}_2$ of H_2. Irreducible representations of H_1 are all one-dimensional and are uniquely specified by a character, irreducible representation of H_2 are characterized by their highest weights.

In the cases we need below we can describe these representations in an easier way. We shall need irreducible representations of the reductive subgroup $G_0 \subset G$, where G is a simple Lie group. Let $\mathfrak{g}$ be the Lie algebra of G and $\mathfrak{h}$ its Cartan subalgebra chosen to be a subalgebra of $\mathfrak{g}_0$. A datum needed to specify such an irreducible representation is an integral weight $\lambda \in \mathfrak{h}^*$, dominant for the semisimple part of G_0.

In the cases below, the center of G_0 will be one-dimensional, generated by the element λ_E corresponding to the grading element E by the isomorphism $\mathfrak{h} \simeq \mathfrak{h}^*$ given the Killing form. The dual of $\mathfrak{h}^*$ can be written as a direct sum of the linear subspace generated by λ_E and its orthogonal complement. It induces the splitting of any $\lambda \in \mathfrak{h}^*$ into $\lambda = w\lambda_E + \lambda'$. The number w will be called the (generalized) conformal weight and λ' is the highest weight of the corresponding representation considered as a representation of semisimple part of G_0. As in the conformal case, there will be a finite number of invariant first order operators acting on fields with values in λ', each one needing a specific value of w to be an invariant operator. Hence the value λ will be, in general, different for these different cases. Note also that irreducible representations of the whole parabolic group P are the same as irreducible representations of the reductive part G_0, because the nilpotent part always acts trivially. To understand better the invariance properties of solutions of an invariant operator D, we can consider fields on subsets of an open flat subspace of M (e.g. on the big cell). The action of the automorphism group G of the homogeneous model M on fields defined on subsets of the big cell is described in the next theorem.

THEOREM 2.3. *Let D be an invariant differential operator acting on fields with values in the bundle associated to an irreducible P-module $(\rho, \mathbb{V})$, where ρ denotes the corresponding representation of P on $\mathbb{V}$. For any $g \in G$, let ϕ_g denote the induced transformation of M given by the left action. Let us consider restrictions of fields and transformations to the big cell $\mathfrak{g}_{-1}$ embedded to M by $x \in \mathfrak{g}_{-1} \to \exp(x) \cdot P \in G/P$. The transformation ϕ_g is well defined on an open dense subset of $\mathfrak{g}_{-1}$ and its differential $d\phi_g$ belongs to G_0 in every point x from this open dense subset. The action of $g \in G$ on the field $f(x)$ is given by*

$$f'(x) \equiv [g \cdot f](x) = \rho(d\phi_g(x)) \cdot [f(\phi_{g^{-1}}(x)].$$

Consequently, if $f(x)$ is a solution of $Df = 0$, then $f'(x)$ is again a solution of the same equation.

In the conformal case, this leads to the well-known transformation properties of solutions under conformal transformation but the same useful information is available also in other types of geometries.

3. Quaternionic geometry

The geometry of n-dimensional quaternionic space is modelled by the homogeneous space $M = \mathbb{P}^n(\mathbb{H})$, the quaternionic projective space. This is the set $\mathbb{H}^{n+1} \backslash \{0\}$ quotiented by the equivalence

$$(q_0, \ldots, q_n)^t \equiv (q_0, \ldots, q_n)^t \cdot q', q' \in \mathbb{H} \backslash \{0\}.$$

The flat subspace $\mathbb{R}^{4n} \equiv \mathbb{H}^n$ is embedded into M by

$$x = (q_1, \ldots, q_n)^t \in \mathbb{H}^n \rightarrow \tilde{x} = \left[(1, q_1, \ldots, q_n)^t \right] \in \mathbb{P}^n(\mathbb{H}).$$

The big group G acting on it is the group $SL(n+1, \mathbb{H})$ of quaternionic regular matrices of determinant 1, P is the parabolic subgroup of all matrices preserving a chosen point in M.

The groups G_0 is the subgroup $S(H^* \times GL(n, \mathbb{H}))$.

The complexification $\mathbb{C}G$ is the group $SL(2n+2, \mathbb{C})$ with $\mathbb{C}G_0 = S(GL(2, \mathbb{C}) \times GL(2n, \mathbb{C})$. Hence the corresponding standard Dynkin diagram for such a complex parabolic pair is $\overset{1}{\bullet}\!\!-\!\!\overset{2}{\times}\!\!-\!\!\overset{3}{\bullet} \cdots \overset{n}{\bullet}\!\!-\!\!\overset{n+1}{\bullet}$. This complex $|1|$-graded case has several real forms; the quaternionic geometry is one of them.

The situation is easily illustrated on the level of the corresponding Lie algebras. We have $\mathfrak{g} = \mathfrak{sl}(n+1, \mathbb{H})$ (all quaternionic matrices with vanishing trace), $\mathfrak{g}_0 = \mathfrak{sl}(1, \mathbb{H}) \oplus \mathfrak{sl}(n, \mathbb{H}) \oplus \mathbb{R}$, $\mathfrak{g}_\pm \simeq \mathbb{H}^n$.

The decomposition of $\mathfrak{g}$ is easily seen in the block matrix form:

$\mathbb{H}$	$\mathbb{H}^n$
$\mathbb{H}^n$	$\mathfrak{gl}(n, \mathbb{H})$

As always in $|1|$-graded parabolic geometries, $\mathfrak{g}_1$ is the model for the cotangent space. The complexification $\mathbb{C}\mathbb{H}$ of the field of quaternions $\mathbb{H}$ is standardly (using the Pauli matrices) identified with $\mathrm{Mat}(2 \times 2, \mathbb{C})$. Hence the complexification $\mathbb{C} \otimes \mathfrak{g}_1 \equiv \mathbb{C}\mathfrak{g}_1$ can be written as $\mathbb{C}\mathbb{H}^n \simeq \mathrm{Mat}(2 \times 2n, \mathbb{C})$. Moreover, breaking the symmetry of $\mathbb{H}$, we can identify $\mathbb{H}$ with $\mathbb{F} = \mathbb{C}^2$.

the characteristic property of $\mathbb{C}\mathfrak{g}_1$ is that $\mathbb{C}\mathfrak{g}_1 \simeq \mathrm{Mat}(2 \times 2n, \mathbb{C}) = \mathbb{F} \otimes \mathbb{E}$, where $\mathbb{E} \simeq (\mathbb{C}^{2n})^t$. Even globally, the complexified cotangent bundle $T_c^* M$ of the homogeneous space M can be written as the tensor product $F \otimes E$ of two complex vector bundles (see e.g. [6]).

To understand this fact on the level of representations, let us note that $\mathfrak{g}_1 \simeq \mathbb{H}^n$ is a G_0-module by the adjoint representation. The block matrix description above makes it clear that if $\alpha \in \mathbb{H}$, $A \in \mathrm{Mat}(n \times n, \mathbb{H})$ are diagonal elements of $g \in G_0$, then the action of g on $x^t \in \mathbb{H}^n$ is given by $g \cdot x^t = \alpha(x^t) A^{-1}$. Hence the relation $\mathbb{C}\mathfrak{g}_1 \simeq \mathbb{E} \otimes \mathbb{F}$ is true also on the level of G_0-modules.

3.1. QUATERNIONIC COMPLEXES

In general there are two types of invariant differential operators on parabolic homogenous models (or on their curved versions), the standard operators and nonstandard operators. The first order operators are all standard. We shall need mostly first order invariant operators (with one exception) so that we shall add only a few comments on standard ones. The family of standard invariant operators splits again into the regular and singular operators.

The regular standard operators come in families which have always the same pattern (depending on the parabolic pair in question) and they fit together into the so called Bernstein-Gelfand-Gelfand (BGG for short) complexes on the homogenous spaces. These are very interesting analogues of the de Rham sequence. There is a whole family of them (numbered by irreducible finite dimensional representations of G). These questions are well understood at present (for more details, see e.g. [7, 22, 17] and references therein).

Much less is known about sequences composed of singular standard operators. In the case of quaternionic geometry discussed in this section, these singular complexes were studied by R.Baston ([6]). He found that singular standard operators in the families should be supplemented by certain nonstandard operators. The tool he was using for the description of such complexes was the Penrose transform (a version of an integral transform, an analogue of the Radon transform) and the relative regular BGG sequences (relative with respect to fibers of one of the two projections in the double fibration characterizing the corresponding integral transform). By such a direct image of the fiberwise BGG sequence, he has found new complexes in quaternionic geometry. Quite surprisingly, these complexes are directly related to the complexes studied for many years in Clifford analysis by the group around D.Struppa ([1, 2, 34]). The purpose of this section is to compare the two types.

The Baston complexes in [6] were formulated on the curved analogues of the homogeneous model M. To show its implications for Clifford analysis, we may consider all fields on the flat space $\mathbb{R}^{4n} \simeq \mathbb{H}^n$ embedded (as above) into M. Instead of sections of associated bundles on M, we shall consider maps with values in certain vector spaces (typically, subspaces of tensor products of tensor powers of $\mathbb{H}$ and $\mathbb{E}$).

To achieve an easier description, we shall consider all maps defined not only on $\mathbb{R}^{4n}$, but we shall consider their holomorphic extensions to $\mathbb{C}^{4n}$. This can certainly be done for polynomial maps considered in complexes in [1, 2, 34], but also for real analytic maps on $\mathbb{R}^{4n}$. (The use of this holomorphic extension is not necessary but it makes the language used simpler). If $\mathbb{V}$ is a fixed (complex) vector space, we shall denote by V the space of holomorphic maps from $\mathbb{C}^{4n}$ to $\mathbb{V}$. Let us denote variables in $\mathbb{C}^{4n}$ by $z_{AA'}, A = 1, 2; A' = 1, \ldots, 2n$. The symbols $\nabla_{AA'}$ will denote derivatives with respect to $z_{AA'}$. Using this notation, we can describe one of Baston's quaternionic complexes (relevant for us) as follows (for details, see [6]):

THEOREM 3.1. *There is an exact complex of invariant differential operators $D_i, i = 1, \ldots 2n - 1$, acting from spaces V_i to V_{i+1}, where*

$$V_1 = H, V_2 = E, V_3 = \Lambda^3(E), V_{j+3} = \odot^j(F) \otimes \Lambda^{j+3}(E); \quad j = 1, \ldots, 2n-3.$$

All operators are of first order with exception of D_2, which is the second order operator. The operator D_2 is described in coordinates by

$$D_2(f_{C'}) = Alt \, (\nabla_{AA'} \nabla_{BB'} f_{C'}),$$

where the symbol 'Alt' denotes antisymmetrization in the pair of indices AB and triple of indices $A'B'C'$ separately. Note that $\Lambda^2(\mathbb{F})$ is one-dimensional (in the Baston notation, it correspond to a power of a line bundle indicated by negative numbers in the squared brackets).

All other first order operators are given by applying the gradient ∇ to $f \in V_k$ (∇f has values in $\mathbb{E} \otimes \mathbb{F} \otimes \mathbb{V}_k$) followed by the projection to $\mathbb{V}_{k+1} \subset \mathbb{E} \otimes \mathbb{F} \otimes \mathbb{V}_k$. In particular, the operator D_1 is given in coordinates by

$$D_1(f_B) = Alt \, (\nabla_{AA'} f_B) = \nabla_{0A'} f_1 - \nabla_{1A'} f_0, A' = 1, \ldots, 2n.$$

It is well known fact ([35]) that in the case $n = 1$, the quaternionic geometry is the four-dimensional conformal geometry and the operator D_1 is the Fueter operator D in one variable (extended by holomorphic continuation to the complexification $\mathbb{CH}$). Similarly, for $n > 1$, D_1 is the vector of Fueter operators acting in n quaternionic variables separately. So this is just the same operator in several quaternionic variables studied by D.Struppa and his coworkers. Hence the quaternionic complex above starts with the same operator as the quaternionic complex (4) described in [34]. Both complexes have the same length and the orders of the differential operators are also the same.

It is easy to compute dimensions (over $\mathbb{C}$) of spaces $V_k : \dim V_1 = 2$, $\dim V_2 = 2n$, $\dim V_3 = \binom{2n}{3}$ and $\dim V_{j+3} = (j + 1) \binom{2n}{j + 3}$, $j =$

$1, \ldots, 2n - 3$. Real dimension are twice as big. We can compare them to dimensions of spaces in the complex from [34] computed there for $n = 3, 4$. They are the same. Hence the complexes should be isomorphic.

3.2. INVARIANT PROPERTIES OF MONOGENIC FUNCTIONS OF SEVERAL QUATERNIONIC VARIABLES

Using the Theroem 2.3 and a discussion of the form of the transformation group G in the quaternionic case, we can deduce that the space of solutions (i.e. monogenic functions of several quaternionic variables) is preserved under the action of G (described in Th.2.3). Note that a general element of G is an $(n + 1) \times (n + 1)$ quaternionic matrix. Diagonal matrices correspond to conformal transformations in individual quaternionic variables, hence G is much bigger transformation group.

4. The projective contact geometry

4.1. A DESCRIPTION OF PROJECTIVE CONTACT GEOMETRY

The case we would like to study now is a $|2|$-graded real Lie algebra $\mathfrak{g}$ of the following form: $\mathfrak{g} = \mathfrak{sp}(2n + 2, \mathbb{R}), \mathfrak{g}_0 = \mathfrak{sp}(2n, \mathbb{R})$ $\mathfrak{g}_\pm \simeq \mathbb{R}^{2n}$ and $\mathfrak{g}_{\pm 2} \simeq \mathbb{R}$. The big cell hence appears there as $\mathfrak{g}_- = \mathbb{R}^{2n} \oplus \mathbb{R}$. The symplectic form defining the algebra $\mathfrak{g}$ induces the symplectic form on the space $\mathfrak{g}_{-1}$, which shall be used below.

It was shown in [38] that any first order differential operator is given by a P-homomorphism, which factorizes through the space of restricted jets. In other words, it means that in the coordinate expression for an invariant first order differential operator, only derivatives in directions of $\mathfrak{g}_{-1}$ will be present. Hence solutions of any invariant first order operator will be arbitrary functions of the last variable from $\mathfrak{g}_{-2} \simeq \mathbb{R}$. It is hence natural to study functions defined only on $\mathfrak{g}_{-1} \simeq \mathbb{R}^{2n}$ with values in irreducible P-modules. For finite dimensional representations, the procedure for construction of such invariant operators is the same as above. We shall need, however, the case of fields with values in an infinite dimensional irreducible representation; in this case a repeated verification of the construction is necessary.

4.2. THE SYMPLECTIC AND THE METAPLECTIC GROUP

Let a symplectic vector space (V, ω) be given. Then the symplectic group $Sp(V, \omega)$ is defined to be the group of all endomorphisms of the space V preserving the bilinear form ω :

$$u \in Sp(V, \omega) \iff \omega((u(v), u(w)) = \omega(v, w); \ v, w \in V.$$

As it is in the case of the orthogonal group, the group $Sp(V, \omega)$ is not simply connected. Its universal simply connected cover, however, is not a finite covering but infinite one. As a quotion of this universal cover, there is a double cover of $Sp(V, \omega)$, which is standardly called *the metaplectic group* and denoted by $Mp(V, \omega)$. As in the orthoghonal case, there is an exact sequence of groups

$$0 \longrightarrow \mathbb{Z}_2 \longrightarrow Mp(V, \omega) \longrightarrow Sp(V, \omega) \longrightarrow 0.$$

In the orthogonal case, the standard realization of the Spin group is inside the orthogonal Clifford algebra. It is possible to elaborate an analogue of such realization in the symplectic case (see [Cr]) but it needs an additional effort caused by the fact that the symplectic Clifford algebra is infinite dimensional, hence the exponential map needs special care to be well defined.

It is useful to note the following simple fact. Every linear map $u \in Sp(V, \omega)$ preserves by definition the form ω. But because we have

$$[u(v), u(w)] = \omega(u(v), u(w)) = \omega(v, w)$$

for all $v, w \in V$, the universal property of the Clifford algebra implies that u extend to a homomorphism of the whole algebra, and, in particular, to a homomorphism of the Lie algebra H.

4.3. THE SPINOR REPRESENTATION

There is a very nice analogue of the basic spinor representation in the case of the symplectic Clifford algebra. The main difference is that in this case, the representation is infinite dimensional. Various names are used for this unusual and many-faceted representation, for example Segal-Shale-Weil representation or (mostly in physics) the Fock space representation.

A specific realisation of the spinor space is, similarly to the orthogonal case, based on a choice of a maximal isotropic subspace in V. Such isotropic subspaces are usually called in the symplectic case *Lagrangean subspaces*.

DEFINITION 4.1. *Let* $\dim V = m = 2n$. *Let* $V = W \oplus W'$ *be a splitting of V into into two maximal isotropic subspace of (V, ω). Let $\{e_i, f_i\}_{i=1}^n$ be a canonical symplectic basis for (V, ω) adapted to the splitting and let $\{x_i, y_i\}_{i=1}^n$ be the corresponding coordinates on V.*

Then the space $S := L_2(W)$ *is the spinor representation of $C(V, \omega)$. The action σ of H on S is given by the basic realisation of the Heisenberg algebra*

$$\sigma(1) = i\mathrm{Id}, \quad \sigma(e_j) = i\, x_j, \quad \sigma(f_j) = \partial/\partial x_j.$$

This representation σ of H extends in a natural way to the whole of $C(V, \omega)$ (which is the universal enveloping algebra of H) and we get in such a way a representation σ of $C(V, \omega)$ on S.

The Clifford multiplication *is then the map $C(V, \omega) \times S \longrightarrow S$ defined by the action of σ* :

$$a \cdot s := \sigma(a)[s], \ a \in C(V, \omega), s \in S.$$

In the orthogonal case, the basic spinor representation of the Clifford algebra induces, by restriction, the spinor representation of the Spin group. Moreover, there is the basic relation between the spin and the orthogonal group (p is the projection of the spin group on the orthogonal group)

$$g \, v \, g^{-1} = p(g)[v], v \in V, \ g \in Spin(V, Q).$$

We do not have a realisation of the metaplectic group in the symplectic Clifford algebra. Hence it is more difficult to define the action of $Mp(V, \omega)$ on S. An analogue of the above relation is used in the following indirect definition. The problem that the metaplectic group is not embedded in the Clifford algebra is avoided by an interpretation of elements of Clifford algebra as well as of the metaplectic group as endomorphisms of the spinor representation. Recall that any linear map $u \in Sp(V, \omega)$ extends to a homomorphism of H.

THEOREM 4.2. (Segal-Shale-Weil). *There is a unique representation ρ of the group $Mp(V, \omega)$ on S characterized by the relation*

$$\rho(g)\sigma(h)\rho^{-1}(g) = \sigma(p(g)[h]), \ g \in Mp(V, \omega), h \in H.$$

Remarks.

1. In the orthogonal as well as in the symplectic case, there is an important subgroup $U(n)$ of these both groups. Its embedding is fixed by a choice of an isotropic subspace W; it is just a subgroup preserving W. The basic spinor representation S is the so called Fock space (it is $\Lambda^*(W)$ for the orthogonal case and $\otimes^*(V)$ in the symplectic one). Elements of W, resp. W' act on S as creation (resp. annihilation) operators. Homogeneous components of S are preserved clearly by the $U(n)$ action, and they are in fact irreducible components of S under the restriction to the $U(n)$ action. In both cases (more precisely, in the symplectic case and in the even dimensional orthogonal case), the spinor representation splits into two pieces $S^{\pm}$, given by the sum of even, resp. odd, powers of W.

2. Finite dimensional representations of $SO(V, Q)$ form a representation ring under the tensor product. The same is true for $Spin(m)$, but this time the representation ring is larger, it also contains the spinor representation, which is finite dimensional. Any power $\otimes^k(S)$ can be decomposed into irreducible pieces and every finite dimensional representation of $Spin(m)$ can be found as a piece in such a decomposition.

In the symplectic case, there is again the ring of all finite dimensional representations of $Sp(V, \omega)$ but this time the space of all finite dimensional representations of $Mp(V, \omega)$ is not bigger and it does not contain S. In an interesting analogy, any power $\otimes^k(S)$ decomposes into irreducible components and it is possible to consider a nice class of all representations of $Mp(V, \omega)$ which can be found in the decomposition of a power $\otimes^k(S)$. They are called *harmonic representations*. The space of all harmonic representations seems to be a good analogue of the space of all finite dimensional representations of the groups $Spin(m)$.

3. The Clifford multiplication commutes with the action of the metaplectic group, as it does in the orthogonal case. More precisely, the Clifford multiplication being denoted by $\cdot$:

$$p(g)[v] \cdot \rho(g)[s] = \rho(g)[v \cdot s]; \quad g \in Mp(V, \omega), v \in V, s \in S.$$

4. The definition of the spinor representation of the metaplectic group was introduced by Kostant in connection with geometrical quantization (see [Ks]). More on the metaplectic group and its spinor representation can be found in [KW].

4.4. THE SYMPLECTIC DIRAC OPERATOR

The setting described above leads in a natural way to the definition of the symplectic Dirac operator. This definition was written down by K.Habermann ([H1],[H2]) in a general situation of a symplectic manifold with a chosen metaplectic structure and a chosen symplectic covariant derivative. We are going to define it in the flat case.

DEFINITION 4.3. *Let ∇ be the gradient acting on the space of spinor-valued maps $C^\infty(\mathbb{R}^m, S)$ with values in the space $C^\infty(\mathbb{R}^m, S \times T^*(\mathbb{R}^m))$. Using the symplectic form (which is supposed to be nondegenerate), we can identify $T^*(\mathbb{R}^m)$ and $T(R^m)$, and the composition of Clifford multiplication with the gradient defines the Dirac operator D.*

We can write it (with respect to a canonical symplectic basis) as

$$D\phi = \sum_{j=1}^{n} \{e_j \nabla_{f_j} - f_j \nabla_{e_j} \phi\}, \ \phi \in C^{\infty}(\mathbb{R}^m, S).$$

Remarks.

1. At present, properties of solutions of the symplectic Dirac equation are not understood. An analogy with the orthogonal case suggests many questions in the corresponding function theory which await an answer. The papers by K.Habermann concentrate on questions arising in differential geometry (how the definition of a symplectic manifold depends on various choices made).

2. A natural question to ask is whether the symplectic Dirac operator introduced above is one of the invariant operators for projective contact geometry described above. The answer is being sought now by L.Kadlčáková in Prague.

References

1. W.Adams, C.Bernstein, P.Loustaunau, I.Sabadini, D.Struppa: Regular functions of several quaternionic variables and the Cauchy-Fueter complex, to appear in J.Geom. Anal.
2. W.Adams, P.Loustaunau, V.Palamodov, D.Struppa: Hartog's phenomenon for polyregular functions and projective dimension of related modules, Ann. Inst. Fourier, 47, 1997, 623-640
3. R.Baston: Verma modules and differential conformal invariants, J.Diff.Geom., 32, 851-898, 1990
4. R.J.Baston, M.G.Eastwood: *The Penrose transform, Its interasction with representation theory*, Oxford University Press, Oxford, 1989
5. R.J.Baston, M.G.Eastwood: Invariant operators, in "Twistors in Mathematics and Physics", LMS Lect. Notes 156, Cambridge University Press, 1990
6. R.J.Baston: Quaternionic complexes, Jour.Geom. and Physics 8, 1992, 29-52
7. I.N. Bernstein, I.M. Gelfand, S.I. Gelfand, Differential operators on the base affine space and a study of g–modules, in "Lie Groups and their Representations" (ed. I.M. Gelfand) Adam Hilger 1975, 21–64
8. Bureš J.: The higher spin Dirac operators, Proc. conf DGA, Brno 1998
9. Bureš, J.: The Rarita-Schwinger operator and spherical monogenic forms, to be published in Complex Variables:Theory and Applications.
10. Bureš, J., Van Lancker, P., Sommen, F., Souček, V.: Symmetric analogies of Rarita-Schwinger equations, submitted for publication
11. Bureš, J., Van Lancker, P., Sommen, F., Souček, V.: Rarita-Schwinger type Operators in Clifford Analysis, submitted for publication
12. F.Brackx, R.Delanghe, F.Sommen: *Clifford analysis*, Pitman, London, 1982

13. Branson T., Olafsson G., Ørsted B. - Spectrum generating operators and intertwining operators for representations induced from a maximal parabolic subgroup, J.Funct. Anal. 135 (1996), 163-205 .

14. T.P.Branson: Conformally covariant equations on differential forms, Comm. in PDE, 7, 392-341, 1982

15. T.P.Branson: Second-order conformal covariants I.,II., preprint, Kobenhaven, 1989

16. J.Bureš: Special invariant operators I, Com.Mat.Univ.Car., 37, 1, 179-198, 1996

17. D.M.J.Calderbank, T.Diemmer: Differential bilinear invariants on curved BGG sequences, preprint MS-99-010, Dept.Math.Stat., Edinbourgh

18. A. Crumeyrolle: *Orthoogonal and symplectic Clifford algebras*, Kluwer, 1990

19. A.Čap, J.Slovák, V.Souček: Invariant operators on manifolds with almost Hermitian symmetric structures, Invariant differentiation Preprint ESI 186, 194, 34 pp., 1994

20. A.Čap, J.Slovák, V.Souček: Invariant operators on manifolds with almost Hermitian symmetric structures, Normal Cartan connections, Preprint ESI 194, 16 pp., 1994

21. A.Čap, J.Slovák, V.Souček: Invariant operators on manifolds with almost Hermitian symmetric structures, Standard operators Preprint ESI 613 1998

22. Čap A.,Slovák J.,Souček V. : Bernstein-Gelfand-Gelfand sequences , to be published .

23. J.Cnops: Hurwits pairs and applications of Moebius transformations, Thesis, Gent, 1994

24. R.Delanghe, F.Sommen, V.Souček: *Clifford algebra and spinor-valued functions*, Kluwer, 1992

25. M.Eastwood: Notes on conformal differential geometry, Proc. of Winter School, Srni, 1995

26. H.D.Fegan: Conformally invariant first order differential equations, Q.J.Math. Oxford, 27, 371-378, 1975

27. J.Gilbert, M.Murray: *Clifford algebras and Dirac operators in harmonic analysis*, Cambridge University Press, 1991

28. K.Haberman: The Dirac operator on symplectic spinors, Ann.Glob.anal. and Geometry, 13 (1995), 155-168

29. K.Habermann: Basic properties of symplectic Dirac operators, Comm.Math. Phys. 184 (1997), 629-652

30. M.Kashiwara, M.Vergne: On the Segal-Shale-Weil representations and harmonic polynomials, Inv.Math. 44 (1978), 1-47

31. B.Kostant: Symplectic spinors, Symposia Mathematica, Vol. XIV, 1974

32. J.Ryan: Conformally covariant operators in Clifford analysis, Jour. for Anal. and its Appl., 14, 4, 677-704, 1995

33. J.Ryan (ed.): Clifford algebras in analysis and related topics, CRC Press, Boca Raton, 1996

34. I.Sabadini, F.Sommen, D.Struppa, P.Van Lancker: Complexes of Dirac operators in Clifford algebras, to be published

35. S.M.Salamon: Differential geometry of quatrenionic manifolds, Ann.Sc.Ec. Norm.Sup. 19, 1986, 31-55

36. J.Slovák: Natural operators on conformal manifolds, Hab. dissertation, Masaryk University, Brno, 1993

37. J.Slovák: Parabolic geometries, Doctoral thesis, Brno, 1997

38.   J.Slovák, V.Souček: Invariant operators of the first order on manifolds with a
      given parabolic structure, to appear in the Proc. of the Conference 'Analyse
      harmonique and analyse sur les varietes', Marseille, June 1999
39.   K. Yamaguchi, Differential systems associated with simple graded Lie algebras,
      Advanced Studies in Pure Mathematics 22 (1993), 413–494

Teodorescu Type Transforms in Applications

Wolfgang Sprößig (sproessig@math.tu-freiberg.de)
Faculty of Mathematics and Informatics, Freiberg University of Mining and Technology

Abstract. This paper points out the importance of the Teodorescu transform for the treatment of some classes of elliptic boundary value problems and eigenvalue estimations. The considerations are restricted to problems in fluid mechanics and the Maxwell equations. Relations to the classical potential theory are stated. The norm of the Teodorescu transform in some pairs of spaces is estimated. A discrete version of a Teodorescu transform is deduced and compared with the continuous one. Exact computations over the unit ball are also obtained.

Keywords: Teodorescu transform, Maxwell equations, fluid mechanics, eigenvalue estimations, electro-magnetic fields

Mathematics Subject Classification: 30G35

Dedicated to Richard Delanghe on the occasion of his 60th birthday.

1. Introduction

This paper is written on the occasion of the 60th birthday of Richard Delanghe who is continuously involved in this field of research. We are deeply indebted to him and his group for valuable discussions and support. Their books [6], [8] have inspired many people to learn Clifford analysis.

In the classical two-dimensional Vekua theory the so-called T-operator plays an essential role. This operator is nothing but a two dimensional weak singular integral operator over a domain in the complex plane, which is right invers to the Cauchy-Riemann operator. It is of the form:

$$(T_G u)(z) = -\frac{1}{2\pi i} \int_G \frac{u(\zeta)}{\zeta - z} d\xi d\eta$$

where $z = \xi + i\eta$ and G is a domain in $\mathbb{C}$. There exist comprehensive treatises on mapping properties and their applications in complex analysis. An excellent reference is the classical book by Ilja N. Vekua: *"Generalized analytic functions"* (1959)[26]). It isn't an exaggeration to say, that the whole theory of complex partial differential equations is

F. Brackx et al. (eds.), *Clifford Analysis and Its Applications*, 341–360.

based on operators of such type. This basic operator has been extended and enables a wide range of further applications in this way (cf. [27] G.C. Wen and H. Begehr). The aim of this article is to identify higher dimensional generalizations and applications of this T-operator, which are called *Teodorescu type transforms* nowadays. One of the first generalizations one can find in [23]. During recent decades we have developed an operator calculus (cf. [12], [13]) together with K. Grlebeck, which is working in real Clifford algebras with special emphasis on Teodorescu type transforms. In cooperation with S. Bernstein ([5], [4], [3]), U. Khler, A. Hommel, F. Kippig [16] and U. Wimmer [15] a lot of special results have been obtained. In recent years it was especially the cooperation with J. Ryan [22], M. Shapiro [17] and H. Malonek [19], that improved and completed this operator calculus. In this paper we are able only to outline the theory and demonstrate this calculus for some selected examples.

2. Preliminaries

Let $\{e_0 = 1, e_1, ..., e_n\}$ be a basis in $\mathbb{R}^{n+1}$. Consider the Clifford algebra $Cl_{0,n}$ with the basis $\{1; e_1, ..., e_n; e_1e_2, e_1e_3, ..., e_{n-1}e_n; ...; e_1e_2...e_n\}$. Note that in our case $e_i^2 = -1$ for $i = 1, 2, ..., n$. By definition the relations $e_ie_j + e_je_i = 0$ are satisfied. With the abbreviation $e_{i_1...i_k} := e_{i_1}...e_{i_k}$ each element $u \in Cl_{0,n}$ has the form

$$u = u_0 + \sum_{k=1}^{n} \sum_{(i_1,...,i_k)} u_{i_1...i_k} e_{i_1...i_k},$$

where $1 \leq i_1 \leq ... \leq i_k \leq n$. Furthermore, we define

$$\overline{e_{i_1...i_k}} := (-1)^{\frac{k(k+1)}{2}} e_{i_1...i_k},$$

and conjugation of u by

$$\bar{u} = \sum_{k=1}^{n} \sum_{(i_1,...,i_k)} u_{i_1...i_k} \overline{e_{i_1...i_k}}.$$

We abbreviate $u_0 =: \text{Sc}\, u$ and $\underline{u} = u_1e_1 + ... + u_ne_n$, denoting the scalar part and the vector part respectively. Further, we assume $G \subset \mathbb{R}^n$ to be a domain with a sufficient smooth boundary Γ and write $x = x_0 + \underline{x}$ with $\underline{x} = \sum_{k=1}^{n} x_ke_k$. For each paravector $x \in \mathbb{R} \oplus \mathbb{R}^n$ we have $x\bar{x} = x_0^2 + x_1^2 + ... + x_n^2$. Let $B(\Omega)$ be a Banach space of functions defined on Ω, an open subset of $\mathbb{R}^{n+1}$. Then the space $B(\Omega) \otimes Cl_{0,n}$ is denoted by $B(\Omega)$, $\Omega \in \{G, \Gamma\}$.

Now we define the *Dirac operator* including the paravector-valued potential b

$$D - b = \sum_{k=1}^{n} e_k(\partial_k - b_k),$$

acting in the vector space $\mathbb{R}^n$, where $\partial_i = \frac{\partial}{\partial x_i}$. Note that

$$D^2 = -\Delta \quad \text{and} \quad \overline{D} = -D.$$

Functions that are defined in G and fulfil the equation

$$(D - b)\, u = 0$$

are called *left Clifford-b-regular*. In this paper we intend to call these functions simply *Clifford-b-regular*. If $b = 0$ this notion coincides with Clifford regularity. Now define the so-called *Cauchy kernel* on $\mathbb{R}^n \backslash \{0\}$ by

$$e(x) = e_0(x) = \frac{-x}{\omega_n |x|^n},$$

where ω_n denotes the surface area of the unit ball in $\mathbb{R}^n$.

3. A class of Teodorescu type transforms

Let G be a domain in $\mathbb{R}^n$ and b a vector in $C\ell_{0,n}$.

Definition 3.1 *We consider T_b to be a linear operator that has the integral representation*

$$(T_b u)(x) := \int_G e_b(x - y)u(y)dy,$$

which is right-inverse to the Dirac operator D including the potential b, e.g. $(D - b)T_b u = u$. The kernel function $e_b(x)$ fulfils for $x \neq 0$ the equation $(D - b)e_b = 0$.

Now, we are going to define the corresponding kernels for several potentials. Each operator T_b is called a *Teodorescu transform*.

We introduce the so-called *Macdonald function* $K_p(t)$. A modification of this function is needed for the construction of the kernel of a corresponding Teodorescu transform. The function $K_p(t)$ is given by

$$K_p(t) := \frac{\pi i}{2} e^{\frac{1}{2}pi\pi} H_p^{(1)}(it),$$

where

$$H_p^{(1)}(t) := J_p(t) + iN_p(t)$$

and for $(p \notin Z)$

$$N_p(t) := \frac{\cos \pi p}{\sin \pi p} J_p(t) - \frac{1}{\sin \pi p} J_{-p}(t),$$

$$J_p(t) := \frac{1}{\Gamma(\frac{1}{2})\Gamma(\frac{1+2p}{2})} \left(\frac{t}{2}\right)^p \int_0^\pi \cos(t \cos \tau)(\sin \tau)^{2p} d\tau.$$

Here J_p denotes a Bessel function of the first kind and N_p the Neumann function. We assume $a_0, p, x, R > 0$.

Now, we are going to define the corresponding kernel of the Teodorescu transform for several potentials.

—   Let $b = 0$, then

$$e(x) = -\frac{1}{\sigma_n} \frac{x}{|x|^n}.$$

—   Let $b = a_0$ and $a_0 \in \mathbb{R}$. We have

$$e_{a_0}(x)$$

$$= -\left(\frac{ia_0}{2\pi}\right)^{n/2} |x|^{1-n/2} \left(K_{n/2}(ia_0|x|)\omega - K_{n/2-1}(ia_0|x|)\right)$$

$$= (D + a_0)\mathcal{K}_{ia_0}(x) \quad (\omega = \frac{x}{|x|}).$$

—   Let $n = 3$. Then the fundamental solution reads:

$$e_{a_0}(x) = e(x) \left(\cos(a_0|x|) - |x|a_0 \sin(a_0|x|)\right) - \frac{a_0 \cos(a_0|x|)}{4\pi|x|}.$$

—   For $b = ia_0$ it follows that

$$e_{ia_0}(x) = (D + ia_0)\mathcal{K}_{a_0}(x).$$

—   Let $b = -a$, then

$$e_a(x) = e^{-(a,x)} e_{-ia_0}(x).$$

−   Let $b = -ia$, then

$$e_{ia}(x) = e^{-(i\underline{a},x)} e_{-a_0}(x).$$

Remark: In [17] a Teodorescu transform with $b \in \mathbb{C}\mathbb{H}$ is introduced and in [28] a TT is defined if b is a vector. In both cases a corresponding Borel-Pompeiu formula is obtained.

4.   Operators and their relations

We have to introduce the so-called Cauchy-Bizadse type operator

$$(F_{\Gamma,b}u)(x) := \int_{\Gamma} e_b(x-y)\underline{n}(y)u(y)d\Gamma_y, x \notin \Gamma,$$

where $\underline{n}(y) = \sum_{k=1}^{n} e_k n_k(y)$ is the unit vector of the outer normal at the point y.

Furthermore, we need to define projections $P_{\Gamma,b}$ and $Q_{\Gamma,b}$. These operators are characterized by the following properties:

(i) $((P_{\Gamma,b})^2 u)(\xi) = (P_{\Gamma,b}u)(\xi), \;\; ((Q_{\Gamma,b})^2 u)(\xi) = (Q_{\Gamma,b}u)(\xi).$

(ii) The subspace $im\, P_{\Gamma,b} \cap C^{0,\alpha}(\Gamma), 0 < \alpha < 1$ describes the space of all $C\ell_{0,n}$-valued functions that are Clifford-b-regular extendable into the domain G

(iii) The subspace $im\, Q_{\Gamma,b} \cap C^{0,\alpha}(\Gamma)$ describes the space of all $C\ell_{0,n}-$ valued functions that are Clifford-b-regular extendable into the domain $R^n \backslash \overline{G}$ and vanish at infinity.

Remark: The images of these projections are called *Hardy spaces*. The projections themselves are called *Pompeiu projections*. If the domain G is a ball, then $P_{\Gamma,b}$ and $Q_{\Gamma,b}$ are just the *Szeg projections*. Usually these projections are generated by the limits (from inside and outside) on the boundary of the Cauchy-Bizadse type operator $F_{\Gamma,b}$.

Proposition 4.1.   (cf.[13]) *Let* $u \in C^1(G) \cap C(\overline{G})$. *Then we can establish the so-called Borel-Pompeiu formula:*

$$(F_{\Gamma,b}u)(x) + (T_b(D-b)\,u)(x) = \begin{cases} u(x), & x \in G \\ 0, & x \in \mathbb{R}^n \backslash \overline{G} \end{cases}.$$

Remark: The operators $F_{\Gamma,b}, P_{\Gamma,b}$ and $Q_{\Gamma,b}$ can be extended to $L_p(\Gamma), 1 < p < \infty$.

5. TT and potentials

In this section we are going to describe the connection of the Teodorescu transform to potential operators.

Definition 5.1. *We introduce the following Newton type potential:*

$$(K_a\cdot)(x) := \int_G \mathcal{K}_{ia_0}(x-y)e^{-(a,x-y)} \cdot dy.$$

Furthermore, we have to define the following single layer potential, too:

$$(V_a\cdot)(x) := -\int_\Gamma \mathcal{K}_{ia_0}(x-y)e^{-(a,x-y)}\underline{n}(y) \cdot d\Gamma_y.$$

Proposition 5.1. *Let $u \in W_2^1(G)$. Then*

$$K_a D_a u + V_a u = T_a u.$$

Proof. The proof can be found in [24]. #

Corollary 5.1. *Let $b = -\bar{a}$. Then in G it can be shown that*

$$F_{\Gamma,a} = -D_b V_a, \quad T_a = -D_b K_a .$$
$$D_a F_{\Gamma,a} = 0, \quad D_a T_a = I, \quad T_a D_a + F_{\Gamma,a} = I.$$

6. The image of orthoprojections with the help of TT

Proposition 6.1. (cf.[13]) *Let $G \subset \mathbb{R}^n$ be a bounded domain with a piecewise smooth boundary. Then the right–linear set*

$$\mathcal{L} := L_2(G) \cap \ker(D - b)$$

is a subspace in $L_2(G)$.

It is important to note the following decomposition of the Hilbert space $L_2(G)$:

$$L_2(G) = \ker(D - b) \cap L_2(G) \oplus (D + b) \overset{\circ}{W}_2^{1}(G).$$

The corresponding projections $\mathcal{P}_b$ (onto $\ker(D - b) \cap L_2(G)$) and $\mathcal{Q}_b$ onto $(D + b) \overset{\circ}{W}_2^1 (G)$ are orthoprojections and $\mathcal{P}_b$ is just a *Bergman type projection*. In case of $b = 0$, we simply write $\mathcal{P}$ and $\mathcal{Q}$. In [24] it is proved (at least in $\mathbb{H}$) that

$$f \in \text{im } \mathcal{Q}_b \cap L_2(G) \Longleftrightarrow tr_\Gamma T_{\bar{b}} f = 0,$$

where tr_Γ denotes the restriction to the boundary Γ. It is shown (cf. [13]) that the Bergman type projection $\mathcal{P}_b$ permits the representation

$$\mathcal{P}_b = F_{\Gamma,b}(tr_\Gamma T_{\bar{b}} F_{\Gamma,b})^{-1} tr_\Gamma T_{\bar{b}}$$

and in the case where $u \in \ker(D - b)$ we have that

$$\mathcal{P}_b = F_{\Gamma,b}(tr_\Gamma V_{\bar{b}})^{-1} tr_\Gamma T_{\bar{b}}.$$

7. Estimations of norms of TT

In this section we are going to give estimations of the norm of versions of the Teodorescu transform in several pairs of Banach spaces. The results are proved in [12] and [13]. Some of them were obtained by S. Bernstein in work preliminary to [13].

7.1. ESTIMATIONS OF THE TRANSFORM $T_0 = T$

We have

1. Let $u \in C(\overline{G})$. Then

$$\|Tu\|_C \leq \frac{1}{\sigma_n} \max_{x \in \overline{G}} \int_G \frac{1}{|x - y|^{n-1}} dy . \|u\|_C.$$

2.

$$|(Tu)(x)| \leq \frac{1}{\sigma_n} \left| \int_g e(x - y) u(y) dy \right| \leq \frac{1}{\sigma_n} \int_G |e(x - y)| \, |u(y)| dy$$

$$\leq \frac{1}{\sigma_n} [dist(x, G)]^{1-n} \|u\|_1.$$

3. The operator

$$\partial_k T : L_p(\mathbb{R}^n) \to L_p(R^n)$$

is continuous and

$$\|\partial_k T\|_{L(L_p(R^n))} \le \left(C\sigma_n^{-\frac{1}{p}} + \frac{1}{n}\right) \quad (1 < p < \infty).$$

Remark: Using the Theorem of CALDERON–ZYGMUND (cf. [20] §3, XI) it can be proved that

$$\left\|\frac{1}{\sigma_n}\int\limits_{\mathbb{R}^n} e(x-y)u(y)dy\right\|_p \le C\left(\int\limits_{S^1}\left(\frac{1}{\sigma_n}\frac{(x-y)}{|x-y|}\right)^q dS_1\right)^{\frac{1}{q}}\|u\|_p,$$

$$\left(\int\limits_{S_1}(\frac{1}{\sigma_n}\frac{(x-y)}{|x-y|})^q dS_1\right)^{\frac{1}{q}} = \frac{1}{\sigma_n}\left(\int\limits_{S_1} dS_1\right)^{\frac{1}{q}} = \sigma_n^{\frac{1}{q}-1} = \sigma_n^{-\frac{1}{p}},$$

where $1/p + 1/q = 1$. In special cases it is possible to compute the value of the constant C. Nevertheless, such a computation remains difficult.

4. Let G be a bounded domain. Then

$$T : L_p(G) \to W_p^1(G)$$

is continuous and it holds

$$\|Tu\|_{p,1} = \left(\|Tu\|_p + \sum_{k=1}^{n}\|\partial_k Tu\|_p\right)^{1/p} \le \|Tu\|_p + \sum_{k=1}^{n}\|\partial_k Tu\|_p$$

$$\le \operatorname{diam} G\,\|u\|_p + \sum_{k=1}^{n}(C\sigma_n^{-\frac{1}{p}} + \frac{1}{n})\|u\|_p$$

$$\le (\operatorname{diam} G + nC\sigma_n^{-\frac{1}{p}} + 1)\|u\|_p.$$

5. Let G be a bounded domain in $\mathbb{R}^n$, $u \in L_p(G)$. Then

(i) $T_G : L_p(G) \to L_\infty(G,)$ for $n < p < \infty$ with

$$\|Tu\|_\infty \le \frac{1}{\sigma_n^{1/p}}\left(\frac{(\operatorname{diam} G)^{n-p(\frac{n-1}{p-1})}}{n - p(\frac{n-1}{p-1})}\right)^{\frac{p-1}{p}}\|u\|_p,$$

(ii) $\|Tu\|_p \le \operatorname{diam} G\,\|u\|_p$ for $1 < p < \infty$.

6. Let $u \in L_p(G), p > n$; then holds

$$\|Tu\|_p \le \frac{1}{\sigma_n}\max_{x \in \bar{G}}\left(\int\limits_G \frac{1}{|x-y|^{(n-1)}}dy\right)^{p/q}|G|^{1/p}\|u\|_p.$$

7. Let $u \in L_p(G), p > n$ and G is a domain with a finite volume. The following estimation is valid:

$$\|T\|_{L[L_p,C]} \le C(n,p)|G|^{\frac{p-n}{p}}$$

with $C(n,p) = \sigma_n^{(\frac{p-n}{p}+\frac{p-1}{p})} \cdot \left(\frac{1}{n(\frac{p-n}{p-1})}\right)^{\frac{p-1}{p}}.$

For $p \to \infty$ we get $\|T\|_{L(L\infty,C)} \le \frac{\sigma_n^2}{n}|G|$.

7.2. A NORM ESTIMATE FOR T_{a_0}

Let $u \in L_\infty(G)$. The norm of the Teodorescu transform T_{a_0} from $L_\infty(G)$ to $L_\infty(G)$ is bounded by

$$c = n\sigma_n \left(\frac{2^{n/2-1}}{a_0^{n/2+1}}\Gamma(n/2) - \frac{\rho^{n/2}}{a_0}K_{n/2}(a_0\rho)\right.$$

$$+ \ \rho^{n/2+1}K_{n/2}(a_0\rho) + a_0\rho\frac{2^{n/2-1}}{a_0^{n/2+1}}\Gamma(n/2)).$$

This result is proved in [1]. We have

$$\|T\|_{L[L_2]} \le \sqrt[n]{\frac{|G|}{|B_1|}}$$

where B_1 is the unit ball in $\mathbb{R}^n$.

Now, we are going to outline the connection of norms with the first eigenvalue of Dirichlet's problem.

8. TT and lower bounds for eigenvalues

Methods of quaternionic and Clifford analysis can be applied successfully to develop a general approach to the estimation of lower bounds for the first eigenvalue for several boundary value problems. Our method uses the close connection between the first eigenvalue of the corresponding elliptic eigenvalue problem and the norm of a Teodorescu transform in a corresponding pair of spaces. Together with results from the previous section we are able to find sufficiently precise lower estimates for eigenvalues. This method works under very weak conditions on the domain G. In [11] two key-relations are proved. They are:

$$\|T\|_{L(im\mathcal{Q},L_2)} = \frac{1}{\sqrt{\lambda_1(G)}}$$

where λ_1 is the first eigenvalue of Dirichlet' s problem and

$$\|T\|_{L(im\mathcal{Q},W_2^1)} = \sqrt{1 + \frac{1}{\lambda_1(G)}}.$$

Modifications of the Teodorescu transform lead to estimations of the first eigenvalue of Neumann's boundary value problem and of mixed boundary value problems, too. As shown in [11] we have to use the operator

$$Tu - \frac{1}{|G|} \int_G (Tu)(x)dx$$

instead of the TT in the case of Neumann's problem. In the case of a mixed boundary value problem for the Laplacian the operator

$$(I - F_\Gamma)^{-1}T$$

has to be used (cf. [11]).

For the first eigenvalue of Dirichlet's problem of the Lam equations of the linear elasticity we get the lower bound

$$\Lambda_1 \geq \frac{m-2}{2(m-1)} \frac{1}{\|T\|_{L(L_2)}\|T\|_{[im\mathcal{Q},L_2]}}.$$

In [24] nonlinear examples are given.

9. TT in a non-linear free convection problem

In this section we are going to show how the methods of Clifford analysis can be used to treat of boundary value problems of partial differential equations. For this reason we have chosen the time-harmonic case of the non-linear stream problem with free convection. The corresponding mathematical description reads as follows:

In the bounded domain G the following equations have to be satisfied:

$$-\Delta \underline{u} + a_1(\underline{u} \cdot grad) \underline{u} + f(u) + a_2 grad\, p + a_3(-e_3)w = F(x)$$
$$div\, \underline{u} = 0$$
$$-\Delta w + a_4(\underline{u} \cdot grad) w = g.$$

On the boundary $\partial G = \Gamma$ the velocity $\underline{u}$ and the temperature w are identically zero, and p denotes the pressure. A well-defined physical meaning can be given to the coefficients. If we denote the density of the fluid by ρ, the viscosity by η, the Grashof number by γ, the temperature conductivity by κ and the Prandl number by m, then they are related to the coefficients a_i ($i=1,2,3,4$) as follows:

$$a_1 = \frac{\rho}{\eta}, \quad a_2 = \frac{1}{\eta}, \quad a_3 = \frac{\gamma}{\eta}, \quad a_4 = \frac{m}{\kappa}.$$

Conditions on the function f are more technical. We will formulate them later.

In order to apply Clifford analysis we first have to transform this problem first. We obtain the following operator integral equation:

$$u = -a_1 TQT[M(u) + a_3 e_3 w] - a_2 TQp$$
$$0 = Sc\{\, a_1 QT[M(u) + a_3 e_3 w] + a_2 Qp\}$$
$$w = a_4 TQTSc\,(uDw) + TQTg \quad,$$

where $M(u) = (u \cdot grad)u + f(u) - F(x)$. Note that e_3 can be identified with the vector $(0,0,1)$. The term u means now $u := \underline{u}$.

Remark: The boundary conditions are fulfilled. In fact, the definition of the orthoprojection Q immediately ensures that

$$tr_\Gamma TQu = tr_\Gamma TDw, \qquad w \in \overset{\circ}{W}{}_2^1(G)\,.$$

Borel-Pompeiu's formula now yields $tr_\Gamma TQu = tr_\Gamma w - tr_\Gamma F_\Gamma w = 0$.

10. An iteration method

The following procedure reduces our problem to the more simple so-called *Stokes problem*:

$$u_n = -a_1 TQT[M(u_{n-1}) + a_3 e_3 w_{n-1}] - a_2 TQp_n \qquad (1)$$
$$0 = Sc\, a_1 QT[M(u_{n-1}) + a_3 e_3 w_{n-1}] + a_2 Sc\, Qp_n \qquad (2)$$
$$w_n = a_4 TQT_G\, Sc\,(u_n Dw_n) + TQTg. \qquad (3)$$

For the computation of w_n we use the "inner" iteration:

$$w_n^{(i)} = a_4 TQT : Sc(u_n Dw_{n-1}^{(i-1)}) + TQTg \quad (i = 1, 2, ...).$$

Theorem 9.1. *1. Let* $u_n \in \overset{\circ}{W}{}_2^1\,(G)$. *If* $a_4 \neq 4$ *and* $\|u_n\| < 1/(a_4 CK)$ *where C is the embedding constant from W_2^1 in L_2 and*

$K = a_1\|T\|_{[im\Omega, W_2^1]}\|T\|_{[L_q, L_2]}$. Then $(w_n^{(i)})$ converges in the sense of $W_2^1(G)$. The norm $\|w_n\|_{2,1}$ can be estimated.

2. Let $F \in L_2(G), g \in L_2(\Gamma)$, $f : W_2^1(G) \mapsto L_2(G)$ with

$$\|f(u) - f(v)\|_2 \leq L\|u - v\|_{2,1}, \quad (L > 0)$$

and $f(0) = 0$. Under the conditions

(i) $a_1\|F\|_2 + a_3K|d|^{-1}\|g\|_2 < 1/(16K^2C)$, $d = (4 - a_4)\kappa$

(ii) $\|g\| < (1 - 1/\sqrt{2})a_2^{-1}d^2/(32K^3Cm)$

(iii) $a_4 < 4$

the iteration procedure (2)-(3)-(4) converges in $W_2^1(G) \times W_2^1(G) \times L_2(G)$ to the unique solution

$$\{u, w, p\} \in \overset{o}{W}{}_2^1(G) \times \overset{o}{W}{}_2^1(G) \times L_2(G)$$

(p is unique up to a constant) of the above formulated boundary value problem.

Remark: The smallness conditions (i) and (ii) can be realized, e.g. for fluids with a sufficient "big" viscosity number. Condition (iii) is easy to check.

We shall now explain how to realize the iteration procedure in practice. There are more or less two ways of doing this within the Clifford operator calculus. First we note that the solution of Stokes problem can be given explicitly. We have:

Theorem 9.2. With the abbreviations

$$h_n := M(u_{n-1}) + a_3e_3w_{n-1} \tag{4}$$

$$Q' := I - Vec\,\mathcal{F}(tr_\Gamma T\,Vec\mathcal{F})^{-1}tr_\Gamma T, \quad (Q'^2 = Q') \tag{5}$$

we get the following representation of the solution $\{u_n, p_n\}$ of (2)-(3):

$$u_n = -a_1TQ'VecT\,h_n$$

$$p_n = \eta a_1\,ScQ'VecT\,h_n, \quad (n = 1, 2, ...).$$

11. TT in electro-magnetic fields

Let $G \subset \mathbb{R}^3$ be a bounded domain symmetric with respect to the origin. An electric charge is distributed with the density $\rho(x)$. Further, let the vectors $\underline{E} = \underline{E}(t, x), \underline{H} = \underline{H}(t, x), \underline{D} = \underline{D}(x, t), \underline{B} = \underline{B}(x, t)$ and $\underline{J} = \underline{J}(x, t)$ describe the *electric field, magnetic field, electric displacements, magnetic field induction* and the *current density vector* respectively. The following relations holds:

$$\operatorname{div} \underline{D} = \rho \qquad \text{(COULOMB's law)}$$
$$\operatorname{div} \underline{B} = 0 \qquad \text{(no free magnetic charge)}$$
$$\operatorname{div} \underline{J} + \partial_t \rho = 0 \qquad \text{(continuity equation)}$$

These equations should be completed by the so-called *constraint equations* reflecting the electro-magnetic properties of the material. For linear and isotropic media the constraint equations read as follows:

$$\underline{D} = \varepsilon \underline{E} \qquad \underline{B} = \mu \underline{H} \qquad \underline{J} = \kappa \underline{E} \qquad \text{where} \qquad \kappa = \frac{1}{\rho}.$$

Here ε denotes the *permittivity*, μ the magnetic *permeability* and κ the *conductivity*. The latter relation is well known as OHM's *law*. In the general case the non-linear relations $\underline{D} = \underline{D}(\underline{E}), \underline{B} = \underline{B}(\underline{H}), \underline{J} = \underline{J}(\underline{E})$ are valid. Sometimes they are written

$$\underline{D} = \varepsilon(\underline{E}), \qquad \underline{B} = \mu(\underline{H}), \qquad \underline{J} = \kappa(\underline{E})$$

We shall now compute the electric field $\underline{E}$ and the magnetic field $\underline{H}$. Assume that time dependence is known, and is time harmonic. Let the electric field and the magnetic field permit the representation as a product of a scalar function that dependent only on time, and a position-dependent vector function. Then we have $\underline{E}(t, x) = E_0(t)\underline{E}_1(x)$ and the magnetic field $\underline{H}(t, x) = H_0(t)\underline{H}_1(x)$. Assume that the permittivity ε, the permeability μ and the electric conductivity κ depend on the position x and the time t, i.e, $\varepsilon = \varepsilon(t, x), \mu = \mu(t, x)$ and $\kappa = \kappa(t, x)$.

Maxwell's equations now read:

$$\operatorname{rot} \underline{E} = \partial_t(\mu \underline{H})$$

$$\operatorname{rot} \underline{H} = \partial_t(\varepsilon \underline{E}) + \kappa \underline{E}$$

$$\operatorname{div}(\varepsilon \underline{E}) = \rho$$

$$\operatorname{div}(\mu \underline{H}) = 0.$$

Using $\underline{E} = E_0(t)\underline{E}_1(x)$ and $\underline{H} = H_0(t)\underline{H}_1(x)$ and inserting these expressions in Maxwell's equations we now obtain

$$E_0 \operatorname{rot} \underline{E}_1 = -\partial_t(\mu H_0)\underline{H}_1$$
$$H_0 \operatorname{rot} \underline{H}_1 = (\partial_t(\varepsilon E_0) + \kappa E_0)\underline{E}_1,$$
$$E_0 \operatorname{grad} \varepsilon \cdot \underline{E}_1 + \varepsilon \operatorname{div} \underline{E}_1) = \rho$$
$$\operatorname{grad} \mu \cdot \underline{H}_1 + \mu \operatorname{div} \underline{H}_1 = 0.$$

In this way we find (with $D = \sum_1^3 \partial_i e_i$)

$$\operatorname{rot} \underline{E}_1 = -\frac{\partial_t(\mu H_0)}{E_0}\underline{H}_1,$$

$$\operatorname{rot} \underline{H}_1 = \frac{\partial_t(\varepsilon E_0) + \kappa E_0}{H_0}\underline{E}_1$$

$$-\operatorname{div} \underline{E}_1 = -\frac{\rho}{\varepsilon E_0} + \left(\frac{D \varepsilon}{\varepsilon}\right) \cdot \underline{E}_1,$$

$$-\operatorname{div} \underline{H}_1 = \left(\frac{D \mu}{\mu}\right) \cdot \underline{H}_1.$$

Note that we denote the usual $\mathbb{R}^3$-scalar product by " $\cdot$ " i.e., $\underline{u} \cdot \underline{v} = \sum_1^3 u_i v_i$. Writing for brevity

$$\underline{a} = \frac{D \mu}{\mu}, \quad \underline{d} = \frac{D \varepsilon}{\varepsilon}, \quad \rho' = -\frac{\rho}{\varepsilon E_0},$$

$$b_0 = \frac{\partial_t(\varepsilon E_0) + \kappa E_0}{H_0}, \quad c_0 = -\frac{\partial_t(\mu H_0)}{E_0}$$

and adding two further trivial equations

$$DE_{10} = 0, \quad DH_{10} = 0$$

we obtain the new system

$$E_1 := E_{10} + \underline{E}_1, \ H_1 = H_{10} + \underline{H}_1, \ d := \underline{d}, \ a := \underline{a}$$

$$D E_1 = c_0 H_1 + d \cdot E_1 + \rho' \qquad D H_1 = a \cdot H_1 + b_0 E_1.$$

Finally, we obtain the matrix equation

$$\begin{pmatrix} D & 0 \\ 0 & D \end{pmatrix}\begin{pmatrix} E_1 \\ H_1 \end{pmatrix} = \begin{pmatrix} d \cdot & c_0 \\ b_0 & a \cdot \end{pmatrix}\begin{pmatrix} E_1 \\ H_1 \end{pmatrix} + \begin{pmatrix} \rho' \\ 0 \end{pmatrix}.$$

Furthermore assuming that $d \times E_1 = 0$ and $a \times H_1 = 0$ it follows that $d \cdot E_1 = -dE_1$ and $a \cdot H_1 = -aH_1$. From above we obtain the matrix equation

$$\begin{pmatrix} D - d & 0 \\ 0 & D - a \end{pmatrix}\begin{pmatrix} E_1 \\ H_1 \end{pmatrix} = \begin{pmatrix} 0 & c_0 \\ b_0 & 0 \end{pmatrix}\begin{pmatrix} E_1 \\ H_1 \end{pmatrix} + \begin{pmatrix} \rho' \\ 0 \end{pmatrix}.$$

Finally we obtain the representation :

$$(D - d)\, E_1 = c_0 H_1 + \rho'$$
$$(D - a)\, H_1 = b_0 E_1.$$

On the boundary the values of H are given by g. For $b_0 \neq 0$ we have

$$(D - d) b_0^{-1} (D - a) H_1 = c_0 \underline{H_1} + \rho'.$$

Now the following theorem can be proved (cf. [13])

Theorem 10.1. *The solution of the boundary value problem of Maxwell's equations formulated above permits the following representation:*

$$
\begin{aligned}
H_1 =\ & F_{-a} g + T_{-a}(F_{-a}(tr_\Gamma T_{-d} b_0 F_{-a})^{-1} Q_{-a} g \\
& + T_{-a} \mathcal{Q}_{(-d)(-a)} b_0 T_{-d} \rho' + T_{-a} \mathcal{Q}_{(-d)(-a)} b_0 T_{-d} c_0 H_1.
\end{aligned}
$$

Remark: If in some sense one of these functions is sufficiently small, a unique solution can be found by Banach's fixed-point theorem. In order to formulate these conditions in a more detailed way we have to study operators of the type T_a. These questions are considered in a more general context. There are known conditions under which it is possible to prove a strong convergence in $W_2^1(G)$ (cf. [24]). The operator $\mathcal{Q}_{(-d)(-a)}$ is definde in ([24]), too.

Sometimes it is more convenient to impose the condition $\underline{a} \cdot \underline{H} = g_0$ on the boundary. This can be easily realized . We then have to replace g by $g(-\underline{n})$, where $g = g_0 + \underline{g}$ and $\underline{n}$ denotes, as previously, the unit vector of the outward pointing normal to Γ.

In the time-harmonic case with constant κ the solution can be represented by:

$$
\tilde{H} = \frac{1}{\mu} F_\Gamma g + \frac{1}{\mu} T F_\Gamma (tr_\Gamma T F_\Gamma)^{-1} Q_\Gamma g - \frac{\kappa}{\varepsilon} T F_\Gamma (tr_\Gamma T F_\Gamma)^{-1} T \rho
$$

$$
+ \frac{\kappa}{\varepsilon} T^2 \rho
$$

$$
\tilde{E} = \frac{1}{\mu^2 \kappa} F_\Gamma (tr_\Gamma T F_\Gamma)^{-1} Q_\Gamma g - \frac{1}{\varepsilon \mu} F_\Gamma (tr_\Gamma T F_\Gamma)^{-1} T \rho - \frac{1}{\mu} T \rho.
$$

Remark: The more general system is solved by using of an iteration method under a smallness condition for a and b (cf. [12]). Far-reaching investigations are made in [21].

12. Action of TT on classes of functions

We will demonstrate in what manner the TT with $b = 0$ acts on classes of very simple functions by some examples. Abbreviate $T_0 =: T$. The domain is taken to be the unit ball.

1. $T(x^{n-1}) =$

$$\begin{cases} -\ln|x| & : \quad n = 0 \\ \frac{-1}{n+2}x^n & : \quad n \text{ odd} \\ \frac{-1}{n}(x^n - 1) & : \quad n \text{ even and integer multiple of 4; } n \neq 0 \\ \frac{-1}{n}(x^n + 1) & : \quad n \text{ even; } n \neq 0 \end{cases}$$

2. $T(|x|^n) =$

$$\frac{-1}{n+3}(x|x|^n), \quad n \neq -3.$$

3. $T(|x|^{\frac{n}{m}}) =$

$$\frac{-m}{n+3m}x|x|^{\frac{n}{m}}$$

4. $T(x|x|^{\frac{n}{m}-2}) =$

$$\begin{cases} \frac{m}{n}(|x|^{\frac{n}{m}} - 1) & : \quad m, n \neq 0 \\ \ln|x| & : \quad m \neq 0, n = 0 \end{cases}$$

5. $T(x^{n-1}|x|^m) =$

$$\begin{cases} -\ln|x| & : \quad n = m = 0 \\ \frac{-1}{n+m+2}|x|^n|x|^m & : \quad n \text{ odd and } n + m \neq -2 \\ \frac{-1}{n+m}(x^n|x|^m - 1) & : \quad n \text{ is a 4-fold, } n + m \neq 0 \\ \frac{-1}{n+m}(x^n|x|^m + 1) & : \quad n \text{ even, } n + m \neq 0, n \neq 0 \end{cases}$$

All of the above formulae are based on the following theorem:

Theorem 11.1. *Let* $G \subset \mathbb{R}^3$ *be a sufficiently smooth bounded domain,* $\{x^{(i)}\}_{i \in \mathbb{N}} \subset \Gamma_e$, $\Gamma_e \subset co\overline{G}$, $x \in G$. *Then,*

$$T\left(\sum_{k=1}^{3} \frac{x_k - x_k^{(i)}}{|x_k - x_k^{(i)}|^3} e_k \right) = -\frac{1}{|x_k - x_k^{(i)}|} + \Phi_{G,i}, \quad \Phi_{G,i} \in \ker D(G)$$

$$T\left(\frac{x_k - x_k^{(i)}}{|x_k - x_k^{(i)}|^3} e_k \right) = \frac{1}{2} \sum_{k=1}^{3} \frac{x_k - x_k^{(i)}}{|x_k - x_k^{(i)}|^3} (x_k^{(i)} - x_k) e_k$$

$$-\frac{1}{2} \frac{1}{|x_k - x_k^{(i)}|} e_0 + \Psi_{G,k}, \quad \text{with} \quad \Psi_{G,k} \in \ker D(G)$$

Proof. Application of D establishes the theorem by a straightforward computation. #

This theorem is one of the basic elements of a corresponding multipole method that is developed in [12][13][14].

13. Discrete versions of TT

As well as the continuous Teodorescu transform a discrete version of the Teodorescu transform can also be constructed. This work has been done in the habilitation paper of K. Gürlebeck. This discrete transform is defined as follows:

$$(T_h^+ u)(x) = \sum_{y \in int G_h \cup \partial G_{h,l}} e_h^+(x - y) u(y) h^3$$

$$- \sum_{\text{"left" edges}} e_j^+(x - y) u(y) h^3 +$$

$$+ \sum_{\text{"left" corner points}} e_h^+(x - y) u(y) h^3.$$

More precisely

$$\partial G_h = \{x \in G_h : \text{dist}(x, co\overline{G}_h) \le \sqrt{3}h\}$$

$$\partial G_{h,l} = \{x \in \partial G_h : \exists i \in \{1, 2, 3\}, V_{i,h}^- x \notin G_h\}$$

$$\partial G_{h,l,i} = \{x \in \partial G_h : V_{i,h}^- \notin G_h\}$$

$$\partial G_{h,l,i,j} = \partial G_{h,l,i} \cap \partial G_{h,l,j}, i \ne j$$

$$\partial G_{h,l,i,j,k} = \partial G_{h,l,i,j} \cap \partial G_{h,l,k}, i \ne j \ne k$$

and

$$(T_h^+ u)(x) = \sum_{y \in \int G_h \cup \partial G_{h,l}} e_h^+(x-y)u(y)h^3$$

$$- \sum_{\substack{k,j=1 \\ j>k}}^{3} \sum_{y \in \partial G_{h,l,j,k}} e_j^+(x-y)u(y)h^3 +$$

$$+ \sum_{\substack{y \in \partial G_{h,l,i,j,k} \\ i \neq j \neq k}} e_h^+(x-y)u(y)h^3 .$$

Here h is the mesh width of the lattice G_h. In order to obtain T_h^- we have to replace "l" by "r". In this construction the basic formulae (Borel-Pompeiu formula, Hilbert space dcomposition) have a corresponding discrete version, too. This enables us to find strong relations between the continuous and the discrete transform under very weak conditions.

We are now going to demonstrate that the discrete Teodorescu transform is really suitably chosen and we are going to estimate the "distance" between the continuous operator and the discrete operator. For more detailed information we refer to [9] and [13].

Theorem 12.1. *Let f be a* RIEMANN-*integrable function that belongs to $L_\infty(G)$. Then*

$$\|T_h^+ f - Tf\|_{C_h(G)} \to 0.$$

If $f \in C^{0,\beta}(\overline{G}), 0 < \beta \leq 1$, we have

$$\|T_h^+ f - Tf\|_{C_h(G)} \leq C(G, \|f\|_{q,h}, \|f\|_q)h^{-2+\frac{3}{p}}|\ln h|$$
$$+ K_{p,G}|h|^\beta \|f\|_{C_h(G)}$$

for $p < 3/2$, $1/p + 1/q = 1$, where $\| \cdot \|_q$ denotes the norm in L_q and $\| \cdot \|_{q,h}$ is the corresponding discrete norm.

Proof. The proof involves a lot of technical details. For this reason we refer to [9] and [13].

It can be shown similarly that corresponding results remain true in L_p-spaces, too.

Theorem 12.2. (cf.[13])*Let $f \in \mathcal{R} \cap L_\infty(G)$. For $p \in (\frac{6}{5}, \frac{3}{2})$ we have*

$$\|T_h^+ f - Tf\|_{p,G} \to 0$$

and under the assumptions that $f \in C^{0,\beta}(\overline{G}), p \in (\frac{6}{5}, \frac{3}{2}), 0 < \beta \leq 1$

$$\|T_h^+ f - Tf\|_{p,G} \leq C\|f\|_{\infty,G} h^{-2+\frac{3}{p}}$$
$$+C_1(\|f\|_{q',h,G}, \|f\|_{q',G}) h^{-2+\frac{3}{p'}} |\ln h|$$
$$+K_{p',G}\|f\|_{C^{0,\beta}}(G) h^{\beta}.$$

Here R describes the class of Riemann integrable functions. As we have seen above, the approximation properties of $T_h^{\pm}$ are as we expect. The proof of these properties requires some work, but there are just technical difficulties. This is very natural in approximating a weakly singular integral by an adapted quadrature formula.

References

1. Bahmann, H., Gürlebeck, K., Shapiro, M. and W. Sprig (2000) On a Modified Teodorescu Transform, submitted for publication.
2. Bernstein S. (1991) Operator calculus for elliptic boundary value problems in unbounded domains. *Zeitschrift f. Analysis u. Anwend.* 10, 4: 447–460.
3. Bernstein S. (1993) Elliptic boundary value problems in unbounded domains. *In: F. Brackx, et al (eds.), Clifford algebras and their applications in mathematical physics, Kluwer, Dordrecht* : 45–53.
4. Bernstein S. (1993) Analytische Untersuchungen in unbeschränkten Gebieten mit Anwendungen auf quaternionische Operatorentheorie und elliptische Randoperatoren. *Dissertation, Freiberg.*
5. Bernstein S. (1996) Fundamental solutions of Dirac type operators. *Banach Center Publications, Banach Center Symposium: Generalizations of Complex Analysis, May 30 – July 1 1994, Warsaw* 37: 159-172.
6. Brackx F., Delanghe R. and Sommen F. (1982) Clifford analysis. *Pitman Research Notes in Math., Boston, London, Melbourne.*
7. Delanghe R. (1970) On regular-analytic functions with values in a Clifford algebra. *Math. Ann.* 185: 91–111.
8. Delanghe R., Sommen F. and Souček V. (1992) Clifford algebra and spinor-valued functions *Kluwer, Dordrecht.*
9. Gürlebeck K. (1988) Grundlagen einer diskreten räumlich verallgemeinerten Funktionentheorie und ihrer Anwendungen. *Habilitationsschrift, TU Karl-Marx-Stadt.*
10. Gürlebeck K. (1990) Approximative solution of the stationary Navier–Stokes equations. *Math. Nachr.* 145: 297–308.
11. Gürlebeck K. and Sprößig W. (1987) A unified approach to estimation of lower bounds for the first eigenvalue of several elliptic boundary value problems. *Math. Nachr.* 131: 183–199.
12. Gürlebeck K. and Sprößig W. (1990) *Quaternionic analysis and elliptic boundary value Problems. Birkhäuser, Basel.*
13. Gürlebeck K. and Sprößig W. (1998) Quaternionic and Clifford Calculus for Physicists and Engineers. *John Wiley& Sons, Chichester, New York, Weinheim, Singapore, Toronto.*

14. Gürlebeck K., Sprößig W. and Tasche M. (1985) Numerical realization of boundary collocation methods. *Int. Ser. of Num. Math. Birkhäuser, Basel* 75: 206-217. 325-334.
15. Gürlebeck K., Sprößig W. and Wimmer U. (1993) Hypercomplex function theory for consideration of non-linear Stokes problems with variable viscosity. *Complex Variables, Theory and Appl.* 22: 195-202.
16. Kippig F. (1996) Untersuchungen zu Randwert- und Anfangswertaufgaben für partielle Differentialgleichungen mit Methoden der Clifford Analysis. *Dissertation, Freiberg*
17. Kravchenko V.V. and M.V. Shapiro (1996) Integral representations for spatial models of mathematical physics. *Pitman Research Notes in Math. Series* 351.
18. Kratzer, A. and Franz, W. (1960) *Transzendente Funktionen.* Mathematik und ihre Anwendungen in Physik und Technik. Reihe A, Bd. 28. Leipzig: Akademische Verlagsgesellschaft Geest & Portig K.-G.. XIII, 375 S. 58.
19. Malonek H.R. (1987) *Zum Holomorphiebegriff in höheren Dimensionen. Habilitationsschrift, Pädagogische Hochschule Halle.*
20. Mikhlin S.G. and Prößdorf S. (1986) *Singular integral operators.* Akademie-Verlag Berlin, Berlin.
21. Mitrea D., Mitrea M. and Pipher J. (1996) Vector potential theory on non-smooth domains in $\mathbb{R}^3$ and applications to electromagnetic scattering. *Preprint of the University of Minneapolis.*
22. Ryan J. (1993) Intertwining operators for iterated Dirac operators over Minkowski-type spaces. *Journal of Mathematical Analysis and Applications* 177: 1-23.
23. Sprößig W. (1978) Räumliches Analogon zum komplexen T-Operator. *Beiträge zur Analysis* 12: (1978), 127-137.
24. Sprößig W. (1995) On decompositions of the Clifford valued Hilbert space and their applications to boundary value problems. *Advances in Applied Clifford Algebras* 5, No. 2: 167-186.
25. Sprößig W. and Gürlebeck K. (1996) On the treatment of fluid problems by methods of Clifford analysis. *In: Deville, Gavrilakis, Ryhming (eds), "Computation of three-dimensional complex flows", Notes on Numerical Fluid Mechanics, Vieweg, Braunschweig, Wiesbaden* 53: 304-310.
26. Vekua I.N. (1963) *Verallgemeinerte analytische Funktionen. Akademie-Verlag, Berlin.*
27. Wen G.C. and H. Begehr, (1997) *Some second order systems in the complex plane, Preprint FU Berlin.*
28. Xu Z. (1991) A function theory for the operator $D - \lambda$. *Complex Variables Theory and Appl.* 16: 27-42.

Combinatorics and Clifford Analysis

Irene Sabadini (sabadini@mate.polimi.it)
Dipartimento di Matematica
Politecnico di Milano

Frank C Sommen (fs@cage.rug.ac.be)*
Department of Mathematical Analysis
Ghent University

Abstract. In this paper we introduce some linear first order systems in the framework of Clifford analysis that are originated by some combinatorial structures. We begin their study by providing their resolutions and we show that most of the systems treated are De Rham like. Being a new field of interest, we also give a list of open problems.

Keywords: combinatorial systems, resolutions, syzygies

Mathematics Subject Classification: 15A66, 30G35, 35A27

Dedicated to Richard Delanghe on the occasion of his 60th birthday.

1. Introduction

In our papers [7] and [9] we have introduced several systems that arise in the framework of Clifford analysis and its developments. In fact, the theory of Dirac or Fueter operators, both in one or several variables, can be considered nowadays a classical topic in Clifford analysis, but there are many other possibilities that can be investigated, e.g. the operators whose nullsolutions are the so-called higher spin monogenic, see [9], [11], [12], [13]. In [7] we started the study of some of the systems we introduced, by looking at the resolution of the associated complex. As it is well known, much of algebraic and analytic information is contained in the resolution, so the resolution is a first global information on how complicated a system is before developing its analysis. Among the systems in [7], we consider particularly original and stimulating those of combinatorial type. In fact we have shown that these systems have, in many cases, a peculiar behaviour since they are De Rham like. The systems of combinatorial type are constructed starting from some incidence structures that are "finite geometries". The "finite geometries" consist of a set of points, also called *tops*, $\{p_1, ..., p_m\}$ and a

---

* Senior Research Associate, FWO, Belgium

F. Brackx et al. (eds.), Clifford Analysis and Its Applications, 361–376.
© 2001 *Kluwer Academic Publishers. Printed in the Netherlands.*

collection of lines or *blocks* $\{b_1, ..., b_n\}$ where every block b_j is a subset of $\{p_1, ..., p_m\}$. The systems we have in mind consist of operators of the type $\sum_{jk} \pm e_j \partial_{x_k}$ where $x_1, \ldots, x_m$ are a set of m scalar coordinates, $e_1, \ldots, e_M$ are generators of the real Clifford algebra $\mathcal{C}_M$, according to the following axioms:

- (A_1) Each operator is of the type $\sum_{jk} \pm e_j \partial_{x_k}$;

- (A_2) Every partial derivative ∂_{x_j} occurs at most once in a given operator (within a term $\pm e_j \partial_{x_k}$);

- (A_3) Also every basis element e_k occurs at most once in a given equation;

- (A_4) Every term $e_j \partial_{x_k}$ occurs at most once in the whole system, preceded by either a plus or minus sign;

- (A_5) the number M of basis elements e_k is minimal.

Then we will associate with each point p_k the partial derivative ∂_{x_k}, so that each block corresponds to set of partial derivatives with which one may form an operator of the above type by attaching to each ∂_{x_j} a well chosen basis element e_k and a signature and taking the sum over all j–indices in the block. A system constructed in this way is not unique and differs from other systems associated with the same finite geometry, either for the assignment of a unit e_j to a given partial derivative ∂_{x_k} in a block, or for the signature of each top of that block. Obviously it is always possible to create such a system if one can choose the units in a set of M elements with M large enough. This is the main reason to explicitly require axiom (A_5).

The construction of a system according to the previous $(A_1), \ldots, (A_5)$ can be translated into the classical problem of colouring the edges of a certain bipartite graph which is obtained as follows (see e.g. [3]):
The points of the graph consist of two disjoint sets: the set of tops of the finite geometry and the set of blocks. The lines of the graph connect a point in the set of tops to a point in the set of blocks if that top belongs to the block. The set of elements $\{e_1, ..., e_M\}$ may be seen as set of colours with which we have to colour the edges of the graph such that all edges issuing from a given point in the graph have different colour and the total number of colours is minimal. This total number is the edge chromatic number for which Konig proved that it equals the highest number of edges per point.
There is another way to form the operators in a system, namely one can consider a finite geometry of incidence structure with tops $\{p_1, ..., p_M\}$ and blocks $\{b_1, ..., b_n\}$ where to each top we assign a basis element p_k

to a unit e_k. Then for each fixed block we have to choose now a certain "colour" ∂_{x_j} to be assigned to each top e_k of the block, and a signature. The axiom (A_5) has to be replaced by

— (A_5') the number m of partial derivatives is minimal.

The systems obtained in this second way are called *"super–dual systems"*, in order to distinguish them by the systems obtained by interchanging the words *"point"* and *"line"* in a given finite geometry.

We insert in this introduction also a short overview of the algebraic treatment of systems of partial differential equations with the basic tools we need in this treatment. For more details we refer the reader to the fundamental book [5] while, for the applications to Clifford analysis we refer to the papers [1], [2], [6], [7], [8], where some of the matrices we will use in the sequel are directly computed.

Let us note that a system of equations of the form

$$(1) \qquad (\sum_{jk} \pm e_j \partial_{x_k})f = g_i$$

where $f, g : U \subseteq \mathbb{R}^m \to \mathcal{C}_M$, can be written in a matrix form by following the next procedure.

Let $\vec{f} = (f_1, \ldots, f_r)$, $r = 2^M$, f_ℓ real differentiable functions on an open set $U \subseteq \mathbb{R}^m$, $\forall \ell = 1, \ldots, r$ and let

$$(2) \qquad \sum_{\ell=1}^{r} P_{i\ell}(D)f_\ell = g_i$$

be a $q \times r$ system of linear constant coefficient partial differential equations. Let $P = [P_{i\ell}]$ be a $q \times r$ matrix of complex polynomials in $\mathbb{C}^m$ and $D = (-i\partial_{x_1}, \ldots, -i\partial_{x_m})$. The polynomial matrix P, which is the symbol of the system, can be obtained from $P(D)$ by replacing (formally) ∂_{x_i} by the complex variable z_i for every $i = 1, \ldots, m$. This procedure is equivalent to taking the Fourier transform of $P(D)$. When we consider a system consisting of equations of the type (1), we take the real components of each equation with respect to each unit in the Clifford algebra $\mathcal{C}_M$ to arrive at a system of the type (2). The transpose matrix P^t of P is an R–homomorphism $R^q \to R^r$ whose cokernel is $\mathcal{M} = R^r/P^t R^q = R^r/<P^t>$, where $R = \mathbb{C}[z_1, \ldots, z_m]$ and $<P^t>$ is the submodule of R^r generated by the columns of P^t. By the Hilbert syzygy theorem, there is a finite free resolution

$$0 \longrightarrow R^{a_s} \xrightarrow{P_s^t} R^{a_s-1} \longrightarrow \ldots \xrightarrow{P_1^t} R^q \xrightarrow{P^t} R^r \longrightarrow \mathcal{M} \longrightarrow 0$$

that together with its transpose

$$0 \longrightarrow R^r \xrightarrow{P} R^q \xrightarrow{P_1} \ldots \longrightarrow R^{a_s-1} \xrightarrow{P_{a_s}} R^{a_s} \longrightarrow 0$$

are key tools for the algebraic analysis of the system (2). Even though there is a lot of information that arises from the resolutions above, we are mainly interested in the point of view of syzygies: every matrix $P_{a_i}^t(D)$ gives the compatibility conditions for the system whose representative polynomial matrix is $P_{a_{i-1}}^t$. In particular, the matrix $P_1(D)$ gives the compatibility conditions that a datum $\vec{g}$ of a inhomogeneous system $P(D)\vec{f} = \vec{g}$ must satisfy to have solutions f. We have computed the resolutions in this paper running CoCoA, version 3.7 on a Digital AlphaServer 4100/600, with 4 CPU and 3 GB RAM.

2. Affine and projective planes

In this section we will study some systems associated with affine or projective spaces over Z_p, p prime.

The affine geometry over Z_3. This geometry has 8 points, 8 lines and 3 points per line, 3 lines through each point such that any two lines have at most one point in common while through every point outside a given line goes exactly one parallel line. In this case, three Clifford basis elements are sufficient to write the following the combinatorial system:

$$(3) \quad \begin{cases} (e_1\partial_{x_1} + e_2\partial_{x_2} + e_3\partial_{x_3})f = g_1 \\ (e_1\partial_{x_4} + e_2\partial_{x_5} + e_3\partial_{x_6})f = g_2 \\ (e_1\partial_{x_8} + e_2\partial_{x_1} + e_3\partial_{x_4})f = g_3 \\ (e_1\partial_{x_6} + e_2\partial_{x_3} + e_3\partial_{x_7})f = g_4 \\ (e_1\partial_{x_5} + e_2\partial_{x_7} + e_3\partial_{x_1})f = g_5 \\ (e_1\partial_{x_3} + e_2\partial_{x_8} + e_3\partial_{x_5})f = g_6 \\ (e_1\partial_{x_7} + e_2\partial_{x_4} + e_3\partial_{x_2})f = g_7 \\ (e_1\partial_{x_2} + e_2\partial_{x_6} + e_3\partial_{x_8})f = g_8. \end{cases}$$

The operators appearing in the system (3) have been applied to functions $f : \mathbb{R}^8 \longrightarrow C_3$. Each equation appearing in the system can be associated with an 8×8 matrix U^i, $i = 1, \ldots, 7$ that can be obtained from (15) by substituting (formally) the P_k with the variables x_j if x_j is multiplied by the unit e_k. For example, the matrix symbol of the first operator can be obtained by setting $P_1 = x_1$, $P_2 = x_2$, $P_3 = x_3$. The matrix associated with the system is the 64×8 matrix $U = [U^1 \ldots U^8]^t$. The resolution is contained in the following

PROPOSITION 2.1. *The resolution of the system (3) is*

$$0 \longrightarrow R^8(-8) \longrightarrow R^{64}(-7) \longrightarrow R^{224}(-6) \longrightarrow R^{448}(-5) \longrightarrow$$

$$\longrightarrow R^{560}(-4) \longrightarrow R^{448}(-3) \longrightarrow R^{224}(-2) \longrightarrow R^{64}(-1) \longrightarrow \mathcal{M} \longrightarrow 0.$$

Moreover it is invariant with respect to changes of signature of the terms.

Proof. The rows of the 64×8 matrix $U = [U^1 \dots U^8]^t$ introduced above generate an R–module M, where $R = \mathbb{C}[x_1, \dots, x_8]$. The resolution was found by CoCoA giving the command Res(M). The invariance up to signature of the terms was checked using CoCoA by putting coefficients in front of the monomials x_j and by giving the command Rand( ) that let the system chose randomly the coefficients in the set $\{-1, +1\}$. ∎

REMARK 2.2. The resolution of the system (3) has the same length and Betti numbers proportional to those ones appearing in the standard De Rham complex (multiplied by 8 i.e. by the dimension of the overall Clifford algebra) $\partial_{x_1}, \dots, \partial_{x_8}$. We will call a resolution of this type a *De Rham like* resolution.

The Fano plane. The Fano plane is the projective plane over $\mathbb{Z}_2$, so it is the smallest projective plane. It consists of 7 points, 7 lines, 3 points per line and 3 lines per point such that every two lines have exactly one point in common. More precisely, we have the following points: $\{x_1 = (0,0,1), x_2 = (1,0,1), x_3 = (0,1,1), x_4 = (1,1,0), x_5 = (0,1,0), x_6 = (1,0,0), x_7 = (1,1,1)\}$ and the lines joining the points x_i, x_j, x_k with

$$(i, j, k) \in \{(1,2,6), (2,3,4), (1,3,5), (2,5,7), (3,6,7), (1,4,7), (4,5,6)\}.$$

Also in this case, a minimal combinatorial Dirac system requires only 3 Clifford basis elements, and an example is given by

(4)
$$\begin{cases} (e_1 \partial_{x_1} + e_2 \partial_{x_6} + e_3 \partial_{x_2})f = g_1 \\ (e_1 \partial_{x_2} + e_2 \partial_{x_4} + e_3 \partial_{x_3})f = g_2 \\ (e_1 \partial_{x_5} + e_2 \partial_{x_3} + e_3 \partial_{x_1})f = g_3 \\ (e_1 \partial_{x_7} + e_2 \partial_{x_2} + e_3 \partial_{x_5})f = g_4 \\ (e_1 \partial_{x_3} + e_2 \partial_{x_7} + e_3 \partial_{x_6})f = g_5 \\ (e_1 \partial_{x_4} + e_2 \partial_{x_1} + e_3 \partial_{x_7})f = g_6 \\ (e_1 \partial_{x_6} + e_2 \partial_{x_5} + e_3 \partial_{x_4})f = g_7. \end{cases}$$

Each equation appearing in the system can be associated with an 8×8 matrix U^i, $i = 1, \dots, 7$, which can be obtained from (15) by substituting (formally) the P_k with the variables x_j, if x_j is multiplied by the unit e_k. For example, the matrix symbol of the first operator can be obtained by setting $P_1 = x_1$, $P_2 = x_6$, $P_3 = x_2$. The matrix associated to the system is then a 56×8 matrix in the monomials $x_1, \dots, x_7$ and its resolution, that was computed with the same technique used in the previous case, is contained in the following

PROPOSITION 2.3. *The resolution of the system (4) is*

$$0 \longrightarrow R^8(-7) \longrightarrow R^{56}(-6) \longrightarrow R^{168}(-5) \longrightarrow R^{280}(-4) \longrightarrow R^{280}(-3)$$

$$\longrightarrow R^{168}(-2) \longrightarrow R^{56}(-1) \longrightarrow M \longrightarrow 0.$$

Moreover the resolution is De Rham like and invariant with respect to changes of signature of the terms in (4).

Projective plane over Z_3. This projective plane has 13 points, 13 lines, 4 lines per each point, 4 points on each line. A possible system one can obtain requires 4 units, and is

(5)
$$
\begin{cases}
(e_1\partial_{x_1} + e_2\partial_{x_2} + e_3\partial_{x_3} + e_4\partial_{x_{12}})f = g_1 \\
(e_1\partial_{x_5} + e_2\partial_{x_6} + e_3\partial_{x_{12}} + e_4\partial_{x_4})f = g_2 \\
(e_1\partial_{x_9} + e_2\partial_{x_{12}} + e_3\partial_{x_7} + e_4\partial_{x_8})f = g_3 \\
(e_1\partial_{x_7} + e_2\partial_{x_4} + e_3\partial_{x_1} + e_4\partial_{x_{10}})f = g_4 \\
(e_1\partial_{x_{10}} + e_2\partial_{x_8} + e_3\partial_{x_2} + e_4\partial_{x_5})f = g_5 \\
(e_1\partial_{x_3} + e_2\partial_{x_{10}} + e_3\partial_{x_6} + e_4\partial_{x_9})f = g_6 \\
(e_1\partial_{x_{11}} + e_2\partial_{x_3} + e_3\partial_{x_5} + e_4\partial_{x_7})f = g_7 \\
(e_1\partial_{x_2} + e_2\partial_{x_9} + e_3\partial_{x_4} + e_4\partial_{x_{11}})f = g_8 \\
(e_1\partial_{x_8} + e_2\partial_{x_1} + e_3\partial_{x_{11}} + e_4\partial_{x_6})f = g_9 \\
(e_1\partial_{x_6} + e_2\partial_{x_7} + e_3\partial_{x_{13}} + e_4\partial_{x_2})f = g_{10} \\
(e_1\partial_{x_{13}} + e_2\partial_{x_5} + e_3\partial_{x_9} + e_4\partial_{x_1})f = g_{11} \\
(e_1\partial_{x_4} + e_2\partial_{x_{13}} + e_3\partial_{x_8} + e_4\partial_{x_3})f = g_{12} \\
(e_1\partial_{x_{12}} + e_2\partial_{x_{11}} + e_3\partial_{x_{10}} + e_4\partial_{x_{13}})f = g_{13}
\end{cases}
$$

The matrix symbol of the system is composed of 13 blocks of the type

(6)
$$V^i = \begin{bmatrix} A^i & B^i \\ -(B^i)^t & C^i \end{bmatrix}$$

where A^i, B^i, C^i are given in the Appendix and where the P_j must be substituted by the monomials x_k that are multiplied by e_j. The matrix associated to the system $V = [V^1 \ldots V^{13}]^t$ gives rise to a module that has 208 independent generators and CoCoA could not compute the whole resolution. We have obtained, at least partially, the resolution by asking CoCoA to compute the first N syzygies by giving the following commands: GB.Start_Res(M); GB.Steps(M,N);
GB.GetBettiNumbers(M): the system will display a table with the Betti numbers and the degree of the syzygies found; what one obtains is

PROPOSITION 2.4. *The system (5) has 78 linear syzygies at the first step, then 286 linear syzygies, and at least 100 linear syzygies at the third step.*

REMARK 2.5. The first two steps suggest that the resolution is De Rham like, so we tried to compute the resolution for the complex $\partial_{x_1}, \ldots, \partial_{x_{13}}$ in the Clifford algebra C_4. Unfortunately, also in this case it was not possible to compute the whole resolution but, with the same procedure illustrated above, we have taken at least some steps. The two complexes coincide at least at the first two steps since both have 78 linear syzygies at the first step and then 286 linear syzygies at the second step, while the second complex has at least 224 linear syzygies at the third step. It is natural to conjecture that the resolutions coincide entirely, so that we can make the following

CONJECTURE 2.6. The resolution of system (5) is De Rham like and coincides with the following

$$0 \longrightarrow R^{16}(-13) \longrightarrow R^{208}(-12) \longrightarrow R^{1248}(-11) \longrightarrow R^{4576}(-10) \longrightarrow$$

$$\longrightarrow R^{12480}(-9) \longrightarrow 0 \longrightarrow R^{20592}(-8) \longrightarrow R^{27456}(-7) \longrightarrow R^{27456}(-6)$$

$$\longrightarrow R^{20592}(-5) \longrightarrow R^{12480}(-4) \longrightarrow R^{4576}(-3) \longrightarrow R^{1248}(-2) \longrightarrow$$

$$\longrightarrow R^{208}(-1) \longrightarrow \mathcal{M} \longrightarrow 0.$$

The results we have obtained so far, have produced examples of De Rham like resolutions. We then tried to deal the case of some other designs. For sake of brevity we do not give the details of the construction of the matrices associated to each system, since they can be easily obtained by the matrices in the Appendix, following the procedure used in the cases we have already treated.

Desargues configuration. Consider 10 points and 10 lines such that every 3 points lie on a line and every point is the intersection of 3 lines. A system we obtained, by using 3 units e_1, e_2, e_3, is the following:

$$(7) \quad \begin{cases} (e_1\partial_{x_1} + e_2\partial_{x_2} + e_3\partial_{x_7})f = g_1 \\ (e_1\partial_{x_9} + e_2\partial_{x_1} + e_3\partial_{x_3})f = g_2 \\ (e_1\partial_{x_8} + e_2\partial_{x_3} + e_3\partial_{x_2})f = g_3 \\ (e_1\partial_{x_4} + e_2\partial_{x_7} + e_3\partial_{x_5})f = g_4 \\ (e_1\partial_{x_6} + e_2\partial_{x_9} + e_3\partial_{x_4})f = g_5 \\ (e_1\partial_{x_5} + e_2\partial_{x_6} + e_3\partial_{x_8})f = g_6 \\ (e_1\partial_{x_{10}} + e_2\partial_{x_4} + e_3\partial_{x_1})f = g_7 \\ (e_1\partial_{x_2} + e_2\partial_{x_5} + e_3\partial_{x_{10}})f = g_8 \\ (e_1\partial_{x_3} + e_2\partial_{x_{10}} + e_3\partial_{x_6})f = g_9 \\ (e_1\partial_{x_7} + e_2\partial_{x_8} + e_3\partial_{x_9})f = g_{10} \end{cases}$$

and the matrix associated with each equation can be obtained by (15).

PROPOSITION 2.7. *The resolution of (7) is*

$$0 \longrightarrow R^8(-10) \longrightarrow R^{80}(-9) \longrightarrow R^{360}(-8) \longrightarrow R^{960}(-7) \longrightarrow R^{1680}(-6)$$
$$\longrightarrow R^{2016}(-5) \longrightarrow R^{1680}(-4) \longrightarrow R^{960}(-3) \longrightarrow R^{360}(-2)$$
$$\longrightarrow R^{80}(-1) \longrightarrow \mathcal{M} \longrightarrow 0.$$

It is De Rham like and invariant with respect to changes of signature of the terms.

Another interesting system that can be obtained with 4 points and 6 lines joining two points at a time, is

(8)
$$\begin{cases} (\partial_{x_1}e_1 + \partial_{x_2}e_2)f = g_1 \\ (\partial_{x_3}e_1 + \partial_{x_4}e_2)f = g_2 \\ (\partial_{x_1}e_2 + \partial_{x_3}e_3)f = g_3 \\ (\partial_{x_2}e_1 + \partial_{x_4}e_3)f = g_4 \\ (\partial_{x_4}e_1 + \partial_{x_1}e_3)f = g_5 \\ (\partial_{x_3}e_2 + \partial_{x_2}e_3)f = g_6, \end{cases}$$

where we have considered $f : \mathbb{R}^4 \to C_3$. The matrix associated with each of the operators appearing in the system can be obtained by (15) and the resolution we have found is contained in the following:

PROPOSITION 2.8. *The system (8) has the De Rham like resolution*

$$0 \longrightarrow R^8(-4) \longrightarrow R^{32}(-3) \longrightarrow R^{48}(-2) \longrightarrow R^{32}(-1) \longrightarrow \mathcal{M} \longrightarrow 0,$$

which is invariant with respect to changes of signature of the terms.

Note that the resolution has R^{32} at the beginning. This means that only 32 rows of the module $\mathcal{M}$ are independent; that means, in terms of Clifford relations, that only 4 equations of the system are independent.

REMARK 2.9. All the resolutions we have obtained so far are De Rham like because of the particular structure of the finite geometry considered. To prove that, in general, the behaviour of a combinatorial system is not De Rham like, we treat the case of two simple finite geometries. We first consider the system associated with the dual (in the geometric sense) of the system (8): it has 6 points, 4 lines, 3 points per line. A system arising from this design is for example

(9)
$$\begin{cases} (e_1\partial_{x_1} + e_2\partial_{x_6} + e_3\partial_{x_2})f = g_1 \\ (e_1\partial_{x_3} + e_2\partial_{x_2} + e_3\partial_{x_4})f = g_2 \\ (e_1\partial_{x_5} + e_2\partial_{x_1} + e_3\partial_{x_3})f = g_3 \\ (e_1\partial_{x_4} + e_2\partial_{x_5} + e_3\partial_{x_6})f = g_4 \end{cases}$$

The resolution is:

$$0 \longrightarrow R^{16}(-7) \longrightarrow R^{88}(-6) \longrightarrow R^{192}(-5) \longrightarrow R^{200}(-4)$$

$$\longrightarrow R^{24}(-2) \oplus R^{80}(-3) \longrightarrow R^{32}(-1) \longrightarrow \mathcal{M} \longrightarrow 0.$$

If we consider a design with 10 points, 5 lines, 4 points per line, 2 lines through a point, we get for example a system of the type

$$\begin{cases} (e_1\partial_{x_1} + e_2\partial_{x_3} + e_3\partial_{x_6} + e_4\partial_{x_9})f = g_1 \\ (e_1\partial_{x_4} + e_2\partial_{x_1} + e_3\partial_{x_7} + e_4\partial_{x_{10}})f = g_2 \\ (e_1\partial_{x_3} + e_2\partial_{x_2} + e_3\partial_{x_4} + e_4\partial_{x_5})f = g_3 \\ (e_1\partial_{x_2} + e_2\partial_{x_6} + e_3\partial_{x_{10}} + e_4\partial_{x_8})f = g_4 \\ (e_1\partial_{x_8} + e_2\partial_{x_9} + e_3\partial_{x_5} + e_4\partial_{x_7})f = g_5 \end{cases}$$

The resolution is very complicated: since the system could not finish the whole resolution, we have computed only the first two steps and we have found that there is 1 linear and 125 quadratic syzygies at the first step, and then 392 quadratic syzygies. We point out that in these two cases, there are many possibility of choosing the indices for the variables and for the units appearing in each equation. We tried several choices of those labels and we found the same resolutions as above. This fact suggests that the resolution depends only on the underlying geometry but not on the choice of the colouring.

3. Platonic bodies

We first consider a tetrahedron. This Platonic body has 4 vertices, 4 faces. Every vertex belongs to 3 faces and every face has 3 vertices. A system associated to it can be written in the following way:

(10)
$$\begin{cases} (e_1\partial_{x_1} + e_2\partial_{x_2} + e_3\partial_{x_3})f = g_1 \\ (e_1\partial_{x_2} + e_2\partial_{x_1} + e_3\partial_{x_4})f = g_2 \\ (e_1\partial_{x_3} + e_2\partial_{x_4} + e_3\partial_{x_2})f = g_3 \\ (e_1\partial_{x_4} + e_2\partial_{x_3} + e_3\partial_{x_1})f = g_4. \end{cases}$$

Consider now a cube and its 8 vertices and 6 faces. Every vertex belongs to 3 faces and every 3 faces intersect at most in one point. A system describing this solid is:

(11)
$$\begin{cases} (e_1\partial_{x_1} + e_2\partial_{x_2} + e_3\partial_{x_3} + e_4\partial_{x_4})f = g_1 \\ (e_1\partial_{x_7} + e_2\partial_{x_8} + e_3\partial_{x_5} + e_4\partial_{x_6})f = g_2 \\ (e_1\partial_{x_5} + e_2\partial_{x_6} + e_3\partial_{x_1} + e_4\partial_{x_2})f = g_3 \\ (e_1\partial_{x_8} + e_2\partial_{x_7} + e_3\partial_{x_4} + e_4\partial_{x_3})f = g_4 \\ (e_1\partial_{x_3} + e_2\partial_{x_5} + e_3\partial_{x_7} + e_4\partial_{x_1})f = g_5 \\ (e_1\partial_{x_6} + e_2\partial_{x_4} + e_3\partial_{x_2} + e_4\partial_{x_8})f = g_6. \end{cases}$$

The octohedron that has 6 vertices and 8 faces. Every face has 3 vertices and every vertex belongs to 4 faces. A system associated with it is

(12)
$$\begin{cases}
(e_1\partial_{x_1} + e_2\partial_{x_2} + e_3\partial_{x_5})f = g_1 \\
(e_1\partial_{x_2} + e_2\partial_{x_1} + e_3\partial_{x_3})f = g_2 \\
(e_1\partial_{x_4} + e_3\partial_{x_1} + e_4\partial_{x_3})f = g_3 \\
(e_2\partial_{x_5} + e_3\partial_{x_4} + e_4\partial_{x_1})f = g_4 \\
(e_1\partial_{x_6} + e_3\partial_{x_2} + e_4\partial_{x_5})f = g_5 \\
(e_1\partial_{x_3} + e_2\partial_{x_6} + e_4\partial_{x_2})f = g_6 \\
(e_2\partial_{x_3} + e_3\partial_{x_6} + e_4\partial_{x_4})f = g_7 \\
(e_1\partial_{x_5} + e_2\partial_{x_4} + e_4\partial_{x_6})f = g_8.
\end{cases}$$

Another Platonic body is the dodecahedron. It has 20 vertices, 12 faces, every face has 5 vertices and every vertex belongs to 3 faces. A system associated with the dodecahedron consists of 12 equations of the form

(13) $$(e_1\partial_{x_i} + e_2\partial_{x_j} + e_3\partial_{x_k} + e_4\partial_{x_\ell} + e_5\partial_{x_n})f = g_\nu$$

where $\nu = 12$, (i, j, k, ℓ, n) (in this order) belongs to

$\{(1,2,3,4,5),\ (5,1,6,14,15),\ (2,12,1,13,14),\ (3,10,2,11,12),$

$(4,5,7,6,8),\ (8,3,4,9,10),\ (6,7,15,19,20),\ (16,17,18,20,19),$

$(17,11,12,16,13),\ (7,8,19,18,9),\ (9,18,11,10,17),\ (13,14,20,15,16)\}.$

Finally, we consider an icosahedron. It has 12 vertices, 20 faces, every face has 3 vertices and each vertex belongs to 5 faces. A system associated with this solid contains 20 equations of the type (13) where $\nu = 20$, (i, j, k, ℓ, n) (in this order) belongs to

$\{(1,2,3,0,0),\ (0,1,2,7,0),\ (0,0,1,3,5),\ (4,0,0,5,3),\ (0,6,5,1,0),$

$(8,9,0,2,0),\ (6,0,7,0,1),\ (0,7,0,8,2),\ (3,0,9,0,4),\ (0,0,6,11,7),$

$(7,8,11,0,0),\ (0,11,0,10,12),\ (12,0,0,6,11),\ (11,0,8,0,10),$

$(0,10,0,9,8),\ (9,4,10,0,0),\ (10,0,4,12,0),\ (2,3,0,0,9),\ (0,5,12,4,0),$

$(5,12,0,0,6)\}.$

The last two cases have been treated in the Clifford algebra $\mathcal{C}_5$. The matrix associated with each system can be obtained by suitable formal substitutions from the matrix (7) in the paper [8]. We obtained the following results:

PROPOSITION 3.1. *The systems associated with the Platonic bodies (10), (11), (12), (13) for $\nu = 20$ have the following resolutions:*

Tetrahedron:

$$0 \longrightarrow R^8(-4) \longrightarrow R^{32}(-3) \longrightarrow R^{48}(-2) \longrightarrow R^{32}(-1) \longrightarrow M \longrightarrow 0$$

Cube:

$$0 \longrightarrow R^{32}(-9) \longrightarrow R^{240}(-8) \longrightarrow R^{768}(-7) \longrightarrow R^{1344}(-6) \longrightarrow$$

$$\longrightarrow R^{48}(-4) \oplus R^{1344}(-5) \longrightarrow R^{160}(-3) \oplus R^{720}(-4) \longrightarrow$$

$$\longrightarrow R^{192}(-2) \oplus R^{160}(-3) \longrightarrow R^{96}(-1) \longrightarrow M \longrightarrow 0.$$

Octohedron:

$$0 \longrightarrow R^{16}(-6) \longrightarrow R^{96}(-5) \longrightarrow R^{240}(-4) \longrightarrow R^{320}(-3) \longrightarrow$$

$$\longrightarrow R^{240}(-2) \longrightarrow R^{96}(-1) \longrightarrow M \longrightarrow 0$$

Icosahedron:

$$0 \longrightarrow R^8(-12) \longrightarrow R^{96}(-11) \longrightarrow R^{528}(-10) \longrightarrow R^{1760}(-9) \longrightarrow$$

$$\longrightarrow R^{3960}(-8) \longrightarrow R^{6336}(-7) \longrightarrow R^{7392}(-6) \longrightarrow R^{6336}(-5) \longrightarrow$$

$$\longrightarrow R^{3960}(-4) \longrightarrow R^{1760}(-3) \longrightarrow R^{528}(-2) \longrightarrow R^{96}(-1) \longrightarrow M \longrightarrow 0$$

REMARK 3.2. The systems (10), (12), (13) for $\nu = 20$ are De Rham like while the system associated with the cube has a more complicated structure. Note also that, despite the fact that the system (13) contains 20 equations, it behaves like a De Rham system in only 12 operators. This fact proves that not all the equations in the system are independent. The case of the dodecahedron is very complicated and exceeds the capabilities of the system. The only fact we were able to prove is that there are linear relations in the first syzygies. As in the case of the cube–system, we are certain that the dodecahedron–system cannot be of De Rham type since the number of equations is less than the number of points, i.e. the total dimension of the system.

PROPOSITION 3.3. *The system (10) associated with a tetrahedron can be reduced to a De Rham system of the type*

$$\begin{cases} \partial_{x_1} f = h_1 \\ \partial_{x_2} f = h_2 \\ \partial_{x_3} f = h_3 \\ \partial_{x_4} f = h_4 \end{cases}$$

Proof. If we consider the second equation of the system (10) multiplied on the left by e_{12} we get that the first pair of equations in the system become

$$(e_1\partial_{x_1} + e_2\partial_{x_2} + e_3\partial_{x_2})f = g_1$$
$$(e_2\partial_{x_2} - e_1\partial_{x_1} + e_{123}\partial_{x_4})f = e_{12}g_2,$$

so that they can be rewritten in the form

(14)
$$2e_1\partial_{x_1}f = g_1 - e_{12}g_2 - e_3\partial_{x_3}f + e_{123}\partial_{x_4}f$$
$$2e_2\partial_{x_2}f = g_1 + e_{12}g_2 - e_3\partial_{x_3}f - e_{123}\partial_{x_4}f.$$

Now we can multiply the third equation by $-e_{23}$ so that we arrive to the new equation

$$e_2\partial_{x_2} = -e_{23}g_3 + e_{123}\partial_{x_3}f + e_3\partial_{x_4}f,$$

which can be combined with the second equation in the system (14) to give

$$(e_3 + 2e_{123})\partial_{x_3}f = g_1 + e_{12}g_2 + 2e_{23}g_3 - (2e_3 - e_{123})\partial_{x_4}f.$$

This allows us to write ∂_{x_3} in terms of ∂_{x_4} since the coefficient $e_3 + 2e_{123}$ admits an inverse. By substituting ∂_{x_1} and ∂_{x_3} in the fourth equation of (10), we can get ∂_{x_4} and, by consequence also ∂_{x_3}. This completes the proof. ∎

REMARK 3.4. We expect that similar reductions may also be established also in the case of the other De Rham systems we have obtained even though the reduction is not obvious, in general.

We finish this section by considering the super–dual of some of the systems we have considered. To that end, it suffices to rewrite the systems already considered by interchanging, formally, ∂_{x_i} with e_i. In the case of dodecahedron and icosahedron, it was not possible to handle the case of super–dual systems since they involve 12 or 20 units respectively, so that the matrices one has to write are too big to be treated. Some of the systems obtained produce an identical resolution, as described in the following propositions.

PROPOSITION 3.5. *The super–dual of systems associated to the Fano plane, the affine geometry over $\mathbf{Z_3}$, the Desargues configuration, and the system (8) have the same De Rham like resolution with respect to three operators, so they have 3 linear first syzygies, then other 3 and 1 linear syzygies.*

Proof. Once one has written the super–dual system in the various cases, one has to find the associated matrix. In the cases of the Fano plane and the affine geometry over $\mathbf{Z}_2$, the matrix is composed of 7 and 8 respectively 16×16 blocks coming from the symbol of the Dirac operator in $\mathcal{C}_8$ via suitable substitutions, and the resolution is

$$0 \longrightarrow R^{16}(-3) \longrightarrow R^{48}(-2) \longrightarrow R^{48}(-1) \longrightarrow \mathcal{M} \longrightarrow 0.$$

In the case of system (8) the matrix consists of 4 blocks, again 16×16, coming from suitable substitutions in the matrix associated with the Dirac operator in $\mathcal{C}_4$; the resolution obtained is the same as above. Finally, in the case of the Desargues configuration, the matrix symbol of the system is made by 10 blocks of size 32×32, which can be obtained by applying the spinor formalism to the original equations which are $\mathcal{C}_{10}$–valued. After the reduction, each equation can be written in $\mathcal{C}_4 \oplus i\mathcal{C}_4$ and once one has the matrix symbol, one can produce with CoCoA the resolution, which is

$$0 \longrightarrow R^{32}(-3) \longrightarrow R^{96}(-2) \longrightarrow R^{96}(-1) \longrightarrow \mathcal{M} \longrightarrow 0.$$

This completes the proof.■

PROPOSITION 3.6. *The super–duals of the systems (10), (11), (12) have the following resolutions*

$$0 \longrightarrow R^8(-4) \longrightarrow R^{32}(-3) \longrightarrow R^{48}(-2) \longrightarrow R^{32}(-1) \longrightarrow \mathcal{M} \longrightarrow 0$$

$$0 \longrightarrow R^{16}(-4) \longrightarrow R^{64}(-3) \longrightarrow R^{96}(-2) \longrightarrow R^{64}(-1) \longrightarrow \mathcal{M} \longrightarrow 0$$

$$0 \longrightarrow R^{16}(-6) \longrightarrow R^{96}(-5) \longrightarrow R^{240}(-4) \longrightarrow R^{320}(-3) \longrightarrow R^{240}(-2)$$
$$\longrightarrow R^{96}(-1) \longrightarrow \mathcal{M} \longrightarrow 0$$

so they are De Rham like.

Proof. The first resolution can be obtained by associating with the system a suitable matrix built with blocks of the type (6). The second system is associated to a matrix built with 6 blocks of size 16×16 that can be obtained from the matrix associated with the Dirac operator in $\mathcal{C}_8$, with suitable substitutions (see [8]). The matrix symbol of the system has 96 rows of which only 64 are independent. The last system is associated with a matrix with blocks of size 8×8, coming from the matrix representing the Dirac operator in $\mathcal{C}_6$. All the resolutions were obtained with CoCoA.■

4. Open problems

The cases we have treated in this paper and in [7] are only a starting point in studying systems of combinatorial type. We think that several problems can be addressed in this field; some of them arise naturally in the combinatorial setting, some others arise from our treatment. Classical problems are for example:

— (P_1) In how many ways can the basis elements e_k be chosen such that all conditions A_i in the introduction are met?

— (P_2) What is the finite group of permutations of the tops leaving the set of blocks invariant which, if combined with a permutation of the elements e_k, leave the set of operators invariant?

  Less classical problems are:

— (P_3) To extend the finite group invariance to a continuous group invariance whereby the continuous extensions of the permutations of the sets $\{\partial_{x_1}, ..., \partial_{x_m}\}$ and $\{e_1, ..., e_M\}$ are subgroups of $SO(m)$ and $SO(M)$ respectively.

— (P_4) In all the cases in which we have tried to choose the signature randomly, we have found that the resolution remains the same. This fact suggests the following question: to what extent is the resolution of the system dependent on the chosen colouring or the choice of the signatures of the terms?

— (P_5) In the same spirit as the previous question we ask: for which finite geometries is the resolution only dependent on the geometry and not on the colouring or choice of signature?

— (P_6) When is the resolution of a system proportional to a De Rham complex? We have found this behaviour in several cases and we conjecture that this happens in the case of all affine and projective geometries over $\mathbf{Z}_p$, p prime, and also in the case of self–dual designs and for designs in which the number of lines is not less than the number of points.

— (P_7) What is the class of geometries that correspond to the same Betti numbers, ignoring the dimension of the Clifford algebra? What geometric information about a finite geometry is included in the sequence of degrees and Betti numbers of a resolution?

5. Appendix I

$$A = \begin{bmatrix} 0 & -P_1 & -P_2 & -P_3 & -P_4 & 0 & 0 & 0 \\ P_1 & 0 & 0 & 0 & 0 & P_2 & P_3 & P_4 \\ P_2 & 0 & 0 & 0 & 0 & -P_1 & 0 & 0 \\ P_3 & 0 & 0 & 0 & 0 & 0 & -P_1 & 0 \\ P_4 & 0 & 0 & 0 & 0 & 0 & 0 & -P_1 \\ 0 & -P_2 & P_1 & 0 & 0 & 0 & 0 & 0 \\ 0 & -P_3 & 0 & P_1 & 0 & 0 & 0 & 0 \\ 0 & -P_4 & 0 & 0 & P_1 & 0 & 0 & 0 \end{bmatrix},$$

$$B = \begin{bmatrix} 0 & 0 & 0 & 0 & 0 & 0 & 0 & 0 \\ 0 & 0 & 0 & 0 & 0 & 0 & 0 & 0 \\ P_3 & P_4 & 0 & 0 & 0 & 0 & 0 & 0 \\ -P_2 & 0 & P_4 & 0 & 0 & 0 & 0 & 0 \\ 0 & -P_2 & -P_3 & 0 & 0 & 0 & 0 & 0 \\ 0 & 0 & 0 & -P_3 & -P_4 & 0 & 0 & 0 \\ 0 & 0 & 0 & P_2 & 0 & -P_4 & 0 & 0 \\ 0 & 0 & 0 & 0 & P_2 & P_3 & 0 & 0 \end{bmatrix},$$

$$C = \begin{bmatrix} 0 & 0 & 0 & -P_1 & 0 & 0 & -P_4 & 0 \\ 0 & 0 & 0 & 0 & -P_1 & 0 & P_3 & 0 \\ 0 & 0 & 0 & 0 & 0 & -P_1 & -P_2 & 0 \\ P_1 & 0 & 0 & 0 & 0 & 0 & 0 & P_4 \\ 0 & P_1 & 0 & 0 & 0 & 0 & 0 & -P_3 \\ 0 & 0 & P_1 & 0 & 0 & 0 & 0 & P_2 \\ P_4 & -P_3 & P_2 & 0 & 0 & 0 & 0 & -P_1 \\ 0 & 0 & 0 & -P_4 & P_3 & -P_2 & P_1 & 0 \end{bmatrix}$$

$$(15) \quad U^i = \begin{bmatrix} 0 & P_1^i & P_2^i & P_3^i & 0 & 0 & 0 & 0 \\ P_1^i & 0 & 0 & 0 & P_2^i & P_3^i & 0 & 0 \\ P_2^i & 0 & 0 & 0 & -P_1^i & 0 & P_3^i & 0 \\ P_3^i & 0 & 0 & 0 & 0 & -P_1^i & -P_2^i & 0 \\ 0 & -P_2^i & P_1^i & 0 & 0 & 0 & 0 & -P_3^i \\ 0 & -P_3^i & 0 & P_1^i & 0 & 0 & 0 & P_2^i \\ 0 & 0 & -P_3^i & P_2^i & 0 & 0 & 0 & -P_1^i \\ 0 & 0 & 0 & 0 & P_3^i & -P_2^i & P_1^i & 0 \end{bmatrix}$$

References

1. W.W. Adams, C.A. Berenstein, P. Loustaunau, I. Sabadini, D.C. Struppa, *Regular functions of several quaternionic variables and the Cauchy–Fueter complex*, J. Geom. Anal. **9**, (1999), 1–16.

2. W.W. Adams, P. Loustaunau, *Analysis of the module determining the properties of regular functions of several quaternionic variables*, to appear in Pacific J.

3. S. Fiorini, R. J. Wilson *Edge colourings of graphs*, Res. Notes in Math. **16**, Pitman 1977,

4. J. Gilbert, M. Murray, *Clifford algebras and Dirac Operators in Harmonic Analysis*, Cambridge, Cambridge Univ. Press n. 26, 1990.

5. V.P. Palamodov, *Linear Differential Operators with Constant Coefficients*, Springer Verlag, New York 1970.

6. I. Sabadini, M. Shapiro, D.C. Struppa, *Algebraic analysis of the Moisil–Theodorescu system*, Compl. Var. **40**, (2000), 333–357.

7. I. Sabadini, F. Sommen, *Special first order systems and resolutions*, preprint, (2000).

8. I. Sabadini, F. Sommen, D.C. Struppa, P. Van Lancker, *Complexes of Dirac operators in Clifford algebras*, to appear in Math. Z.

9. F. Sommen, *Special first order systems in Clifford Analysis*, Proc. Symp. Anal. Num. Meth. Quat. Clif. Anal., Seiffen 1996, 169–274.

10. F. Sommen, *Monogenic Functions of Higher Spin*, Z. A. A. **15**, (1996), n.2, 279–282.

11. F. Sommen, *Clifford analysis on super-space*, to appear in Proc. Cetraro Conf. 1998, Adv. Clif. Alg and Appl.

12. F. Sommen, G. Bernardes, *Multivariable monogenic functions of higher spin*, to appear in J. Nat. Geom.

13. V. Souček, *Clifford analysis for higher spins*, Proc. third Intern. Conf. Clif. Alg. Appl. Math. Phys., Deinze 1993, Kluwer Acad. Publ. (1993), 223–232.

Double Covers of Pseudo-orthogonal Groups

Andrzej Trautman (`andrzej.trautman@fuw.edu.pl`)
Instytut Fizyki Teoretycznej, Uniwersytet Warszawski

Abstract. For every pair (m, n) of non-negative integers one defines $\mathsf{E}_{m,n}$ to be the group of equivalence classes of central extensions of the pseudo-orthogonal group $\mathsf{O}_{m,n}$ by $\mathbf{Z}_2$. The isomorphism $k : \mathsf{E}_{m,n} \to \mathsf{H}^2(B\mathsf{O}_{m,n}, \mathbf{Z}_2)$ is used to show that $\mathsf{E}_{m,n}$ is isomorphic to the group $\mathbf{Z}_2^{l(m,n)}$, where $l(0,0) = 0$, $l(1,0) = 1$, $l(m,0) = 2$, $l(1,1) = 3$, $l(1,n) = 4$ and $l(m,n) = 5$ for $m, n > 1$. If M is a manifold with a metric tensor g of signature (m, n) and f is a smooth map from M to the classifying space $B\mathsf{O}_{m,n}$ inducing the principal $\mathsf{O}_{m,n}$-bundle P of orthonormal frames defined by g, then the bundle P can be reduced to an element H of $\mathsf{E}_{m,n}$—i.e. to a double cover of $\mathsf{O}_{m,n}$—if, and only if, the element $f^* k(H)$ of $\mathsf{H}^2(M, \mathbf{Z}_2)$ vanishes. This generalizes the classical topological condition for the existence of a pin structure on a pseudo-Riemannian manifold. The set of all $32 = 2^5$ inequivalent double covers of $\mathsf{O}_m \times \mathsf{O}_n$, the maximal compact subgroup of $\mathsf{O}_{m,n}$, $m, n > 1$, is described explicitly.

Keywords: Pin and spin groups, extensions of pseudo-orthogonal groups by $\mathbf{Z}_2$, generalized pin structures, topological obstructions

Mathematics Subject Classifications 2000: Primary 11E88, 55R40; Secondary 20C35, 22E43.

Dedicated to Richard Delanghe on the occasion of his 60th birthday.

1. Introduction

Ever since the discovery of the spin of the electron, spinors have played a major role in theoretical physics; see the chapters by Jost and van der Waerden in Fierz and Weisskopf (1960) for the early history of the subject and an account of the major role of Pauli. At first, spinors were considered only in the context of flat spaces; they were defined in terms of suitable representations of the groups SU_2 and $\mathsf{SL}_2(\mathbb{C})$, providing the unique non-trivial double covers of SO_3 and of the connected component of the Lorentz group, respectively. The discovery of parity non-conservation in weak interactions (see the chapter by Wu in *loc. cit.*) induced an increased interest in the behaviour of spinors under reflections. Mathematicians have coined the name of 'pin groups' to denote double covers of the full orthogonal groups that reduce to spin groups upon restriction to the connected component (Atiyah et al., 1964). The development of general relativity forced physicists to consider spinor fields on pseudo-Riemannian manifolds (see Schrödinger (1932) and

377

F. Brackx et al. (eds.), Clifford Analysis and Its Applications, 377–388.

the references to earlier work given there). A precise definition of 'spin structures' on manifolds was possible only after the notion of a fibre bundle had been introduced; Haefliger (1956) found the topological obstruction to the existence of a spin structure on an orientable, Riemannian manifold and Karoubi (1968) extended this result to the non-orientable and pseudo-Riemannian cases.

In the late 1950s, Shirokov (1960) (see also the references to earlier papers given there) pointed out that the full Lorentz group $O_{1,3}$ may have 8 inequivalent double covers. His argument, later taken up and extended to $O_{m,n}$ by Dąbrowski (1988), is as follows. Consider a double cover H — i.e. a central extension by $\mathbb{Z}_2$ — of $O_{1,3}$. Every element of the Lorentz group is covered by two elements, say h and $-h$ of H. In particular, let $\pm s$ and $\pm t$ be elements of H covering commuting space and time reflections, respectively. Since the square of a reflection is the identity, one has $s^2 = a1$, $t^2 = b1$ and $(st)^2 = c1$, where 1 is the unit of H and $a, b, c \in \{+, -\}$. Since there are 8 different combinations of the signs a, b and c, one expects to have 8 inequivalent double covers. For example, $a = b = c = +$ may be realized as the trivial extension $O_{1,3} \times \mathbb{Z}_2$, whereas the two cases when $a \neq b$ and $c = +$ correspond to the two pin groups.

Shirokov discussed the possible relevance of the different covers of the Lorentz group to the description of elementary particles. Similar ideas, in a modern setting and in connection with strings, have been put forward by Carlip and DeWitt-Morette (1988) and DeWitt-Morette and DeWitt (1990). Chamblin (1994) determined the topological obstructions to the reductions of an $O_{m,n}$-bundle to the 8 double covers considered by Dąbrowski. None of these authors have shown that there are precisely 8 such double covers. The only double covers that have been described explicitly, besides the trivial one, are the two corresponding to $\mathrm{Pin}_{m,n}$ and $\mathrm{Pin}_{n,m}$.

In this paper, I determine the group $E_{m,n}$ consisting of inequivalent central extensions by $\mathbb{Z}_2$ of the group $O_{m,n}$ for $m, n \geqslant 0$. The order of this group is the number of inequivalent double covers of $O_{m,n}$. It turns out that the Lorentz group $O_{1,3}$ has 16 such double covers; only 8 among them are 'Cliffordian' in the sense of being generated, as elements of $E_{1,3}$, by the pin groups that are subsets of the complex Clifford algebra $\mathrm{Cliff}_4(\mathbb{C})$. For $m, n > 1$, the group $O_{m,n}$ has as many as 32 inequivalent double covers. The group $O_{1,1}$ has 8 double covers; all of them are Cliffordian. The isomorphism $E_{m,n} \to H^2(BO_{m,n}, \mathbb{Z}_2)$ gives rise to a simple and general formulation of the topological condition for the existence of a generalized pin structure, i.e. of a reduction of the bundle of orthonormal frames of a pseudo-Riemannian manifold to a double cover of $O_{m,n}$.

2. Notation

For every $n \in \mathbb{N}$, one has the orthogonal group O_n, its connected component SO_n, the spin group $Spin_n$, two pin groups Pin_n^+ and Pin_n^-: if $u \in \mathbb{R}^n \cap Pin_n^\pm$, then $u^2 = \pm 1$ and one says that u is a unit vector. For $m, n \in \mathbb{N}$, one has the group $O_{m,n} \subset GL_{m+n}(\mathbb{R})$ of automorphisms of the quadratic form

$$(1) \qquad x_1^2 + \ldots + x_m^2 - x_{m+1}^2 - \ldots - x_{m+n}^2.$$

It is understood that O_0 and $O_{0,0}$ are both the trivial group. For $mn \neq 0$, the connected component $SO_{m,n}^0$ of the group $O_{m,n}$ is a proper subgroup of $SO_{m,n} = \{a \in O_{m,n} | \det a = 1\}$; a similar notation is used for the spin and pin groups. The real Clifford algebra associated with the quadratic form (1) is denoted by $Cliff_{m,n}$. The set $\mathbb{Z}_2 = \{0, 1\}$ is considered, depending on the context, either as a group (with respect to addition mod 2) or as a ring and span X denotes the linear span over $\mathbb{Z}_2$ of the elements of the set X.

3. Generalities on central extensions of topological groups

Recall that a *central extension* of a topological group G by an Abelian discrete group A is an exact sequence of continuous group homomorphisms,

$$(2) \qquad A \xrightarrow{i} H \xrightarrow{p} G,$$

such that i is injective, p is surjective and $i(A)$ is contained in the centre of H; see, e.g., Ch. I § 6 in Bourbaki (1970) for the algebraic aspect.

Another extension

$$A \xrightarrow{i'} H' \xrightarrow{p'} G$$

is said to be *equivalent* to (2) whenever there is an isomorphism of topological groups $j : H \to H'$ such that $j \circ i = i'$ and $p' \circ j = p$. There is always the extension

$$(3) \qquad A \to G \times A \xrightarrow{pr_1} G.$$

An extension equivalent to (3) is said to be *trivial*.

It is often convenient to abuse the language by saying that the group H, appearing in (2), is the extension. This is a real abuse: for example, the dihedral group D_4 and the group $\mathbb{Z}_2 \times \mathbb{Z}_4$ provide each 3 inequivalent extensions of $\mathbb{Z}_2 \times \mathbb{Z}_2$ by $\mathbb{Z}_2$. This being kept in mind, it is possible to use the simplified notation without running into trouble. The trivial extension is then denoted by **O**.

Given extensions by A of two groups,

$$A \xrightarrow{i_\alpha} H_\alpha \xrightarrow{p_\alpha} G_\alpha, \quad \alpha = 1, 2,$$

one defines an extension of their direct product,

(4) $$A \xrightarrow{i} H_1 \diamond H_2 \xrightarrow{p} G_1 \times G_2,$$

by putting $H_1 \diamond H_2 = (H_1 \times H_2)/A$, $i(a) = [(1, a)]$, $p([(h_1, h_2)]) = (p_1(h_1), p_2(h_2))$, etc.

The set $E(G, A)$ of equivalence classes of such extensions of G by A has the structure of an Abelian group; to describe it, consider two extensions

$$A \xrightarrow{i_\alpha} H_\alpha \xrightarrow{p_\alpha} G, \quad \alpha = 1, 2.$$

One defines their *sum* to be the extension $A \xrightarrow{i} H \xrightarrow{p} G$ given as follows. Let $H' = \{(a_1, a_2) \in H_1 \times H_2 | p_1(a_1) = p_2(a_2)\}$. The injection $A \to H'$, $a \mapsto (a, a^{-1})$, makes A into a normal subgroup of H'; let H be the resulting quotient group: $[(a_1, a_2)] = [(a'_1, a'_2)] \in H$ whenever there is $a \in A$ such that $a'_1 = aa_1$ and $a'_2 = a^{-1}a_2$. The map $p : H \to G$ given by $p([a_1, a_2]) = p_1(a_1)$ is a surjective homomorphism and its kernel is the group $\{[(a, 1)] = i(a) \in H | a \in A\}$, isomorphic to A; therefore $H \in E(G, A)$. This extension is written as $H_1 + H_2$; one easily checks that $H_2 + H_1$ is an extension equivalent to $H_1 + H_2$, that the so defined sum of (equivalence classes of) extensions is associative and the trivial extension is the neutral element of the group $E(G, A)$.

4. Cohomology of classifying spaces

How can one find the group $E(G, A)$? There is a well-known isomorphism of the group of all central extensions of G by A with the algebraic (in the sense of Eilenberg and Mac Lane (1942)) second cohomology group $H^2_{\text{alg}}(G, A)$ defined without any reference to the topology of G; see, e.g., Ch. IV in MacLane (1963). With a topological group G there is associated its *classifying space* BG whose singular cohomology with coefficients in A, relatively easy to compute, is closely related to the Eilenberg-Mac Lane cohomology of G and to $E(G, A)$.

Recall that, for every *topological* group G, there is a *universal* principal G-bundle $EG \to BG$ such that every principal G-bundle over a paracompact space M is obtained as the bundle induced by a map from M to the classifying space BG. Two maps from M to BG induce the same bundle if, and only if, they are homotopy equivalent (Husemoller, 1966).

For a topological group G one can consider the group G^δ that has the same underlying set of elements, but is endowed with the discrete topology. The natural homomorphism $G^\delta \to G$ is continuous and lifts to a continuous map $BG^\delta \to BG$. Following a suggestion of Friedlander, Milnor (1983) conjectured that, for a Lie group G and a finite Abelian group A, this map induces an isomorphism of the corresponding cohomology rings with coefficients in A.

Jackowski (2000) informs me that, independently of Milnor's conjecture, for a locally simply connected G—therefore, in particular, for a Lie group— and a discrete A, there holds the isomorphism

(5) $$E(G, A) \cong H^2(BG, A).$$

This paper is based on the validity of the isomorphism of groups (5) for $G = O_{m,n}$ and $A = \mathbb{Z}_2$.

5. Cohomology of the classifying spaces of pseudo-orthogonal groups

From now on assume $A = \mathbb{Z}_2$ so that (2) reads $\mathbb{Z}_2 \to H \to G$ and, for every extension H of G, one has $H + H = O$. The composition of elements in G and H is denoted multiplicatively, one writes $i(0) = 1$, $i(1) = -1$ and $H^*(BG)$ instead of $H^*(BG, \mathbb{Z}_2)$. One refers now to the extensions as 'double covers'.

The cohomology rings mod 2 of the classifying spaces of the groups O_n and SO_n are well-known; see, e.g., Milnor and Stasheff (1974). These rings are polynomials over $\mathbb{Z}_2$ in the Stiefel-Whitney classes,

$$H^*(BO_n) = \mathbb{Z}_2[w_1, w_2, \ldots, w_n]$$

and

$$H^*(BSO_n) = \mathbb{Z}_2[w_2, \ldots, w_n],$$

where $\deg w_k = k$. In particular, the second cohomology groups are $H^2(BO_1) = \mathrm{span}\{w_1^2\}$ and

$$H^2(BO_n) = \mathrm{span}\{w_2, w_1^2\} \quad \text{for } n > 1.$$

Since $O_{n'} \times O_{n''}$ is a maximal compact subgroup of $O_{n',n''}$, the quotient $O_{n',n''}/(O_{n'} \times O_{n''})$ is contractible and so the spaces $B(O_{n'} \times O_{n''})$ and $BO_{n',n''}$ are homotopy equivalent. Using now the Künneth theorem one obtains

$$H^*(BO_{n',n''}) = H^*(BO_{n'}) \otimes_{\mathbb{Z}_2} H^*(BO_{n''}).$$

This implies, in a self-explanatory notation,

$$H^2(BO_{1,1}) = \mathrm{span}\{w_1'^2, w_1''^2, w_1'w_1''\},$$

$$H^2(BO_{1,n''}) = \mathrm{span}\{w_2'', w_1'^2, w_1''^2, w_1'w_1''\} \quad \text{for } n'' > 1,$$

and

(6) $\quad H^2(BO_{n',n''}) = \mathrm{span}\{w_2', w_2'', w_1'^2, w_1''^2, w_1'w_1''\} \quad \text{for } n', n'' > 1.$

Therefore, the groups O_n, $O_{1,1}$, $O_{1,n}$ and $O_{m,n}$ have (for $m, n > 1$) 4, 8, 16 and 32 double covers, respectively. In particular, the set of extensions of $O_{1,1}$ is in a natural, bijective correspondence with the set of extensions of $\mathbb{Z}_2 \times \mathbb{Z}_2 = O_1 \times O_1$ by $\mathbb{Z}_2$.

Similarly,

$$H^2(BSO_{1,n''}^0) = \mathrm{span}\{w_2''\} \quad \text{for } n'' > 1,$$

$$H^2(BSO_{n',n''}) = \mathrm{span}\{w_2', w_2'', w_1'^2\} \quad \text{for } n', n'' > 1$$

and

$$H^2(BSO_{n',n''}^0) = \mathrm{span}\{w_2', w_2''\} \quad \text{for } n', n'' > 1.$$

Clearly, by symmetry, $w_2' + w_2''$ corresponds to the cover of $SO_{n',n''}^0$ by $\mathrm{Spin}_{n',n''}^0$. But it is not evident what corresponds to w_2'. This problem is at the core of the difficulties in constructing explicitly more than 8 double covers of the Lorentz group $O_{1,n}$.

The groups $E(O_n, \mathbb{Z}_2)$ and $E(O_{m,n}, \mathbb{Z}_2)$ are from now on denoted by E_n and $E_{m,n}$, respectively.

6. The isomorphism $E_{m,n} \to H^2(BO_{m,n})$

What is the correspondence between the double covers of $O_{n',n''}$ and the elements of (6)? By inspection of the obstructions to the classical pin structures (Karoubi, 1968), one expects that it can be described as follows. The group $O_{n',n''}$ is known to be generated by reflections in hyperplanes orthogonal to non-isotropic vectors (Cartan-Dieudonné). Every reflection is an involution; reflections in hyperplanes associated with orthogonal vectors commute. Therefore, the square of an element of H covering a reflection is either 1 or -1. Two elements of H that cover two commuting and distinct reflections either commute or anticommute. Let $V = \mathbb{R}^{n'+n''} \subset \mathrm{Cliff}_{n',n''}$. Consider the spaces of unit vectors

$$U' = \{u \in V | u^2 = 1\} \quad \text{and} \quad U'' = \{u \in V | u^2 = -1\}.$$

Reflections associated with elements of U' (resp., U'') are called time (resp., space) reflections. Let

$$(7) \qquad k(H) = \lambda' w_2' + \lambda'' w_2'' + \mu w_1' w_1'' + \nu' w_1'^2 + \nu'' w_1''^2,$$

where $\lambda', \lambda'', \mu, \nu', \nu'' \in \mathbb{Z}_2$, be the element of (6) corresponding to the double cover H.

PROPOSITION 1. *The isomorphism* $k : E_{n',n''} \to H^2(BO_{n',n''})$ *is as follows:*

$\lambda' = 0$ *(resp.,* $\lambda' = 1$*) if every two elements of H covering reflections associated with orthogonal elements of U' commute (resp., anticommute); similarly for λ'' and U'';*

$\mu = 0$ *(resp.,* $\mu = 1$*) if every two elements of H covering reflections associated with orthogonal elements, one in U' and the other in U'', commute (resp., anticommute);*

$\nu' = 0$ *(resp.,* $\nu' = 1$*) if the square of every element of H covering a reflection associated with an element of U' is 1 (resp., -1); similarly for ν'' and U''.*

Moreover, to characterize the double cover of $SO_{n',n''}^0$, obtained by restriction of the one corresponding to (7), one puts $\mu = \nu' = \nu'' = 0$ in (7).

Proof. To justify the interpretation of ν', ν'' look at double covers of O_1. Similarly, the significance of μ is obtained from $O_{1,1}$. By considering two extensions H_1 and H_2, and two pairs of elements $(u_1, u_2) \in H_1 \times H_2$ and $(u_1', u_2') \in H_1 \times H_2$ such that $p_1(u_1) = p_2(u_2)$ and $p_1(u_1') = p_2(u_2')$ are reflections in hyperplanes determined by orthogonal vectors, one easily checks that if $u_1 u_1' + u_1' u_1 = 0$ and $u_2 u_2' + u_2' u_2 = 0$, then the elements $[(u_1, u_2)]$ and $[(u_1', u_2')]$ of $H_1 + H_2$ commute, etc. □

7. Construction of double covers

7.1. THE 4 DOUBLE COVERS OF O_n

This case is the simplest and best known: there are two generators of the group E_n,

Pin_n^+ corresponding to w_2 and Pin_n^- corresponding to $w_2 + w_1^2$.

Their sum $Pin_n^+ + Pin_n^-$ is a non-trivial double cover that trivializes upon restriction to SO_n. It corresponds to w_1^2.

It is worth-while to describe in some detail the extension

(8) $\mathbb{Z}_2 \to \mathrm{Pin}_n^+ + \mathrm{Pin}_n^- \to O_n;$

all sums of double covers of the groups $O_{m,n}$ are constructed in a similar manner; I refer to them as being given *explicitly* in terms of the summands. Consider

$$\mathbb{Z}_2 \to \mathrm{Pin}_n^\pm \xrightarrow{p_\pm} O_n$$

and put $H' = \{(a,b) \in \mathrm{Pin}_n^+ \times \mathrm{Pin}_n^- | p_+(a) = p_-(b)\}$, as in Section 3. The group $\mathrm{Pin}_n^+ + \mathrm{Pin}_n^-$ is the quotient of H' by the equivalence relation $(a,b) \equiv (a',b') \Leftrightarrow a = a'$ and $b = b'$ or $a = -a'$ and $b = -b'$. Let $[(a,b)]$ denote the corresponding equivalence class. The injection $i : \mathbb{Z}_2 \to \mathrm{Pin}_n^+ + \mathrm{Pin}_n^-$ is given by $i(1) = [(1,-1)]$. For every unit vector u one has $[(u,u)]^2 = [(1,-1)]$. Therefore, the extension (8) does not split. If u and v are orthogonal unit vectors, then the elements $[(u,u)]$ and $[(v,v)]$ commute.

7.2. EXPLICIT CONSTRUCTION OF 8 DOUBLE COVERS OF $O_{m,n}$ FOR $m,n \geqslant 1$

For every $m,n \geqslant 1$ one constructs 8 inequivalent double covers by giving a set of 3 generators in the group $E_{m,n}$.

Let $i = \sqrt{-1}$. There are 4 pin groups, defined as subsets of the *complexified* Clifford algebra $\mathrm{Cliff}_{n'+n''}(\mathbb{C}) = \mathbb{C} \otimes \mathrm{Cliff}_{n',n''}$, generated multiplicatively:

(9) $\mathrm{Pin}_{n',n''}^{\nu',\nu''}$ is generated by $i^{\nu'} U' \cup i^{1+\nu''} U'',$

where $\nu', \nu'' \in \mathbb{Z}_2$. In the traditional notation, one usually writes $\mathrm{Pin}_{n',n''}$ instead of $\mathrm{Pin}_{n',n''}^{0,1}$ and $\mathrm{Pin}_{n'',n'}$ instead of $\mathrm{Pin}_{n',n''}^{1,0}$. In the positive definite case, $U'' = \varnothing$, one has $\mathrm{Pin}_n^+ = \mathrm{Pin}_{n,0}^{0,*}$ and $\mathrm{Pin}_n^- = \mathrm{Pin}_{n,0}^{1,*}$; there is no need for the elaborate notation. The epimorphism $p : \mathrm{Pin}_{n',n''}^{\nu',\nu''} \to O_{n',n''}$ is as in the classical case, $p(a)v = \alpha(a)va^{-1}$, where $v \in V$ and α is the main (grading) automorphism of $\mathrm{Cliff}_{n'+n''}(\mathbb{C})$. The 'new' groups $\mathrm{Pin}_{n',n''}^{0,0}$ and $\mathrm{Pin}_{n',n''}^{1,1}$ also provide double covers of $O_{n',n''}$. (They should not be confused with the compact groups $\mathrm{Pin}_{n'+n''}^+$ and $\mathrm{Pin}_{n'+n''}^-$.) No two extensions among the four are equivalent, but only 3 among them are independent, as elements of the group $E_{n',n''}$; the following relation is easy to check:

$$\mathrm{Pin}_{n',n''}^{0,0} + \mathrm{Pin}_{n',n''}^{1,0} + \mathrm{Pin}_{n',n''}^{0,1} + \mathrm{Pin}_{n',n''}^{1,1} = \mathbf{O}.$$

It is convenient to introduce the 'total 2nd Stiefel-Whitney class',

$$w_2 = w_2' + w_2'' + w_1'w_1''.$$

It follows from Prop. 1 that there is the following correspondence between the double covers of $O_{n',n''}$ described in (9) and the elements of $H^2(BO_{n',n''})$ given by (6):

$$(10) \qquad k(\mathrm{Pin}_{n',n''}^{\nu',\nu''}) = w_2 + \nu'w_1'^2 + \nu''w_1''^2.$$

To describe all the 32 double covers of $O_{n',n''}$, $n', n'' > 1$, one would need two more independent generators, for example, those corresponding to w_2' and w_2''. For the (generalized) Lorentz group $O_{1,n}$, $n > 1$, one extra generator suffices. I do not know how to construct them.

Among the 32 double covers only the 8 corresponding to

$$w_2 + \bar{\mu}w_1'w_1'' + \nu'w_1'^2 + \nu''w_1''^2, \quad \text{where } \bar{\mu}, \nu', \nu'' \in \mathbb{Z}_2,$$

are 'spinorial' in the sense that, restricted to the connected component, they reduce to the classical double cover $\mathrm{Spin}_{n',n''}^0 \to SO_{n',n''}^0$. Among those 8, there are the 4 given in (10); I do not know how to describe the remaining 4 which are characterized by $\bar{\mu} = 1$.

There are also 8 double covers (corresponding to $\mu w_1'w_1'' + \nu'w_1'^2 + \nu''w_1''^2$) that trivialize upon restriction to the connected component. Among them, 4 (those with $\mu = 0$) are generated from (10).

The following table summarizes the relations between the notation of Dabrowski (1988) and Chamblin (1994), briefly described here in Section 1, and the present one. It covers only the cases when $\lambda' = \lambda'' = 1$. Here Q is the quaternion group. The dimensions n' and n'' are omitted.

a	b	c	μ	ν'	ν''	group H	its subgroup generated by $\{-1, s, t\}$
+	+	+	0	0	0		$\mathbb{Z}_2 \times \mathbb{Z}_2 \times \mathbb{Z}_2$
+	−	−	0	1	0		$\mathbb{Z}_2 \times \mathbb{Z}_4$
−	+	−	0	0	1		$\mathbb{Z}_2 \times \mathbb{Z}_4$
−	−	+	0	1	1		$\mathbb{Z}_2 \times \mathbb{Z}_4$
+	+	−	1	0	0	$\mathrm{Pin}^{0,0}$	D_4
+	−	+	1	1	0	$\mathrm{Pin}^{0,1}$	D_4
−	+	+	1	0	1	$\mathrm{Pin}^{1,0}$	D_4
−	−	−	1	1	1	$\mathrm{Pin}^{1,1}$	Q

7.3. ALL THE DOUBLE COVERS OF $O_m \times O_n$

It is easy to describe explicitly all the 32 double covers of the group $O_{n'} \times O_{n''}$, $n', n'' > 1$. Recall first that there are 4 double covers of O_n,

$$H_n^0 = O_n \times \mathbb{Z}_2, \quad H_n^1 = \mathrm{Pin}_n^+,$$

$$H_n^2 = \mathrm{Pin}_n^+ + \mathrm{Pin}_n^-, \quad H_n^3 = \mathrm{Pin}_n^-.$$

The 16 groups (cf. (4)) $H_{n'}^\alpha \diamond H_{n''}^\beta$, $\alpha, \beta = 0, \ldots, 3$, correspond to the elements of the form (7) with $\mu = 0$.

To describe the remaining ones, one uses a graded ('supersymmetric') construction that appears already in Karoubi (1968). Every double cover $p : H_n^\alpha \to O_n$ is $\mathbb{Z}_2$-graded by putting, for $h \in H_n^\alpha$, $\det p(h) = (-1)^{\deg h}$. In the Cartesian product $H_{n'}^\alpha \times H_{n''}^\beta$ one defines a group structure by

$$(11) \qquad (h_1', h_1'')(h_2', h_2'') = (h_1' h_2', (-1)^{\deg h_1'' \deg h_2'} h_1'' h_2''),$$

where $h_1', h_2' \in H_{n'}^\alpha$, $h_1'', h_2'' \in H_{n''}^\beta$. The groups

$$H_{n'}^\alpha \diamond^{\mathrm{gr}} H_{n''}^\beta = (H_{n'}^\alpha \times H_{n''}^\beta)/\mathbb{Z}_2, \quad \alpha, \beta = 0, \ldots, 3,$$

with a composition induced from (11), provide the 16 double covers with $\mu = 1$.

7.4. COMPLEX DOUBLE COVERS

In the complex domain, it is natural to consider the conformal group,

$$CO_n(\mathbb{C}) = \{A \in GL_n(\mathbb{C}) | AA^t = \lambda \mathrm{id}, \lambda \in \mathbb{C} \smallsetminus \{0\}\},$$

and the corresponding Clifford (=conformal spin) group $\mathrm{CPin}_n(\mathbb{C})$ generated multiplicatively in $\mathrm{Cliff}_n(\mathbb{C})$ by all non-isotropic elements of $\mathbb{C}^n$. There is the exact sequence

$$\mathrm{CPin}_n(\mathbb{C}) \overset{\rho}{\longrightarrow} CO_n(\mathbb{C}) \to 1,$$

where $\rho(a)v = \alpha(a)v\beta(a)$, $v \in \mathbb{C}^n$. The kernel of ρ is of order 4 (see, e.g., (Robinson and Trautman, 1993); here α and β are the grading automorphism and the main anti-automorphism of $\mathrm{Cliff}_n(\mathbb{C})$, respectively; $\beta(ab) = \beta(b)\beta(a)$, etc.).

One can define a *complex* double cover of $O_{n',n''}$ as a double cover $p : H \to O_{n',n''}$ such that there exists a cover $p_\mathbb{C} : H_\mathbb{C} \to CO_{n'+n''}(\mathbb{C})$ and a monomorphism j such that $p_\mathbb{C} \circ j = \mathrm{inj} \circ p$, where inj is the injection of $O_{n',n''}$ into $CO_{n'+n''}(\mathbb{C})$.

The 4 extensions of the compact group O_n are complex.

Conjecture. A double cover $p : H \to O_{n',n''}$ is complex if, and only if,

$$k(H) = \lambda w_2 + \nu' w_1'^2 + \nu'' w_1''^2, \quad \text{where } \lambda, \nu', \nu'' \in \mathbb{Z}_2.$$

In other words, the conjecture says that, for every $n', n'' \geqslant 1$, there are 8 complex double covers of $O_{n',n''}$, generated be $\text{Pin}_{n',n''}^{\nu',\nu''}$.

8. Topological obstructions

Given a pseudo-Riemannian manifold M with a metric tensor of signature (m, n) and a double cover $H \to O_{m,n}$, one can consider the reduction of the bundle P of all orthonormal frames of M to the group H. The corresponding topological obstruction can be determined in a way similar to the one used for (s)pin structures; see § 26.5 in Borel and Hirzebruch (1959), Milnor (1963), Prop. 1.1.26 in Karoubi (1968), and § 3.5 in Ward and Wells Jr (1990). In fact, from an obstruction theory argument (Spanier, 1966) one obtains

PROPOSITION 2. *Let* $f : M \to BO_{m,n}$ *be the map inducing the bundle* $P \to M$. *The pull-back* $f^*k(H)$ *is an element of* $H^2(M, \mathbb{Z}_2)$ *that vanishes if, and only if, there is a reduction of* P *to* H.

As an application of Prop. 2, consider the real projective space $\mathbb{R}P_5$. It is orientable, but its second Stiefel-Whitney class is $\neq 0$; therefore, it has no spin structure, but its bundle of all orthonormal frames can be reduced to the group $\text{Pin}_5^+ + \text{Pin}_5^-$. A less trivial example would be provided by a non-orientable Riemannian space with $w_2 \neq 0$ and $w_1^2 = 0$.

Acknowledgements

I thank Professor Stefan Jackowski for enlightening discussions on the subject of group extensions and cohomology. Sections 4 and 5 have been written under his influence, but I am solely responsible for the shortcomings of the present text.

Work on this paper was supported in part by the Polish Committee for Scientific Research (KBN) under grant no. 2 P03B 060 17.

References

Atiyah, M. F., R. Bott, and A. Shapiro: 1964, 'Clifford modules'. *Topology* **3**, Suppl. 1, 3–38.

Borel, A. and F. Hirzebruch: 1959, 'Characteristic classes and homogeneous spaces, II'. *Amer. J. Math.* **81**, 315–382.

Bourbaki, N.: 1970, *Éléments de Mathématiques: Algèbre, Ch. 1 à 3.* Paris: Diffusion C.C.L.S.

Carlip, S. and C. DeWitt-Morette: 1988, 'Where the sign of the metric makes a difference'. *Phys. Rev. Lett.* **60**, 1599–1601.

Chamblin, A.: 1994, 'On the obstructions to non-Cliffordian pin structures'. *Comm. Math. Phys.* **164**, 65–85.

Dąbrowski, L.: 1988, *Group Actions on Spinors.* Napoli: Bibliopolis.

DeWitt-Morette, C. and B. S. DeWitt: 1990, 'Pin groups in physics'. *Phys. Rev. D* **41**, 1901–07.

Eilenberg, S. and S. MacLane: 1942, 'Group extensions and homology'. *Ann. Math.* **43**, 757–831.

Fierz, M. and V. F. Weisskopf: 1960, *Theoretical Physics in the Twentieth Century: A Memorial Volume to Wolfgang Pauli.* New York: Interscience.

Haefliger, A.: 1956, 'Sur l'extension du groupe structural d'un espace fibré'. *C. R. Acad. Sci. (Paris)* **243**, 558–560.

Husemoller, D.: 1966, *Fibre Bundles.* New York: McGraw-Hill.

Jackowski, S.: 2000, private communication.

Karoubi, M.: 1968, 'Algèbres de Clifford et K-théorie'. *Ann. Sci. Éc. Norm. Sup.,* *4ème sér.* **1**, 161–270.

MacLane, S.: 1963, *Homology.* Berlin: Springer.

Milnor, J.: 1963, 'Spin structures on manifolds'. *Enseign. Math.* **9**, 198–203.

Milnor, J.: 1983, 'On the homology of Lie groups made discrete'. *Comment. Math. Helv.* **58**, 72–85. MR **85b**:57050 by E. Friedlander.

Milnor, J. W. and J. D. Stasheff: 1974, *Characteristic Classes.* Princeton: Princeton University Press.

Robinson, I. and A. Trautman: 1993, 'The conformal geometry of complex quadrics and the fractional-linear form of Möbius transformations'. *J. Math. Phys.* **34**, 5391–5406.

Schrödinger, E.: 1932, 'Diracsches Elektron im Schwerefeld I'. *Sitzungsber. Preuss. Akad. Wiss., Phys.-Math. Kl.* **XI**, 105–28.

Shirokov, Y. M.: 1960, 'Space and time reflections in relativistic theory'. *Nuclear Physics* **15**, 1–12.

Spanier, E. H.: 1966, *Algebraic Topology.* New York: McGraw-Hill.

Ward, R. S. and R. O. Wells, Jr.: 1990, *Twistor Geometry and Field Theory.* Cambridge: Cambridge University Press.

Higher Spin Fields on Smooth Domains

Peter Van Lancker (pvl@cage.rug.ac.be)
Department of Mathematical Analysis, Ghent University

Abstract. This paper deals with a generalization of the classical Rarita-Schwinger equations for spin 3/2-fields to the case of functions taking values in irreducible representation spaces with weight $k+1/2$ (realised as functions taking values in spaces of spherical monogenics earlier considered in [SVA2]). In this paper we consider such fields on smooth domains in $\mathbb{R}^m$.

Keywords: Clifford analysis, Dirac operator, Rarita-Schwinger operators, Boundary values.

Mathematics Subject Classification: 30G35, 58G20

Dedicated to Richard Delanghe on the occasion of his 60th birthday.

1. Introduction

Let $\mathbb{C}_m$ be the complex 2^m-dimensional Clifford algebra over $\mathbb{R}^m$ generated by the relations $e_i e_j + e_j e_i = -2\delta_{ij}$. Vectors $x \in \mathbb{R}^m$ are identified with $x = \sum_j x_j e_j$ and are often referred to as vector variables. Conjugation on $\mathbb{C}_m$ is the anti-involution on $\mathbb{C}_m$ given by $\bar{a} = \sum_{A \subset M} \bar{a}_A \bar{e}_A$ where $\bar{e}_A = \bar{e}_{\alpha_h} \ldots \bar{e}_{\alpha_1}$ and $\bar{e}_j = -e_j$. The Dirac operator on $\mathbb{R}^m$ is given by $\partial_x = \sum_j e_j \partial_{x_j}$. Spaces of homogeneous monogenic polynomials of degree k (spherical monogenics of degree k) (see also [BDS], [DSS]) are denoted by $\mathcal{M}_k$. Monogenic operators are special operators which transform polynomials of type $u^s P_k(u)$ into polynomials of similar type $u^{s'} P_{k'}(u)$, where P_k and $P_{k'}$ are spherical monogenics. Any Clifford valued differential operator $P(u, \partial_u)$ decomposes into monogenic operators (see [SVA1]). In particular the Dirac operator ∂_x acting on functions $f(x, u)$ depending on the variable x and u (spherical monogenic of degree k in u) decomposes as $\partial_x = R_k(\partial_x) + T_k^*(\partial_x)$ where

$$R_k(\partial_x) f(x, u) = \sum_j R_k(e_j) \partial_{x_j} f(x, u) = \left(\frac{u \partial_u}{2k + m - 2} + 1 \right) \partial_x f(x, u)$$

is our generalized version of the Rarita-Schwinger operator and

$$T_k^*(\partial_x) f(x, u) = \sum_j T_k^*(e_j) \partial_{x_j} f(x, u) = -\frac{1}{2k + m - 2} u \partial_u \partial_x f(x, u)$$

F. Brackx et al. (eds.), Clifford Analysis and Its Applications, 389–398.
© 2001 *Kluwer Academic Publishers. Printed in the Netherlands.*

is called the dual twistor operator. In other words, the operator $R_k(\partial_x)$ transforms $\mathcal{M}_k$-valued functions $f : x \mapsto (u \mapsto f(x,u))$ into $\mathcal{M}_k$-valued functions. These polynomial valued monogenics are one way of introducing generalized Rarita-Schwinger type operators which is very natural from the function theory point of view (see [SVA2] and [BSoSoVL]). For functions of two vector variables, see e.g. [So1]. In general, we use the notation $R_k(a) = (\frac{u\partial_u}{2k+m-2} + 1)a$ and dotted operators act on dotted variables. From a geometry and physics point of view one may define these generalizations starting from functions $f(x)$ with values in an abstract irreducible representation space of $Spin(m)$. In the course of this paper we restrict ourselves to the irreducible representation spaces with weight $(k + \frac{1}{2}, \frac{1}{2}, \ldots, \pm\frac{1}{2})$ which can be exactly realised as spaces of spherical monogenics. One could also consider such operators acting on functions with values in even more complicated representations (simplicial monogenics)(see [VLSC]) which requires an algebraic machinery of functions of at least three vector variables. The Rarita-Schwinger operator is then defined as the unique conformally invariant operator acting on functions with values in such spaces. The conformal invariance using Vahlen matrices is expressed as follows (see e.g. [S])

THEOREM 1. *Suppose that $f(x,u)$ ($\mathcal{M}_k$-valued) is a null solution of $R_k(\partial_x)$; then so is the transformed function*

$$F(x,u) = |cx + d|^{-m+1} L((cx + d)^*/|cx + d|) f((ax + b)(cx + d)^{-1}, \dot{u}),$$

where the L-representation $g \mapsto sg(\tilde{s}us)$ of $Pin(m)$ acts on the dotted variable.

Associated to the fundamental solution of $R_k(\partial_x)$ we introduce the Cauchy and Hilbert transforms C_k and $\mathcal{H}_k$ (initially defined on C^∞-functions) on the Hilbert module of $\mathcal{M}_k$-valued square integrable functions on $\partial\Omega$. We will restrict our attention to the case where $\overline{\Omega}$ is a C^∞-submanifold-with-boundary of $\mathbb{R}^m$ (referred to as smooth domain). By $L_2(\partial\Omega, \mathcal{M}_k)$ we denote the space of $\mathcal{M}_k$-valued square integrable functions on $\partial\Omega$ provided with the $\mathbb{C}_m$-valued inner product:

$$\langle f, g \rangle_{\partial\Omega} = \int_{\partial\Omega} (f(x,u), g(x,u))_u \, ds(x)$$

where ds denotes Lebesgue measure on $\partial\Omega$ and $(\ ,\ )_u$ denotes the Fischer inner product with respect to the variable u. This inner product can be turned into a complex inner product by putting $(f,g)_{\partial\Omega} = [\langle f, g \rangle_{\partial\Omega}]_0$, $f, g \in L_2(\partial\Omega, \mathcal{M}_k)$. In this way $L_2(\partial\Omega, \mathcal{M}_k)$ can be considered either as a right Hilbert module $(L_2(\partial\Omega, \mathcal{M}_k), \langle, \rangle_{\partial\Omega})$ over $\mathbb{C}_m$ or as a complex Hilbert space $(L_2(\partial\Omega, \mathcal{M}_k), (,)_{\partial\Omega})$. A norm on $L_2(\partial\Omega, \mathcal{M}_k)$ is given by $\| f \|_{\partial\Omega}^2 = (f, f)_{\partial\Omega}$. Obviously $L_2(\partial\Omega, \mathcal{M}_k)$ can be regarded as the completion of $C^\infty(\partial\Omega)$ with respect to the above inner product. The definition of certain Hardy spaces leads to several splittings of $L_2(\partial\Omega, \mathcal{M}_k)$ and an analogue of the Kerzman-Stein theorem is formulated (see also [VL1,VL2]). For the case of the classical Dirac operator on manifolds with boundary (see e.g. [C]). Finally we give a geometric characterization of the self-adjointness of the Cauchy transform C_k.

2. The fundamental solution of the operator $R_k(\partial_x)$

The identity of $End(\mathcal{M}_k)$ can be represented by the reproducing kernel $K_k(u, v)$ for the inner spherical monogenics of degree k. This so called zonal spherical monogenic satisfies

$$P_k(v) = (K_k(u, v), P_k(u))_u = \frac{1}{A_m} \int_{S^{m-1}} \overline{K_k(u, v)} P_k(u) dS(u).$$

The fundamental solution can be constructed using the conformal invariance (see [BSoSoVL]). Put $C_k = -\frac{1}{A_m} \frac{m-2+2k}{m-2}$, A_m being the surface area of the sphere S^{m-1}, and define

$$E_k(x; u, v) = C_k \frac{1}{|x|^{m-1}} \dot{L}(\frac{x}{|x|}) K_k(\dot{u}, v) = C_k \frac{x}{|x|^{m+2k}} K_k(xux, v).$$

Let y, v be fixed, then the Cauchy kernel $E_k(x-y; u, v)$ (with singularity in y) for $R_k(\partial_x)$ satisfies the following relations:

$$R_k(\partial_x) E_k(x - y; u, v) = E_k(x - y; u, v) R_k(\partial_x) = \delta(x - y) K_k(u, v).$$

Each null solution of $R_k(\partial_x)$ is a null solution of the operator $\triangle^{k+1}$ (see [BVLSoSo, BSoSoVL]). As a matter of fact the Rarita-Schwinger operator $R_k(\partial_x)$ factorizes a specific power of the Laplacian $\triangle$. This is stated in the following theorem.

THEOREM 2. *There exists a differential operator* D_{2k+1} *of order* $2k+1$ *acting between spaces of* $\mathcal{M}_k$-*valued functions such that:*

$$R_k(\partial_x) D_{2k+1} = \triangle^{k+1}.$$

An explicit formula for the operator D_{2k+1} follows from the following basic identity.

2.1. The minimal polynomial of $R_k(e_1)$

Let $P_k(u)$ be a spherical monogenic of degree k. With respect to the orthogonal splitting $\mathbb{R}^m = \mathbb{R}e_1 \oplus \mathbb{R}^{m-1}$ each vector u can be written as $u = u_1 e_1 + \underline{u}$. Let us denote by $\Gamma_{\underline{u}}$ and $\Delta_{\underline{u}}^S$ the spherical Dirac operator and the Laplace Beltrami operator in $\mathbb{R}^{m-1}$. Then we have

$$R_k^2(e_1) = \frac{4}{(2k+m-2)^2}(\Delta_{\underline{u}}^S + \Gamma_{\underline{u}} - (\frac{m-2}{2})^2)$$

$$= \frac{4}{(2k+m-2)^2}(C(\underline{L}) + \frac{2m-3}{4})$$

where $C(\underline{L})$ is the Casimir operator corresponding to the usual L-representation of $Spin(m)$ restricted to the subgroup $Spin(m-1)$ fixing e_1. By branching rules the minimal polynomial is given by

$$\prod_{j=0}^{k}(R_k^2(e_1) + (\frac{m-2+2j}{m-2+2k})^2)P_k(u) \equiv 0.$$

Put $\lambda_j = \frac{m-2+2k}{m-4+2j}$, $j = 1, \ldots, k$ and let σ_l be the l-th elementary symmetric polynomial in k variables. Then for any vector variable x,

$$R_k^2(x)\left(|x|^{2k} + (|x|^2 + R_k^2(x))\sum_{l=0}^{k-1}\sigma_{l+1}(\lambda_1^2, \ldots, \lambda_k^2)|x|^{2k-2l-2}R_k^{2l}(x)\right) =$$

$$-|x|^{2k+2}.$$

Replacing x by ∂_x leads then to the factorization of the Laplacian raised to the power $k+1$.

2.2. Basic integral formulae

Put $dx = dx_1 \wedge \cdots \wedge dx_m$ and $d\sigma = \partial_x \lrcorner dx = \sum_{k=1}^{m}(-1)^{j-1}e_j d\hat{x}_j$ where $d\hat{x}_j = dx_1 \wedge \cdots \wedge [dx_j] \wedge \cdots \wedge dx_m$.

THEOREM 3. Let $\Omega' \subset \mathbb{R}^m$ and $\Omega \subset \Omega'$. Then for $f, g \in C^1(\Omega', \mathcal{M}_k)$:

(i) (Stokes' Theorem)

$$\int_{\Omega}[-(R_k(\partial_x)g(x,u), f(x,u))_u + (g(x,u), R_k(\partial_x)f(x,u))_u]\,dx$$

$$= \int_{\partial\Omega}(g(x,u), \dot{R}_k(d\sigma_x)\,f(x,\dot{u}))_u$$

(ii) (Cauchy-Pompeiu Theorem) Let $y \in \Omega$. Then

$$-f(y,v) + \int_{\Omega} (E_k(x-y;u,v), R_k(\partial_x)f(x,u))_u \, dx$$

$$= \int_{\partial\Omega} (E_k(x-y;u,v), \dot{R}_k(d\sigma_x) f(x,\dot{u}))_u$$

(iii) (Cauchy integral formula). If $R_k(\partial_x)f = 0$, then for $y \in \Omega$:

$$f(y,v) = -\int_{\partial\Omega} (E_k(x-y;u,v), \dot{R}_k(d\sigma_x) f(x,\dot{u}))_u,$$

where $\dot{R}_k(d\sigma_x) f(x,\dot{u})$ is an $\mathcal{M}_k$-valued $(m-1)$-form.

3. The Cauchy and Hilbert transforms on $L_2(\partial\Omega, \mathcal{M}_k)$

Definition. Let $\overline{\Omega} \subset \mathbb{R}^m$ be a smooth domain and let h be a C^∞-smooth function on $\partial\Omega$ taking values in $\mathcal{M}_k$. Then the Cauchy transform $C_k h$ of the function h with respect to the operator $R_k(\partial_x)$ is defined by:

$$(C_k h)(y,v) = -\int_{\partial\Omega} (E_k(x-y;u,v), \dot{R}_k(d\sigma_x) h(x,\dot{u}))_u$$

where n is the outer unit normal field to $\partial\Omega$ with respect to Ω. The space of null solutions of $R_k(\partial_y)$ in Ω is denoted by $\mathcal{R}_k(\Omega)$. If Ω is unbounded, we also need the extra vanishing condition at ∞. Clearly, when $h \in C^\infty(\partial\Omega)$, the function $C_k h$ belongs to $\mathcal{R}_k(\Omega)$. Notice that we can also take the Cauchy transform of an element h in $C^\infty(\partial\Omega)$ with respect to $\mathbb{R}^m \backslash \overline{\Omega}$, thus creating an element of $\mathcal{R}_k(\mathbb{R}^m \backslash \overline{\Omega})$. Before considering the behaviour of the Cauchy transform C_k on the space $C^\infty(\partial\Omega)$ we introduce the following definition:

Definition. The submodule $\mathcal{R}_k(\Omega) \cap C(\overline{\Omega})$ of $\mathcal{R}_k(\Omega)$ will be denoted by $\mathcal{R}_k^\infty(\Omega)$.

THEOREM 4. *The Cauchy transform C_k maps $C^\infty(\partial\Omega)$ onto $\mathcal{R}_k^\infty(\Omega)$.*

By the reproducing property of Cauchy's integral formula it is clear that each element of $\mathcal{R}_k^\infty(\Omega)$ is uniquely determined by its restriction to $\partial\Omega$. Let now $\mathcal{R}_k^\infty(\partial\Omega)$ be the subspace of $C^\infty(\partial\Omega)$ obtained by taking the restriction to $\partial\Omega$ of the elements belonging to $\mathcal{R}_k^\infty(\Omega)$. From now on identifying the spaces $\mathcal{R}_k^\infty(\Omega)$ and $\mathcal{R}_k^\infty(\partial\Omega)$, the above theorem says that the Cauchy transform C_k can be regarded as a map from $C^\infty(\partial\Omega)$ to $\mathcal{R}_k^\infty(\partial\Omega)$. We thus have the following:

COROLLARY 5. *The Cauchy transform C_k is a linear map between the right modules $C^\infty(\partial\Omega)$ and $\mathcal{R}_k^\infty(\partial\Omega)$ (or $\mathcal{R}_k^\infty(\Omega)$) satisfying $C_k^2 = C_k$.*

Now let $f, g \in C^\infty(\partial\Omega)$; then $C_k f \in C^\infty(\partial\Omega)$. The formal adjoint C_k^* of C_k is then defined to be the linear map from $C^\infty(\partial\Omega)$ into itself satisfying the property $\langle C_k f, g \rangle_{\partial\Omega} = \langle f, C_k^* g \rangle_{\partial\Omega}$ for all $f, g \in C^\infty(\partial\Omega)$.

LEMMA 6. *Let $\overline{\Omega} \subset \mathbb{R}^m$ be a smooth domain and let n be the outer unit normal field to $\partial\Omega$ with respect to Ω. Define $C_k^* g = g - R_k(n) C_k R_k^{-1}(n) g$, $g \in C^\infty(\partial\Omega)$. Then C_k^* is formally adjoint to C_k.*

Definition. Let $h \in C^\infty(\partial\Omega)$. Then the *Hilbert transform* $\mathcal{H}_k$ on $C^\infty(\partial\Omega)$ is defined via the singular integral

$$(\mathcal{H}_k h)(y, v) = -2 \lim_{\delta \to 0} \int_{\partial\Omega \setminus B(y, \delta)} (E_k(x - y; u, v), \dot{R}_k(d\sigma_x) h(x, \dot{u}))_u$$

$$= -2 \, p.v. \int_{\partial\Omega} (E_k(x - y; u, v), \dot{R}_k(d\sigma_x) h(x, \dot{u}))_u, \, y \in \partial\Omega.$$

The fact that $\mathcal{H}_k h$ makes sense follows from the relationship between the Hilbert and Cauchy transforms expressed in the following

THEOREM 7. *(Plemelj formula)*
Let $h \in C^\infty(\partial\Omega)$. Then $C_k h = \frac{1}{2}(h + \mathcal{H}_k h)$.

4. Hardy spaces and the Kerzman-Stein Theorem

Consider the space $\mathcal{R}_k(\Omega)$ of null solutions of the operator $\mathcal{R}_k$ in Ω which extend to C^∞-functions on $\overline{\Omega}$. Restricting elements of this space to the boundary of Ω establishes an isomorphism between the subspace $\mathcal{R}_k^\infty(\partial\Omega)$ of $L_2(\partial\Omega, \mathcal{M}_k)$ and $\mathcal{R}_k^\infty(\Omega)$. We now introduce the following definitions:

Definitions.

(i) The *Hardy Space* $\mathcal{HS}_k(\Omega)$ is defined as the closure of $\mathcal{R}_k^\infty(\partial\Omega)$ in $L_2(\partial\Omega, \mathcal{M}_k)$

(ii) The *Szegö projection* S_k is the orthogonal projection of $L_2(\partial\Omega, \mathcal{M}_k)$ on $\mathcal{HS}_k(\Omega)$.

The notation $\mathcal{HS}_k(\Omega)$ emphasises that $\mathcal{HS}_k(\Omega)$ is precisely the completion of the space of boundary value elements belonging to $\mathcal{R}_k^\infty(\Omega)$ (instead of $\mathcal{R}_k^\infty(\mathbb{R}^m\setminus\overline{\Omega})$). We will now state the analogue of the Kerzman-Stein theorem in the plane (see [KS]).

LEMMA 8. *(Kerzman-Stein Formula)*
Let $\overline{\Omega}\subset\mathbb{R}^m$ be a smooth domain. Then

$$C_k = S_k(1 + C_k - C_k^*) \text{ on } C^\infty(\partial\Omega).$$

We will now give a singular integral representation for the operator $C_k - C_k^*$ on $C^\infty(\partial\Omega)$. To this end we first introduce the Kerzman-Stein kernel A_k.

Definition. The Kerzman-Stein kernel $A_k(x,y;u,v)$ is defined as

$$A_k(x,y;u,v) = \dot{R}_k(n(x))E_k(x-y;\dot{u},v) + \overline{\dot{R}_k(n(y))E_k(x-y;\dot{v},u)},$$

$x,y \in (\partial\Omega \times \partial\Omega)\setminus L$, L being the diagonal of $\partial\Omega \times \partial\Omega$.

Let $f \in C^\infty(\partial\Omega)$. In view of the definition of C_k^* and the relation between the Hilbert and Cauchy transforms,

$$(C_k - C_k^*)f = C_k f - f + R_k(n)C_k R_k^{-1}(n)f$$
$$= \frac{1}{2}\mathcal{H}_k f + \frac{1}{2}R_k(n)\mathcal{H}_k R_k^{-1}(n)f.$$

Therefore the singular integral operator $C_k - C_k^*$ is given by

$$((C_k - C_k^*)f)(y,v) = p.v.\int_{\partial\Omega}(\dot{R}_k(n(x))E_k(x-y;\dot{u},v), f(x,u))_u ds(x)$$

$$+ p.v.\int_{\partial\Omega}\overline{(\dot{R}_k(n(y))E_k(x-y;\dot{v},u)}, f(x,u))_u ds(x)$$

$$= p.v.\int_{\partial\Omega}(A_k(x,y;u,v), f(x,u))_u ds(x). \qquad (1)$$

The fact that the principal value in relation (1) can actually be omitted follows from the following theorem. This means that the operator $C_k - C_k^*$ turns out to be an integral operator on $C^\infty(\partial\Omega)$.

THEOREM 9. *(The Kerzman-Stein Theorem for Rarita-Schwinger operators)*

(i) The integral operator $\mathcal{A}_k : L_2(\partial\Omega, \mathcal{M}_k) \to L_2(\partial\Omega, \mathcal{M}_k)$ defined by

$$(\mathcal{A}_k f)(y,v) = \int_{\partial\Omega}(A_k(x,y;u,v), f(x,u))_u ds(x), y \in \partial\Omega$$

is compact and $i\mathcal{A}_k$ is in addition self-adjoint.

(ii) *The Kerzman-Stein formula $C_k = S_k(1 + C_k - C_k^*)$ extends to $L_2(\partial\Omega, \mathcal{M}_k)$.*

An immediate consequence of this theorem is the following

THEOREM 10.

(i) *The Cauchy transform C_k extends to a bounded linear operator from $L_2(\partial\Omega, \mathcal{M}_k)$ onto $\mathcal{HS}_k(\Omega)$ satisfying $C_k^2 = C_k$.*

(ii) *The Hilbert transform $\mathcal{H}_k$ extends to a bounded linear operator on $L_2(\partial\Omega, \mathcal{M}_k)$, mapping $C^\infty(\partial\Omega)$ into itself, and is inverted via the relation $\mathcal{H}_k^2 = 1$.*

5. Some decompositions of $L_2(\partial\Omega, \mathcal{M}_k)$

Let us recall that if $\overline{\Omega}$ is a proper smooth domain of $\mathbb{R}^m$, then so is $\mathrm{Co}\Omega = \mathbb{R}^m \backslash \Omega$. Therefore the $(m-1)$-dimensional submanifold $\partial\Omega$ of $\mathbb{R}^m$ may be regarded either as the boundary of Ω or as the boundary of $\mathbb{R}^m \backslash \overline{\Omega}$. In this way one can also consider the Cauchy transform and other relevant operators from the point of view of $\mathbb{R}^m \backslash \overline{\Omega}$ and thus reformulate all results obtained in the previous sections in terms of the manifold- with-boundary $\mathbb{R}^m \backslash \Omega$. This will lead to (in general) *different* decompositions of $L_2(\partial\Omega, \mathcal{M}_k)$.

THEOREM 11. *Let $\overline{\Omega} \subset \mathbb{R}^m$ be a proper smooth domain and let n be the outer unit normal field to $\partial\Omega$ with respect to Ω. Then*

(i)

$$L_2(\partial\Omega, \mathcal{M}_k) = \mathcal{HS}_k(\Omega) \oplus_\perp R_k(n)\mathcal{HS}_k(\Omega),$$

(ii)

$$L_2(\partial\Omega, \mathcal{M}_k) = \mathcal{HS}_k(\Omega) \oplus \mathcal{HS}_k(\mathrm{Co}\overline{\Omega})$$

where the decomposition (i) is orthogonal while the mixed decomposition (ii) is direct but in general not orthogonal .

6. The self-adjointness of the Cauchy transform

Since $C_k - C_k^*$ corresponds to an integral operator having as kernel the Kerzman-Stein kernel $A_k(x, y; u, v)$; $C_k = C_k^*$ iff $A_k(x, y; u, v)$ vanishes on $\partial\Omega \times \partial\Omega$. In view of the Kerzman-Stein formula $C_k = S_k(1 + C_k - C_k^*)$ this is also equivalent to the following problem: when do the Cauchy transform C_k and the Szegö projection S_k coincide or when is the mixed decomposition in theorem 11 orthogonal? In this case the decompositions formulated in theorem 11 coincide. To answer this question we need the following geometric characterization of hyperspheres and hyperplanes in $\mathbb{R}^m$ (a similar result in the plane is proved in [KS]).

LEMMA 12. *(Characterization of hyperspheres and hyperplanes in $\mathbb{R}^m$ in terms of the unit normal field)*

Hyperspheres and hyperplanes are the only $(m-1)$-dimensional manifolds in $\mathbb{R}^m$ such that, given any two vectors x, y on them, the outer unit normals (induced by their embedding in $\mathbb{R}^m$) in the corresponding points satisfy the relation:

$$(y - x)n(x) + n(y)(y - x) = 0 .$$

THEOREM 13. *(Characterisation of the self-adjointness of the Cauchy transform)*
Let $\overline{\Omega} \subset \mathbb{R}^m$ be a smooth domain. Then the Cauchy transform C_k induced on $\partial\Omega$ is self-adjoint iff $\partial\Omega$ is a $(m-1)$-dimensional sphere or hyperplane in $\mathbb{R}^m$.

REFERENCES

[BDS] F. Brackx, R. Delanghe and F. Sommen: *Clifford Analysis*, Pitman, London, 1982.

[BSoSoVL] J. Bures, F. Sommen, V. Souček, P. Van Lancker: *Rarita-Schwinger operators in Clifford Analysis*, Preprint.

[BVLSoSo] J. Bures, P. Van Lancker, F. Sommen, V. Souček: *Symmetric analogues of Rarita-Schwinger equations*, Preprint.

[C] D. M. J. Calderbank: *Clifford analysis for Dirac operators on manifolds with boundary*, Report MPI 96-131.

[DSS] R. Delanghe, F. Sommen, V. Souček: *Clifford analysis and spinor valued functions*, Kluwer Acad. Publ., Dordrecht, 1992.

[GM] J. Gilbert and M. Murray: *Clifford algebras and Dirac operators in harmonic analysis*, Cambridge University Press, 1991.

[KS] N. Kerzman, E. M. Stein: *The Cauchy kernel, the Szegö kernel, and the Riemann mapping function*, Math. Ann. 236, 1978, pp. 85-93.

[So1] F. Sommen: *Clifford analysis in two and several vector variables*, Appl. Anal. Vol. 73 (1-2), pp. 225-253.

[S] V. Souček: *Higher Spins and Conformal Invariance in Clifford Analysis*, Proc. Conf. Seiffen, 1996.

[SVA1] F. Sommen, N. Van Acker: *Monogenic Differential Operators*, Results in Math. 22, 1992, pp. 781-798.

[SVA2] F. Sommen, N. Van Acker: *Invariant differential operators on polynomial valued functions*, F. Brackx et al. (eds.), Clifford Algebras and their Applications in Mathematical Physics, Kluwer Acad. Publ., 1993, pp. 203-212.

[VL1] P. Van Lancker: *Clifford Analysis on the Unit Sphere*, Ph. D. thesis, University of Gent, 1996.

[VL2] P. Van Lancker: *The Kerzman-Stein Theorem on the Sphere*, to appear in Complex Variables.

[VLSC] P. Van Lancker, F. Sommen, D. Constales: Models for irreducible representations of $Spin(m)$, to appear in the Proc. of Conf. on Dirac Operators, Cetraro, 1998.

Bergman Type Spaces on the Unit Disk *

Nikolai Vasilevski

Departamento de Matemáticas, CINVESTAV del I.P.N., México

Abstract. The decomposition of $L_2(\mathbb{D})$ into Bergman and Bergman type spaces on the unit disk $\mathbb{D}$ is studied. Connections between the Bergman and the Hardy spaces, as well as between the Bergman and Szegö projections are established.

Keywords: Bergman space, Hardy space, Bergman projection, unit disk

Mathematics Subject Classification: 30H05, 46E20, 47B32

Dedicated to Richard Delanghe on the occasion of his 60th birthday.

1. Introduction

Let $\mathbb{D}$ be the unit disk in $\mathbb{C}$. Rearranging the basis of $L_2(\mathbb{D})$ L. Peng, R. Rochberg and Z. Wu [1] proved that the space $L_2(\mathbb{D})$ can be decomposed onto a direct sum of Bergman type spaces

$$L_2(\mathbb{D}) = \bigoplus_{n=1}^{\infty} \mathcal{A}^2_{(n)} \oplus \bigoplus_{n=1}^{\infty} \tilde{\mathcal{A}}^2_{(n)}, \tag{1.1}$$

where $\mathcal{A}^2_{(n)} = \ker(\bar{z}\,\partial/\partial\bar{z})^n \ominus \ker(\bar{z}\,\partial/\partial\bar{z})^{n-1}$, $\tilde{\mathcal{A}}^2_{(n)} = \ker(z\,\partial/\partial z)^n \ominus \ker(z\,\partial/\partial z)^{n-1}$. Studying this question we follow the ideas of [2], which allows us to obtain more information. In particular, we construct a unitary operator W on the space $L_2(\mathbb{D})$, which *simultaneously* reduces all the pieces of decomposition (1.1) to a simple transparent form

$$W : \mathcal{A}^2_{(n)} \longrightarrow L_{n-1} \otimes H^2(\mathbb{D}), \qquad W : \tilde{\mathcal{A}}^2_{(n)} \longrightarrow L_{n-1} \otimes H^2(\mathbb{C} \setminus \overline{\mathbb{D}}),$$

where $H^2(\Omega)$ is the Hardy space in Ω, and L_n is the one-dimensional subspace of $L_2((0,1), r\,dr)$ generated by the function $\ell_n(r)$ of the form (3.1). Several connections between the Bergman and the Hardy spaces, as well as between the corresponding Bergman and Szegö projections are established. In particular, we construct a partial isometry $\tilde{R}$: $L_2(\mathbb{D}) \to L_2(S^1)$, for which we have the following isometric isomorphisms between the Bergman $\mathcal{A}^2(\mathbb{D})$ and the Hardy $H^2(\mathbb{D})$ spaces:

$$\tilde{R}|_{\mathcal{A}^2(\mathbb{D})} : \mathcal{A}^2(\mathbb{D}) \longrightarrow H^2(\mathbb{D}), \qquad \tilde{R}^*|_{H^2(\mathbb{D})} : H^2(\mathbb{D}) \longrightarrow \mathcal{A}^2(\mathbb{D}).$$

* This work was partially supported by CONACYT Project 27934-E, México.

399

F. Brackx et al. (eds.), Clifford Analysis and Its Applications, 399–409.
© *2001 Kluwer Academic Publishers. Printed in the Netherlands.*

Furthermore, the operators $\tilde{R}$ and $\tilde{R}^*$ provide the following decomposition of the Bergman $B_{\mathbb{D}}$ and the Szegö P_{S^1} projections:

$$\tilde{R}^*\tilde{R} = B_{\mathbb{D}} : L_2(\mathbb{D}) \longrightarrow \mathcal{A}^2(\mathbb{D}), \quad \tilde{R}\tilde{R}^* = P_{S^1} : L_2(\mathbb{R}^n) \longrightarrow H^2(\mathbb{D}).$$

2. Bergman space and Bergman projection

Let $\mathbb{D}$ be the unit disk in $\mathbb{C}$, introduce the space $L_2(\mathbb{D})$ with the usual Lebesgue plane measure $d\mu(z) = dxdy$, $z = x + iy$, and its Bergman subspace $\mathcal{A}^2(\mathbb{D})$ consisting of all functions analytic in $\mathbb{D}$. It is well known that the orthogonal Bergman projection $B_{\mathbb{D}}$ of $L_2(\mathbb{D})$ onto $\mathcal{A}^2(\mathbb{D})$ is given by

$$(B_{\mathbb{D}}\varphi)(z) = \frac{1}{\pi} \int_{\mathbb{D}} \frac{\varphi(\zeta)}{(1 - z\bar{\zeta})^2} \, d\mu(\zeta).$$

Note, that the Bergman space $\mathcal{A}^2(\mathbb{D})$ can be described alternatively as the (closed) subspace of $L_2(\mathbb{D})$, which consists of all functions satisfying the equation

$$\frac{\partial}{\partial \bar{z}} \varphi = \frac{1}{2} \left(\frac{\partial}{\partial x} + i \frac{\partial}{\partial y} \right) \varphi = 0.$$

Passing to polar coordinates we have

$$L_2(\mathbb{D}) = L_2(\mathbb{D}, d\mu(z)) = L_2((0,1), rdr) \otimes L_2([0, 2\pi), d\alpha)$$
$$= L_2((0,1), rdr) \otimes L_2(S^1, \frac{dt}{it}) = L_2((0,1), rdr) \otimes L_2(S^1),$$

where S^1 is the unit circle, and $\frac{dt}{it} = |dt| = d\alpha$ is the element of length; in addition

$$\frac{\partial}{\partial \bar{z}} = \frac{\cos\alpha + i\sin\alpha}{2} \left(\frac{\partial}{\partial r} + i\frac{1}{r} \frac{\partial}{\partial \alpha} \right) = \frac{t}{2} \left(\frac{\partial}{\partial r} - \frac{t}{r} \frac{\partial}{\partial t} \right).$$

Introduce the unitary operator

$$U_1 = I \otimes \mathcal{F} : L_2((0,1), rdr) \otimes L_2(S^1) \longrightarrow L_2((0,1), rdr) \otimes l_2,$$

where the Fourier transform $\mathcal{F} : L_2(S^1) \to l_2$ is given by

$$\mathcal{F} : f \longmapsto c_n = \frac{1}{\sqrt{2\pi}} \int_{S^1} f(t) t^{-n} \frac{dt}{it}, \quad n \in \mathbb{Z}.$$

It is easy to see that

$$(I \otimes \mathcal{F})\frac{t}{2} \left(\frac{\partial}{\partial r} - \frac{t}{r} \frac{\partial}{\partial t} \right) (I \otimes \mathcal{F}^{-1})\{c_n(r)\}_{n \in \mathbb{Z}} =$$

$$\left\{ \frac{1}{2} \left(\frac{\partial}{\partial r} - \frac{n-1}{r} \right) c_{n-1}(r) \right\}_{n \in \mathbb{Z}}.$$

Thus the image of the Bergman space $\mathcal{A}_1^2 = U_1(\mathcal{A}^2(\mathbb{D}))$ can be described as the (closed) subspace of $L_2((0,1), r dr) \otimes l_2 = l_2(L_2((0,1), r dr))$ which consists of all sequences $\{c_n(r)\}_{n \in \mathbb{Z}}$ satisfying the equations

$$\frac{1}{2} \left(\frac{\partial}{\partial r} - \frac{n}{r} \right) c_n(r) = 0, \quad n \in \mathbb{Z}. \tag{2.1}$$

The equations (2.1) are easy to solve, and their general solutions have the form

$$c_n(r) = c_n' r^n = \sqrt{2(|n|+1)} \, c_n r^n, \quad n \in \mathbb{Z}.$$

But each function $c_n(r) = \sqrt{2(|n|+1)} \, c_n r^n$ has to be in $L_2((0,1), r dr)$, which implies that $c_n(r) \equiv 0$, for each $n < 0$. Thus the space $\mathcal{A}_1^2$ ($\subset L_2((0,1), r dr) \otimes l_2 = l_2(L_2((0,1), r dr))$) coincides with the space of all two-sided sequences $\{c_n(r)\}_{n \in \mathbb{Z}}$ with

$$c_n(r) = \begin{cases} \sqrt{2(n+1)} \, c_n r^n, & \text{if } n \in \mathbb{Z}_+ \\ 0, & \text{if } n \in \mathbb{Z}_- \end{cases},$$

where $\mathbb{Z}_+ = \{0\} \cup \mathbb{N}$, $\mathbb{Z}_- = \mathbb{Z} \setminus \mathbb{Z}_+$, and

$$\|\{c_n(r)\}_{n \in \mathbb{Z}}\| = \left(\sum_{n \in \mathbb{Z}_+} |c_n|^2 \right)^{1/2} = \|\{c_n\}_{n \in \mathbb{Z}_+}\|_{l_2}.$$

For each $n \in \mathbb{Z}_+$ introduce the unitary operator $u_n : L_2((0,1), r dr) \longrightarrow L_2((0,1), r dr)$ by the rule

$$(u_n f)(r) = \frac{1}{\sqrt{n+1}} r^{-\frac{n}{n+1}} f(r^{\frac{1}{n+1}}),$$

then the inverse operator $u_n^{-1} = u_n^* : L_2((0,1), r dr) \longrightarrow L_2((0,1), r dr)$ is given by

$$(u_n^{-1} f)(r) = \sqrt{n+1} \, r^n f(r^{n+1}).$$

Finally, define the unitary operator

$$U_2 : l_2(L_2((0,1), r dr)) \longrightarrow l_2(L_2((0,1), r dr)) = L_2((0,1), r dr) \otimes l_2$$

as follows:

$$U_2 : \{c_n(r)\}_{n \in \mathbb{Z}} \longmapsto \{(u_{|n|} c_n)(r)\}_{n \in \mathbb{Z}}.$$

Then the space $\mathcal{A}_2^2 = U_2(\mathcal{A}_1^2)$ coincides with the space of all sequences $\{d_n(r)\}_{n \in \mathbb{Z}}$, where

$$d_n = u_n(\sqrt{2(n+1)} \, c_n r^n) = \sqrt{2} \, c_n,$$

for $n \in \mathbb{Z}_+$, and $d_n(r) \equiv 0$, for $n \in \mathbb{Z}_-$.

We introduce some notation. Letting $\ell_0(r) = \sqrt{2}$, we have $\ell_0(r) \in L_2((0,1), r dr)$ and $\|\ell_0(r)\| = 1$. Denote by L_0 the one-dimensional subspace of $L_2((0,1), r dr)$ generated by $\ell_0(r)$, then the one-dimensional projection P_0 of $L_2((0,1), r dr)$ onto L_0 has the form

$$(P_0 f)(r) = \langle f, \ell_0 \rangle \cdot \ell_0 = \sqrt{2} \int_0^1 f(\rho) \sqrt{2} \rho \, d\rho. \qquad (2.2)$$

Denote by l_2^+ (l_2^-) the subspace of (two-sided) l_2, consisting of all sequences $\{c_n\}_{n \in \mathbb{Z}}$, such that $c_n = 0$ for all $n \in \mathbb{Z}_-$ ($n \in \mathbb{Z}_+$). Then $l_2 = l_2^+ \oplus l_2^-$, and denote by p^+ (p^-) the orthogonal projection of l_2 onto l_2^+ (l_2^-). Introduce the sequences $\chi_\pm = \{\chi_\pm(n)\}_{n \in \mathbb{Z}} \in l_\infty$, where $\chi_\pm(n) = 1$ for $n \in \mathbb{Z}_\pm$, and $\chi_\pm(n) = 0$ for $n \in \mathbb{Z}_\mp$. Then, $p^\pm = \chi_\pm I$.

Now $\mathcal{A}_2^2 = L_0 \otimes l_2^+$, and the orthogonal projection B_2 of $L_2((0,1), r dr) \otimes l_2$ onto $\mathcal{A}_2^2$ obviously has the form $B_2 = P_0 \otimes p^+$. Thus we arrive at the following theorem.

THEOREM 2.1. *The unitary operator $U = U_2 U_1$ is an isometric isomorphism of the space $L_2(\mathbb{D})$ onto $L_2((0,1), r dr) \otimes l_2$ under which*

1. *the Bergman space $\mathcal{A}^2(\mathbb{D})$ is mapped onto $L_0 \otimes l_2^+$*

$$U : \mathcal{A}^2(\mathbb{D}) \longrightarrow L_0 \otimes l_2^+,$$

   where L_0 is the one-dimensional subspace of $L_2((0,1), r dr)$, generated by $\ell_0(r) = \sqrt{2}$,

2. *the Bergman projection $B_\mathbb{D}$ is unitary equivalent to the following,*

$$U B_\mathbb{D} U^{-1} = P_0 \otimes p^+,$$

   where P_0 is the one-dimensional projection (2.2) of $L_2((0,1), r dr)$ onto L_0.

REMARK 2.2. The above result describes the structure of the Bergman space inside $L_2(\mathbb{D})$, and is a "Bergman version" of the following well known "Hardy version" result: *the Fourier transform $\mathcal{F}$ gives an isometric isomorphism of the space $L_2(S^1)$ onto two-sided l_2 under which the Hardy space $H^2(\mathbb{D})$ is mapped onto l_2^+: $\mathcal{F} : H^2(\mathbb{D}) \longrightarrow l_2^+$, and the Szegö projection $P_{S^1}^+ : L_2(S^1) \longrightarrow H_+^2(\mathbb{D})$ is unitary equivalent to $\mathcal{F} P_{S^1}^+ \mathcal{F}^{-1} = p^+$.*

Introduce the isometric imbedding $R_0 : l_2^+ \longrightarrow L_2((0,1), r dr) \otimes l_2$ by the rule

$$R_0 : \{c_n\}_{n \in \mathbb{Z}_+} \longmapsto \ell_0(r) \{\chi_+(n) c_n\}_{n \in \mathbb{Z}}.$$

The image of R_0 obviously coincides with the space $\mathcal{A}_2^2$. The adjoint operator $R_0^* : L_2((0,1), r\,dr) \otimes l_2 \to l_2^+$ is given by

$$R_0^* : \{c_n(r)\}_{n \in \mathbb{Z}} \longmapsto \left\{ \chi_+(n) \int_0^1 c_n(\rho) \sqrt{2}\, \rho\, d\rho \right\}_{n \in \mathbb{Z}_+}.$$

Now the operator $R = R_0^* U$ maps the space $L_2(\mathbb{D})$ onto l_2^+, and the restriction

$$R|_{\mathcal{A}^2(\mathbb{D})} : \mathcal{A}^2(\mathbb{D}) \longrightarrow l_2^+$$

is an isometric isomorphism. The adjoint operator

$$R^* = U^* R_0 : l_2^+ \longrightarrow \mathcal{A}^2(\mathbb{D}) \subset L_2(\mathbb{D})$$

is an isometric isomorphism of l_2^+ onto the subspace $\mathcal{A}^2(\mathbb{D})$ of the space $L_2(\mathbb{D})$.

REMARK 2.3. We have

$$R R^* = I : l_2^+ \longrightarrow l_2^+, \quad R^* R = B_\mathbb{D} : L_2(\mathbb{D}) \longrightarrow \mathcal{A}^2(\mathbb{D}).$$

THEOREM 2.4. *The isometric isomorphism* $R^* = U^* R_0 : l_2^+ \longrightarrow \mathcal{A}^2(\mathbb{D})$ *is given by*

$$R^* : \{c_n\}_{n \in \mathbb{Z}_+} \longmapsto \frac{1}{\sqrt{2\pi}} \sum_{n \in \mathbb{Z}_+} \sqrt{2(n+1)}\, c_n\, z^n.$$

COROLLARY 2.5. *The inverse isomorphism* $R : \mathcal{A}^2(\mathbb{D}) \longrightarrow l_2^+$ *is given by*

$$R : \varphi(z) \longmapsto \left\{ \frac{\sqrt{2(n+1)}}{\sqrt{2\pi}} \int_\mathbb{D} \varphi(z)\, \bar{z}^n\, d\mu(z) \right\}_{n \in \mathbb{Z}_+}.$$

Introduce the operator $\widetilde{R} : L_2(\mathbb{D}) \to L_2(S^1)$ as follows: $\widetilde{R} = \mathcal{F}^{-1} R$.

COROLLARY 2.6. *We have the following isometric isomorphisms between the Bergman* $\mathcal{A}^2(\mathbb{D})$ *and the Hardy* $H^2(\mathbb{D})$ *spaces:*

$$\widetilde{R}|_{\mathcal{A}^2(\mathbb{D})} : \mathcal{A}^2(\mathbb{D}) \longrightarrow H^2(\mathbb{D}), \quad \widetilde{R}^*|_{H^2(\mathbb{D})} : H^2(\mathbb{D}) \longrightarrow \mathcal{A}^2(\mathbb{D}).$$

The operators $\widetilde{R}$ and $\widetilde{R}^*$ provide the following decomposition of the Bergman $B_\mathbb{D}$ and the Szegö P_{S^1} projections:

$$\widetilde{R}^* \widetilde{R} = B_\mathbb{D} : L_2(\mathbb{D}) \longrightarrow \mathcal{A}^2(\mathbb{D}), \quad \widetilde{R} \widetilde{R}^* = P_{S^1} : L_2(S^1) \longrightarrow H^2(\mathbb{D}).$$

Another connection between the Bergman and the Hardy spaces, and
between the corresponding projections is given by the following

THEOREM 2.7.   *The unitary operator* $W = (I \otimes \mathcal{F}^{-1})U_2(I \otimes \mathcal{F})$ *gives
an isometric isomorphism of the space*
$L_2(\mathbb{D}) = L_2((0,1), r dr) \otimes L_2(S^1, \frac{dt}{it})$ *under which*

1. *the Bergman* $\mathcal{A}^2(\mathbb{D})$ *and the Hardy* $H^2(\mathbb{D})$ *spaces are connected by
   the formula*
   $$W(\mathcal{A}^2(\mathbb{D})) = L_0 \otimes H^2(\mathbb{D}),$$

2. *the Bergman* $B_\mathbb{D}$ *and the Szegö* $P_{S^1}^+$ *projections are connected by
   the formula*
   $$W B_\mathbb{D} W^{-1} = P_0 \otimes P_{S^1}^+,$$

where P_0 *is the one-dimensional projection (2.2) of* $L_2((0,1), r dr)$ *onto
one-dimensional space* L_0 *generated by* $\ell_0(r) = \sqrt{2} \in L_2((0,1), r dr)$.

In addition to the Bergman space $\mathcal{A}^2(\mathbb{D})$ of functions analytic in $\mathbb{D}$,
introduce the space $\widetilde{\mathcal{A}}^2(\mathbb{D})$ as the (closed) subspace of $L_2(\mathbb{D})$ consisting
of all functions which are anti-analytic in $\mathbb{D}$ and have the zero value at
the point $0 \in \mathbb{D}$ (otherwise the spaces $\mathcal{A}^2(\mathbb{D})$ and $\widetilde{\mathcal{A}}^2(\mathbb{D})$ would intersect
in the constants). Denote by $\widetilde{B}_\mathbb{D}$ the orthogonal Bergman projection of
$L_2(\mathbb{D})$ onto $\widetilde{\mathcal{A}}^2(\mathbb{D})$

THEOREM 2.8.   *The unitary operator* $U = U_2 U_1$ *is an isometric
isomorphism of the space* $L_2(\mathbb{D})$ *onto* $L_2((0,1), r dr) \otimes l_2$ *under which*

1. *the space* $\widetilde{\mathcal{A}}^2(\mathbb{D})$ *is mapped onto* $L_0 \otimes l_2^-$
   $$U : \widetilde{\mathcal{A}}^2(\mathbb{D}) \longrightarrow L_0 \otimes l_2^-,$$

   where L_0 *is the one-dimensional subspace of* $L_2((0,1), r dr)$, *generated by* $\ell_0(r) = \sqrt{2}$,

2. *the projection* $\widetilde{B}_\mathbb{D}$ *is unitary equivalent to the following one*
   $$U \widetilde{B}_\mathbb{D} U^{-1} = P_0 \otimes p^-,$$

   where $p^- = \chi_- I$ *is the orthogonal projection of (two-sided)* l_2 *onto
   l_2^-, and* P_0 *is the one-dimensional projection (2.2) of* $L_2((0,1), r dr)$
   onto L_0.

PROOF.   The space $\widetilde{\mathcal{A}}^2(\mathbb{D})$ can be described alternatively as the (closed)
subspace of $L_2(\mathbb{D})$ which consists of all functions satisfying the equation

$$D\varphi = z\frac{\partial}{\partial z}\varphi = \frac{z}{2}\left(\frac{\partial}{\partial x} - i\frac{\partial}{\partial y}\right)\varphi = 0.$$

Passing to polar coordinates, it is easy to see that

$$U_2 U_1 D U_1^{-1} U_2^{-1} = U_2 \left\{ \frac{r}{2} \left(\frac{\partial}{\partial r} + \frac{n}{r} \right) c_n(r) \right\}_{n \in \mathbb{Z}} U_2^{-1} = \{D_n\}_{n \in \mathbb{Z}},$$

where

$$D_n = \begin{cases} \frac{n+1}{2} \left(\frac{2n}{n+1} + r \frac{d}{dr} \right), & \text{if } n \in \mathbb{Z}_+ \\ \frac{|n|+1}{2} r \frac{d}{dr}, & \text{if } n \in \mathbb{Z}_- \end{cases} \tag{2.3}$$

The general solution of the equations $D_n c_n(r) = 0$, $n \in \mathbb{Z}$, has the form

$$c_n(r) = c_n \, r^{-\frac{2n}{n+1}}, \quad \text{for } n \in \mathbb{Z}_+,$$
$$c_n(r) = \sqrt{2}\, c_n, \quad \text{for } n \in \mathbb{Z}_-.$$

All functions from $\tilde{A}^2(\mathbb{D})$ have the value zero at origin, which implies that $c_0(r) = c_0 = 0$. Again, each $c_n(r)$, $n \in \mathbb{N}$, has to be in $L_2((0,1), r dr)$, which implies that $c_n = 0$, for all $n \in \mathbb{N}$.

Then the space $\tilde{A}_2^2 = U(\tilde{A}^2(\mathbb{D}))$ coincides with the space of all sequences $\{c_n(r)\}_{n \in \mathbb{Z}}$, where

$$c_n(r) = \begin{cases} 0, & \text{if } n \in \mathbb{Z}_+ \\ \sqrt{2}\, c_n, & \text{if } n \in \mathbb{Z}_- \end{cases},$$

and $\|\{c_n(r)\}\|_{l_2(L_2((0,1), r dr))} = \|\{c_n\}\|_{l_2^-}$.

Thus $\tilde{A}_2^2 = L_0 \otimes l_2^-$, and the orthogonal projection $\tilde{B}_2$ of $l_2(L_2((0,1), r dr)) = L_2((0,1), r dr) \otimes l_2$ onto $\tilde{A}_2^2$ has the form $\tilde{B}_2 = P_0 \otimes p^-$. $\qquad \square$

Introduce now the Hardy space $H^2(\mathbb{D}^-)$ as the subspace of $L_2(S^1)$ consisting of the functions having analytic extension into $\mathbb{C} \setminus \overline{\mathbb{D}}$ and vanishing at infinity. Denote by $P_{S^1}^-$ is the orthogonal (Szegö) projection of $L_2(S^1)$ onto $H^2(\mathbb{D}^-)$.

COROLLARY 2.9. *The unitary operator* $W = (I \otimes \mathcal{F}^{-1}) U_2 (I \otimes \mathcal{F})$ *gives an isometric isomorphism of the space* $L_2(\mathbb{D}) = L_2((0,1), r dr) \otimes L_2(S^1, \frac{dt}{it})$ *under which*

1. the spaces $\tilde{A}^2(\mathbb{D})$ *and* $H^2(\mathbb{D}^-)$ *are connected by the formula*

$$W(\tilde{A}^2(\mathbb{D})) = L_0 \otimes H^2(\mathbb{D}^-),$$

2. the projections $\tilde{B}_{\mathbb{D}}$ *and* $P_{S^1}^-$ *are connected by the formula*

$$W \tilde{B}_{\mathbb{D}} W^{-1} = P_0 \otimes P_{S^1}^-,$$

where P_0 is the one-dimensional projection (2.2) of $L_2((0,1), rdr)$ onto one-dimensional space L_0 generated by $\ell_0(r) = \sqrt{2} \in L_2((0,1), rdr)$.

3. Poly-Bergman type spaces, decomposition of $L_2(\mathbb{D})$

Besides the operator

$$D = z\frac{\partial}{\partial z} = \frac{r}{2}\left(\frac{\partial}{\partial r} + \frac{t}{r}\frac{\partial}{\partial t}\right)$$

introduce

$$\overline{D} = \bar{z}\frac{\partial}{\partial \bar{z}} = \frac{r}{2}\left(\frac{\partial}{\partial r} - \frac{t}{r}\frac{\partial}{\partial t}\right).$$

Analogously to the Bergman spaces $\mathcal{A}^2(\mathbb{D})$ and $\tilde{\mathcal{A}}^2(\mathbb{D})$, which can be treated as the L_2–kernels of $\overline{D}$ and D respectively, following [1] introduce the spaces of poly-$\overline{D}$-analytic and poly-D-analytic functions, the poly-Bergman type spaces.

Define the space $\mathcal{A}_n^2(\mathbb{D})$ of n-$\overline{D}$-analytic functions as the (closed) subspace of $L_2(\mathbb{D})$ of all functions $\varphi = \varphi(z, \bar{z}) = \varphi(x, y)$, which satisfy the equation

$$\overline{D}^n\varphi = \left(\bar{z}\frac{\partial}{\partial \bar{z}}\right)^n \varphi = 0.$$

Similarly, define the space $\tilde{\mathcal{A}}_n^2(\mathbb{D})$ of n-D-analytic functions as the (closed) subspace of $L_2(\mathbb{D})$ of all functions $\varphi = \varphi(z, \bar{z}) = \varphi(x, y)$, which satisfy the equation

$$D^n\varphi = \left(z\frac{\partial}{\partial z}\right)^n \varphi = 0,$$

and have the value zero at origin.

Of course, we have $\mathcal{A}_1^2(\mathbb{D}) = \mathcal{A}^2(\mathbb{D})$ and $\tilde{\mathcal{A}}_1^2(\mathbb{D}) = \tilde{\mathcal{A}}^2(\mathbb{D})$, for $n = 1$, as well as $\mathcal{A}_n^2(\mathbb{D}) \subset \mathcal{A}_{n+1}^2(\mathbb{D})$ and $\tilde{\mathcal{A}}_n^2(\mathbb{D}) \subset \tilde{\mathcal{A}}_{n+1}^2(\mathbb{D})$, for each $n \in \mathbb{N}$. It is known [1] that the system of functions

$$\ell_n(r) = \sqrt{2}\sum_{k=0}^{n}\frac{n!}{k!(n-k)!}\frac{2^k}{k!}(\log r)^k, \quad r \in (0,1), \quad n \in \mathbb{Z}_+ \quad (3.1)$$

forms an orthonornal base in $L_2((0,1), rdr)$. Denote by L_n, $n \in \mathbb{Z}_+$, the one-dimensional subspace of $L_2((0,1), rdr)$, generated by the function $\ell_n(r)$. Note, that for $n = 0$ this definition gives exactly the previously defined space L_0. Let finally $L_n^\oplus = \bigoplus_{k=0}^{n} L_k$ be the direct sum of the first $(n+1)$ spaces.

THEOREM 3.1. *The unitary operator* $U : L_2(\mathbb{D}) \to L_2((0,1), r\,dr) \otimes l_2$
maps the space $\mathcal{A}_n^2(\mathbb{D})$ *of* n-$\overline{D}$-*analytic functions onto* $L_{n-1}^{\oplus} \otimes l_2^+$.

PROOF. The space $U(\mathcal{A}_n^2(\mathbb{D}))$ obviously coincides with the set of all
sequences from $l_2(L_2((0,1), r\,dr)) = L_2((0,1), r\,dr) \otimes l_2$, which satisfy
the equation

$$U\overline{D}^n U^{-1}\{c_k(r)\}_{k\in\mathbb{Z}} = U_2\left\{\left[\frac{r}{2}\left(\frac{\partial}{\partial r} - \frac{k}{r}\right)\right]^n\right\}_{k\in\mathbb{Z}} U_2^{-1}\{c_k(r)\}_{k\in\mathbb{Z}}$$
$$= \{\overline{D}_k^n c_k(r)\}_{k\in\mathbb{Z}} = 0,$$

where

$$\overline{D}_k = \begin{cases} \frac{k+1}{2} r \frac{d}{dr}, & \text{if } k \in \mathbb{Z}_+ \\ \frac{|k|+1}{2}\left(\frac{2|k|}{|k|+1} + r\frac{d}{dr}\right), & \text{if } k \in \mathbb{Z}_- \end{cases}.$$

It is easy to see that the intersection of the general solution of this equation with the space $l_2(L_2((0,1), r\,dr)) = L_2((0,1), r\,dr) \otimes l_2$ coincides
with the set of all sequences of the form

$$\sum_{m=0}^{n-1} (\log r)^m \{\chi_+(k) d_k^{(m)}\}_{k\in\mathbb{Z}},$$

where $\{d_k^{(m)}\}_{k\in\mathbb{Z}} \in l_2$, for all $m = \overline{0, n-1}$, or, rearranging polynomials
on $\log r$, with the set of all sequences

$$\sum_{m=0}^{n-1} \ell_m(r)\{c_k^{(m)}\}_{k\in\mathbb{Z}_+},$$

where $\{c_k^{(m)}\}_{k\in\mathbb{Z}_+} \in l_2^+$, for all $m = \overline{0, n-1}$. $\qquad\square$

Introduce the space $\mathcal{A}_{(n)}^2$ of true-n-$\overline{D}$-analytic functions by

$$\mathcal{A}_{(n)}^2 = \mathcal{A}_n^2 \ominus \mathcal{A}_{n-1}^2,$$

for $n > 1$, and by $\mathcal{A}_{(1)}^2 = \mathcal{A}_1^2$, for $n = 1$, then $\mathcal{A}_n^2 = \bigoplus_{k=1}^n \mathcal{A}_{(k)}^2$.

COROLLARY 3.2. *The unitary operator* $U : L_2(\mathbb{D}) \to L_2((0,1), r\,dr) \otimes$
l_2 *maps the space* $\mathcal{A}_{(n)}^2(\mathbb{D})$ *of true-*n-$\overline{D}$-*analytic functions onto the space*
$L_{n-1} \otimes l_2^+$.

Analogously we have the following results:

THEOREM 3.3. *The unitary operator* $U : L_2(\mathbb{D}) \to L_2((0,1), r\,dr) \otimes l_2$
maps the space $\tilde{\mathcal{A}}_n^2(\mathbb{D})$ *of* n-D-*analytic functions onto* $L_{n-1}^{\oplus} \otimes l_2^-$.

Symmetrically, introduce the space $\widetilde{\mathcal{A}}_{(n)}^2$ of true-n-D-analytic functions by $\widetilde{\mathcal{A}}_{(n)}^2 = \widetilde{\mathcal{A}}_n^2 \ominus \widetilde{\mathcal{A}}_{n-1}^2$, for $n > 1$, and by $\widetilde{\mathcal{A}}_{(1)}^2 = \widetilde{\mathcal{A}}_1^2$, for $n = 1$, analogously, $\widetilde{\mathcal{A}}_n^2 = \bigoplus_{k=1}^n \widetilde{\mathcal{A}}_{(k)}^2$.

COROLLARY 3.4. *The unitary operator* $U : L_2(\mathbb{D}) \to L_2((0,1), r dr) \otimes l_2$ *maps the space* $\widetilde{\mathcal{A}}_{(n)}^2(\mathbb{D})$ *of true-n-D-analytic functions onto the space* $L_{n-1} \otimes l_2^-$.

The above results lead to the following theorem.

THEOREM 3.5. *We have the following isometric isomorphisms and decompositions of the spaces:*

1. *Isomorphic images of poly-$\overline{D}$-analytic spaces,*

$$W : \mathcal{A}_{(n)}^2(\mathbb{D}) \longrightarrow L_{n-1} \otimes H^2(\mathbb{D}),$$

$$W : \mathcal{A}_n^2(\mathbb{D}) \longrightarrow \bigoplus_{k=0}^{n-1} L_k \otimes H^2(\mathbb{D}),$$

$$W : \bigoplus_{k=1}^{\infty} \mathcal{A}_{(k)}^2(\mathbb{D}) \longrightarrow L_2((0,1), r dr) \otimes H^2(\mathbb{D}).$$

2. *Isomorphic images of poly-D-analytic spaces,*

$$W : \widetilde{\mathcal{A}}_{(n)}^2(\mathbb{D}) \longrightarrow L_{n-1} \otimes H^2(\mathbb{D}^-),$$

$$W : \widetilde{\mathcal{A}}_n^2(\mathbb{D}) \longrightarrow \bigoplus_{k=0}^{n-1} L_k \otimes H^2(\mathbb{D}^-),$$

$$W : \bigoplus_{k=1}^{\infty} \widetilde{\mathcal{A}}_{(k)}^2(\mathbb{D}) \longrightarrow L_2((0,1), r dr) \otimes H^2(\mathbb{D}^-).$$

3. *Decomposition of the space* $L_2(\mathbb{D})$,

$$L_2(\mathbb{D}) = \bigoplus_{k=1}^{\infty} \mathcal{A}_{(k)}^2(\mathbb{D}) \oplus \bigoplus_{k=1}^{\infty} \widetilde{\mathcal{A}}_{(k)}^2(\mathbb{D}).$$

Here L_n is the one-dimensional subspace of $L_2((0,1), r dr)$, generated by the function $\ell_n(r)$ of the form (3.1).

REFERENCES

1. L. Peng, R. Rochberg, and Z. Wu.
   Orthogonal polynomials and middle Hankel operators on Bergman
   spaces, *Studia Math.*, 102(1):57–75, 1992.

2. N. L. Vasilevski.
   On the structure of Bergman and poly–Bergman spaces, *Integr.
   Equat. Oper. Th.*, 33:471–488, 1999.

List of participants

S Bernstein
Bauhaus-Universität Weimar, Institut für Mathematik und Physik,
Coudraystr. 13b, 99421 Weimar, Deutschland
e-mail: bernstei@tigger.scc.uni-weimar.de

F Brackx
Department of Mathematical Analysis, Ghent University, Galglaan 2,
Gent, B 9000, Belgium
e-mail: fb@cage.rug.ac.be

T Branson
Department of Mathematics, University of Iowa, Iowa City IA 52242,
USA
e-mail: branson@math.uiowa.edu

D Bryukhov
19 Mira av., apt. 225, Fryazo Moscow region 141190, Russia
e-mail: bryukhov@mail.ru

J Bureš
Mathematical Institute of Charles University, Sokolovská 83, 186 75
Prague, Czech Republic
e-mail: jbures@karlin.mff.cuni.cz

D Calderbank
Department of Mathematics and Statistics, University of Edinburgh,
King's Buildings, Mayfield Road, Edinburgh EH9 3JZ, United Kingdom
e-mail: davidmjc@maths.ed.ac.uk

P Cerejeiras
Universidade de Aveiro, Departamento de Matemática, Universidade
de Aveiro, P-3810-193 Aveiro, Portugal
e-mail: pceres@mat.ua.pt

AA Chernitskii
St.-Petersburg Electrotechnical University,
Prof. Popov str. 5, St.-Petersburg, 197376, Russia
e-mail: aa@cher.etu.spb.ru

JSR Chisholm
Wheelwright Cottage, Tile Kiln Hill, Blean, Canterbury, Kent CT2
9EB, United Kingdom
e-mail: ChisholmRandM@wcb.u-net.com

R Delanghe
Department of Mathematical Analysis, Ghent University, Galglaan 2,
Gent, B 9000, Belgium
e-mail: rd@cage.rug.ac.be

S-L Eriksson-Bique
Department of Mathematics, University of Joensuu, PO Box 111,
FIN-80101 Joensuu, Finland
e-mail: Sirkka-Liisa.Eriksson-Bique@joensuu.fi

K Gürlebeck
Bauhaus-Universität Weimar, Institut für Mathematik und Physik,
Coudraystr.13, 99421 Weimar, Germany
e-mail: guerlebe@fossi.uni-weimar.de

K Habetha
RWTH, Aachen, Germany
e-mail: habetha@math2.RWTH-Aachen.de

J Hirvonen
Department of Mathematics, University of Joensuu,
FIN-80101 Joensuu, Finland
e-mail: hirvonen@cc.joensuu.fi

B Jancewicz
Institute of Theoretical Physics, University of Wrocław, pl Maksa Borna
9, PL-50-204 Wrocław, Poland
e-mail: bjan@ift.uni.wroc.pl

L Kadlčáková
Mathematical Institute of Charles University, Sokolovská 83, 186 75
Prague, Czech Republic
e-mail: kadlcak@karlin.mff.cuni.cz

U Kähler
Universidade de Aveiro, Departamento de Matemática, Universidade
de Aveiro, P-3810-193 Aveiro, Portugal
e-mail: uwek@mat.ua.pt

G Kaiser
Virginia Center for Signals and Waves, 1921 Kings Road, Glen Allen
VA 23060, USA
e-mail: kaiser@wavelets.com

J Kettunen
Department of Mathematics, University of Joensuu,
FIN-80101 Joensuu, Finland
e-mail: jkettune@cc.joensuu.fi

G Khimshiashvili
Department of Theoretical Physics, University of Łodz,
Pomorska str.149/153, Łodz 90-236, Poland
e-mail: khimsh@krysia.uni.lodz.pl

V Kisil
School of Mathematics, University of Leeds, Leeds LS2 9JT, United
Kingdom
e-mail: kisilv@e-math.ams.org

V Kravchenko
ESIME del Instituto Politecnico Nacional, Dept. de Telecom, Unidad
Zacatenco, C.p. 07738, D.F., Mexico
e-mail: vkravche@maya.esimez.ipn.mx

L Krump
Mathematical Institute of Charles University, Sokolovská 83, 186 75
Prague, Czech Republic
e-mail: krump@karlin.mff.cuni.cz

K Kwásniewski
Institute of Computer science, University of Białystok, ul. Sosnowa 64,
PL 15-887 Białystok, Poland
e-mail: kwandr@noc.uwb.edu.pl

V Labunets
Signal Processing Laboratory, Tampere University of Technology,
P.O.Box-533, FIN-33101, Tampere, Finland
e-mail: lab@cs.tut.fi

L Lanzani
Univeristy of Arkansas, Dept. of Mathematics, Fayetteville, AR 72701,
USA
e-mail: lanzani@comp.uark.edu

414

G Laville
Le Laboratoire SDAD, Université de Caen, Campus II, B.P. 5186, 14032
Caen Cedex, France
e-mail: guy.laville@math.unicaen.fr

H Leutwiler
Mathematisches Institut, Universität Erlangen - Nurenberg,
Bismarckstraße 1 1/2, D-91054 Erlangen, Deutschland
e-mail: leutwil@mi.uni-erlangen.de

HR Malonek
Universidade de Aveiro, Departamento de Matemática, Universidade
de Aveiro, P-3810-193 Aveiro, Portugal
e-mail: hrmalon@mat.ua.pt

N Marchuk
Steklov Mathematical Institute, Gubkina st. 8, Moscow 117966, Russia
e-mail: nikolai@marchuk.mian.su

A McIntosh
Centre for Mathematics and Applications,
Australian National University, Canberra ACT, Australia
e-mail: alan@wintermute.anu.edu.au

A Meskhi
A. Razmadze Mathematical Institute, Georgian Academy of Sciences
1, M. Aleksidze St., 380093 Tbilisi, Georgia
e-mail: meskhi@rmi.acnet.ge

O Pokorná
Technical Faculty CZU, Kamýcká, Praha 6, Suchdol, Czech Republic
e-mail: Pokorna@tf.czu.cz

J Ryan
Department of Mathematics, University of Arkansas, Fayetteville,
AR 72701, USA
e-mail: jryan@comp.uark.edu

I Sabadini
Dipartimento di Matematica, Politecnico di Milano, Via Bonardi, 9,
20133 Milano, Italy
e-mail: sabadini@mate.polimi.it

H Schaeben
Mathematische Geologie und Geoinformatik, Institut für Geologie,
TU Bergakademie Freiberg,
Gustav-Zeuner-Str. 12, D 09596 Freiberg, Germany
e-mail: schaeben@geo.tu-freiberg.de

P Somberg
Mathematical Institute of Charles University, Sokolovská 83, 186 75
Prague, Czech Republic
e-mail: somberg@karlin.mff.cuni.cz

FC Sommen
Dept. of Math. Analysis, Ghent University, Galglaan 2, Gent, B 9000,
Belgium
e-mail: fs@cage.rug.ac.be

V Souček
Mathematical Institute of Charles University, Sokolovská 83, Prague,
Czech Republic
e-mail: soucek@karlin.mff.cuni.cz

W Sprößig
Freiberg University of Mining and Technology, Faculty of Mathematics
and Informatics,
Bernhard-von-Cotta-Str. 2, D-09599 Freiberg, Germany
e-mail: sproessig@math.tu-freiberg.de

DC Struppa
Department of Mathematical Sciences, George Mason University, 4400
University Drive, Fairfax, WA 22030, USA
e-mail: dstruppa@osf1.gmu.edu

A Trautman
Instytut Fizyki Teoretycznej, Uniwersytet Warszawski, Hoza 69,
PL-00-681, Warszawa, Poland
e-mail: Andrzej.Trautman@fuw.edu.pl

O Váňa
Mathematical Institute of Charles University, Sokolovská 83, 186 75
Prague, Czech Republic
e-mail: vana@karlin.mff.cuni.cz

P Van Lancker
Department of Mathematical Analysis, Ghent University, Galglaan 2,
Gent, B 9000, Belgium
e-mail: pvl@cage.rug.ac.be

N Vasilevski
Departamento de Matemáticas, CINVESTAV del I.P.N., Apartado Postal
14-740, 07000 Mexico, D.F., Mexico
e-mail: nvasilev@math.cinvestav.mx